Shock Wave and High Pressure Phenomena

More information about this series at http://www.springer.com/series/1774

Shock Wave and High Pressure Phenomena

L.L. Altgilbers, M.D.J. Brown, I. Grishnaev, B.M. Novac, I.R. Smith, I. Tkach, and Y. Tkach: Magnetocumulative Generators

T. Antoun, D.R. Curran, G.I. Kanel, S.V. Razorenov, and A.V. Utkin: Spall Fracture

J. Asay and M. Shahinpoor (Eds.): High-Pressure Shock Compression of Solids

S.S. Batsanov: Effects of Explosion on Materials: Modification and Synthesis Under High-Pressure Shock Compression

G. Ben-Dor: Shock Wave Reflection Phenomena

L.C. Chhabildas, L. Davison, and Y. Horie (Eds.): High-Pressure Shock Compression of Solids VIII

L. Davison: Fundamentals of Shock Wave Propagation in Solids

L. Davison, Y. Horie, and T. Sekine (Eds.): High-Pressure Shock Compression of Solids

V.L. Davison and M. Shahinpoor (Eds.): High-Pressure Shock Compression of Solids III

R.P. Drake: High-Energy-Density Physics

A.N. Dremin: Toward Detonation Theory

J.W. Forbes: Shock Wave Compression of Condensed Matter

V.E. Fortov, L.V. Altshuler, R.F. Trunin, and A.I. Funtikov: High-Pressure Shock Compression of Solids VII

B.E. Gelfand, M.V. Silnikov, S.P. Medvedev, and S.V. Khomik: Thermo-Gas Dynamics of Hydrogen Combustion and Explosion

D. Grady: Fragmentation of Rings and Shells

Y. Horie, L. Davison, and N.N. Thadhani (Eds.): High-Pressure Shock Compression of Solids VI

J. N. Johnson and R. Cherét (Eds.): Classic Papers in Shock Compression Science

V.K. Kedrinskii: Hydrodynamics of Explosion

C.E. Needham: Blast Waves

V.F. Nesterenko: Dynamics of Heterogeneous Materials

S.M. Peiris and G.J. Piermarini (Eds.): Static Compression of Energetic Materials

M. Sućeska: Test Methods of Explosives

M.V. Zhernokletov and B.L. Glushak (Eds.): Material Properties under Intensive Dynamic Loading

J.A. Zukas and W.P. Walters (Eds.): Explosive Effects and Applications

Tracy J. Vogler • D. Anthony Fredenburg
Editors

Shock Phenomena in Granular and Porous Materials

Springer

Editors
Tracy J. Vogler
Sandia National Laboratories
Livermore, CA, USA

D. Anthony Fredenburg
Los Alamos National Laboratory
Los Alamos, NM, USA

ISSN 2197-9529 ISSN 2197-9537 (electronic)
Shock Wave and High Pressure Phenomena
ISBN 978-3-030-23001-2 ISBN 978-3-030-23002-9 (eBook)
https://doi.org/10.1007/978-3-030-23002-9

This Springer imprint is published by the registered company Springer Nature Switzerland AG.
The registered company address is: Gewerbestrasse 11, 6330 Cham, Switzerland

Preface

The study of porous materials under dynamic loading conditions is a field rich in physics that are important to a wide range of applications. While investigators have been active in this field for over 70 years, recent improvements in experimental and computational sciences have furthered our interpretation and understanding of these shock phenomena. In concert with advances in the underlying sciences, new applications and areas of study have emerged, while others have faded.

In this volume, we focus on materials and applications where the porosity has an effect on the behavior that is comparable to or greater than other material properties. Thus, the porosity is a critical contributor to the behavior rather than an aspect that perturbs the behavior modestly. In general, this implies a porosity level of the order of 50% or greater. The materials with this level of porosity are likely to be foams, aerogels, additively manufactured trusses, or loose or semi-packed powders. While partially sintered bodies can have porosities of this level, their behavior under dynamic loading has not been studied extensively and are thus not the focus of the present volume. Reflecting the developments in both the fundamental and applied sciences of shock consolidation, this book is arranged into three parts. The first part provides a look at the fundamental science of porous materials including the high-pressure equation of state response, the low-pressure compaction response, and the strength under dynamic loading. The second part considers application of the fundamental sciences to the areas of penetration, planetary impact, and reactive munitions. The third part concludes with a look at emergent technologies, developments and future avenues for research using advanced light sources, classical molecular dynamics, and additive manufacturing.

The editors intend this volume to serve as an update to capture the significant advances that have occurred since the release of the previous Springer books on the topic of shock compression science for initially porous materials, *High-Pressure Shock Compression of Solids IV: Response of Highly Porous Solids to Shock Loading*, 1997, and *Dynamics of Heterogeneous Materials*, 2001. In the roughly 20 years since those books were published, the field has advanced significantly while also moving in new directions. In some cases, the new directions represent completely new areas (e.g., additively manufactured materials), and in others, they represent increased emphasis (e.g., penetration). The developments in computational capabilities, experimental platforms, and diagnostics that are important across the broader field of shock physics have also had a strong impact on the study of porous materials.

The editors have tried to collect individual chapter contributions that captured these advancements across a diverse set of fundamental and applied areas of interest. As such, this work is meant as a representative snapshot of present-day capabilities and is by no means a completely comprehensive description of this extremely diverse field. We hope the reader finds this book both stimulating and useful, and encourage those active in the field to continue their pursuits toward advancing the area of shock compression science of porous materials.

The book begins with Sjostrom's chapter "Equation of State Modeling for Porous the fundamentals for developing wide-ranging representations of a material's pressure, density, and energy response under shock loading conditions. While, strictly speaking, equations of state are developed for materials in the absence of porosity, Sjostrom successfully demonstrates the utility of using initially porous Hugoniot data as one of the few means of validating the thermal models underlying an equation of state, particularly in the absence of direct temperature measurements under shock loading. Also prevalent in this chapter is the use of quantum molecular dynamics in helping constrain material response models in the warm dense matter regime, an area historically sparse on data. At these elevated temperatures and pressures, the robustness of quantum molecular dynamics calculations is clearly demonstrated with several examples showing agreement between experiment and theory.

At the opposite end of the pressure and temperature spectrum are the following two chapters "Low-Pressure Dynamic Compression Response of Porous Materials" by Vogler and Fredenburg and "Continuum Modeling of Partially Saturated Soils" by Banerjee and Brannon. In the former, Vogler and Fredenburg provide an experimentally focused look at the shock response of porous materials in the region of incomplete compaction. The authors begin by presenting aspects of target configurations, instrumentation considerations, and analysis methods that are unique to measuring the response of highly heterogeneous initially porous materials. Subsequently, a brief look is given to several of the more prevalent empirically derived engineering-level models used to computationally represent compaction in shock physics hydrocodes at the continuum, followed by the discussion of mesoscale, or particle-level modeling methods. The authors conclude by providing insights into several phenomena that have been observed in a variety of initially porous materials subjected to low to moderate shock loading conditions.

Where Vogler and Fredenburg cover experimental aspects associated with the region of incomplete compaction, the following chapter "Continuum Modeling of Partially Saturated Soils" by Banerjee and Brannon explores this region from a largely theoretical point of view. Within the context of dry and saturated soils, the authors develop a three-component (solid, fluid, gas) continuum model, ARENA, aimed largely at capturing the initial loading and unloading behavior experienced under impulsive loading conditions. The elastic and plastic constitutive responses are described in detail, as are the calibration procedures, which include a mix of quasi-static and intermediate strain rate compression tests. The attractiveness of this model stems from its relatively straightforward calibration procedures and its ability to be scaled, once calibrated, across multiple initial densities.

The next part of this book covers applications of shock waves in granular materials, and begins with the chapter "Planetary Impact Processes in Porous Materials" by Collins et al., who focus on impact and penetration at planetary length scales. It begins by highlighting the role played by initial porosity in the transient formation and final resting configurations of impact craters, providing a set of general scaling laws governing the relationship between properties of the semi-infinite granular body and resultant crater size. The discussion continues for collisions where the impacting and target bodies are of comparable length scales, where the presence of porosity is shown to significantly increase the ability of an impacted body to withstand violent collisions. Building upon these theories, the authors develop the concepts of mass ejection and momentum transfer within the context of an asteroid deflection scenario, demonstrating that the same shock-absorbing properties that enhance the ability to withstand collisions also create challenges in deflecting highly porous asteroids. The focus then turns toward localized collisional effects in particulate materials, where extents of compaction, phase change, and melting are discussed in terms of initial porosity.

The following chapter "Recent Insights into Penetration of Sand and Similar Granular Materials" by Omidvar, Bless, and Iskander provides an experimental look at the penetration of rigid bodies into dry and saturated sand. Here, the authors couple high-speed digital photography and optical velocimetry techniques to characterize the strain fields surrounding the nose of blunt and conical penetrators and the deceleration properties of spherical projectiles into granular beds. Through these experimental observations, the authors propose a four-stage, velocity-delimited framework for describing the dominant mechanisms controlling the behavior of penetrating bodies into granular materials, as well as develop a set of engineering constants to parameterize a set of penetration equations for dry sand over a wide range of initial densities and impact velocities.

Also touching on penetration but from a different perspective is the chapter "Applications of Reactive Materials in Munitions" by Peiris and Bolden-Frazier, which focuses on the use of reactive powders for munitions applications. The authors begin by providing a review of several of the more prevalent methods of producing reactive powders, including cold rolling and extrusion laminate methods, electrospraying, and ball milling. Using these reactive materials as feedstock, the discussion then turns to their potential use in advanced munitions concepts such as chemical and biological agent defeat, high-energy yielding explosives, directed energy munitions, and solid rocket propellants. Finally, the use of powder metallurgy techniques to produce high-density structural components is covered with specific emphasis on forming reactive structural munitions components that could replace inert structural components, such as steel, to enhance blast, fragmentation, and penetration properties.

The final part of this book explores advanced concepts in the area of granular and porous materials, beginning with the chapter "X-Ray Phase Contrast Imaging of Granular Systems" from Jensen et al. on the use of advanced light sources to interrogate particle-level physics under shock loading conditions. It begins by describing the theoretical foundation of X-ray phase contrast imaging and how these images can be used to determine the in situ areal density of dynamically impacted materials. The authors then show how the multi-frame X-ray phase contrast imaging system developed for the Advanced Photon Source at Argonne National Laboratory can be used to interrogate the deformation and fracture response of granular materials under impact loading. These high-resolution images are then used to inform and validate mesoscale modeling approaches, a necessary step in linking the physics between the mesoscale and continuum.

The following chapter "Shock Compression of Porous Materials and Foams Using Classical Molecular Dynamics" by Lane reviews historical and recent advancements in the application of molecular dynamics techniques to interrogate the shock response of porous materials at local and global length scales. This fresh look at classical molecular dynamics offers first an introduction to the principles underlying the method, which is followed by an overview of the different means by which porosity is incorporated into molecular dynamic simulations. From this point, the authors then introduce several techniques by which predictions of the continuum response can be made from simulations of materials and voids at the nanoscale. The techniques are then applied to several porous materials systems, highlighting molecular dynamics applications to metals, high-strength solids, and soft energetic materials. The chapter concludes by offering a look at potential growth areas for molecular dynamics investigations, commensurate with the expected increases in accessible length and time scales.

The final chapter "Additively Manufactured Cellular Materials" by Winter and McShane provides a look at porous structures within the context of additively manufactured cellular materials, a field that has progressed significantly with the increased performance and availability of advanced manufacturing techniques. The chapter begins with a brief review of the different methods available for producing cellular structures, following which, the authors focus on the characterization of the compressive response of a cellular 316L stainless steel manufactured using the selective laser melting technique. The results from quasi-static and intermediate strain rate compression tests over a broad range of initial lattice dimension and relative density conditions are used to show how strain rate sensitivities in the stress-strain response can be attributed to properties of the underlying steel as well as those due to buckling of the cellular structure. The focus then shifts to the experimental and simulated compression response of these same cellular structures at higher strain rates commensurate with impact velocities of several hundreds of meters per second, again highlighting the evolution in energy-absorbing processes from cavity collapse to material jetting with increasing strain rate. With the previous study as motivation, the authors conclude with a computational investigation into how changes to the initial structure geometry influence the energy-absorbing capability, demonstrating the feasibility of using additively manufactured cellular structures to achieve an optimal impact response for a given application.

From this collection of individual contributions, it is the editors hope that this volume will serve as a useful desk reference, capturing a snapshot of the current state of the art for shock loading of porous materials. However, as in all growing fields, we expect that the technologies, theories, and capabilities will continue to evolve, and we look forward to see where these advancements take the field in the coming years.

Best wishes and fruitful discoveries.

Livermore, CA, USA — Tracy J. Vogler
Los Alamos, NM, USA — D. Anthony Fredenburg

Contents

Contents

Part I
Fundamental Aspects

Equation of State Modeling for Porous Materials

Travis Sjostrom

1 Introduction

The equation of state (EOS) is a representation of the thermodynamic properties of a material under a range of physical conditions, such as the pressure, temperature, and volume. Knowledge of the equation of state is vital in all manner of modeling systems which can include astrophysics applications including planetary core modeling, simulation of shock physics experiments including the inertial confinement fusion campaign, and in weapons physics design. Given these applications, we are interested in EOS with very broad ranging pressure and temperature conditions. This means for the density, we are concerned with densities approaching zero where we have a gas or plasma state to tens or even hundreds of times the ambient solid density. For the temperatures we can require temperature from 0 to tens of millions degrees Kelvin or up to the keV range ($1\,\mathrm{eV} \approx 11{,}605\,\mathrm{K}$) [1]. While EOS modeling has a long history and extensive literature, this chapter is not intended to be all encompassing but rather to ensure that the reader has one solid perspective on how EOS is treated for porous materials with pointers to references where other perspectives may exist. Additionally, that perspective is the state-of-the-art EOS model currently in use by a select theoretical group at Los Alamos National Laboratory.

Porous materials are a particular state for which we need understanding through an EOS. The EOS itself is for a pure material and does not have knowledge of the porosity [2]. For example, a powder is made of small crystals of solid density and the voids between them. So the ambient energy inside a particle is just the solid energy at ambient pressure and temperature. There may be a surface energy, but this

T. Sjostrom (✉)
Theoretical Division, Los Alamos National Laboratory, Los Alamos, NM, USA
e-mail: sjostrom@lanl.gov

T. J. Vogler, D. A. Fredenburg (eds.), *Shock Phenomena in Granular and Porous Materials*, Shock Wave and High Pressure Phenomena,
https://doi.org/10.1007/978-3-030-23002-9_1

is typically outside the scope of the EOS. The issues for EOS arise when we want to shock a porous material. Initially, compaction plays a significant role, but beyond some threshold shock strength the resulting state should be described by the EOS alone. There are certain benefits to shocking porous materials, for example, metals such as iron and copper [3], and that is to access states of higher internal energy and temperature. Additionally, as in the case of some metal-oxides, a material may only be readily available in non-crystal form [4]. Study of porous aerogels is also an important area as these materials can be made at very low density and are used as a shock standard to study the shock release of materials [5].

In order to construct an EOS, whether for a porous material or not, a certain degree of constraining data are needed [6]. Limiting theoretical cases such as the ideal gas limit are one example. These can be built into the EOS models, but they will not lead to a high accuracy EOS on their own. Static experimental data can be very useful, this includes diamond anvil cell data which can isothermally compress materials up to 2–3 Mbar at low temperatures. Also, ambient pressure data may be available, including measurements of the thermal expansion and heat capacity. But if one wants to go to extreme conditions of pressure and temperature the only experimental approach is dynamic shock experiments. These generally provide a pressure versus density Hugoniot curve; however, a mostly absent set of experimental data for EOS is that of temperature measurements along the shock Hugoniot. Even in the relatively low temperature condensed matter regime this has proven difficult to obtain in dynamic experiments [7]. The importance for the temperature constraint is that it is possible to generate an EOS that captures the pressure–density Hugoniot, but does not have the right temperature, meaning the EOS is inaccurate, and when used, for example, in a hydrodynamic simulation the results may be far from reality.

Shock experiments on porous materials tend to raise the temperature much more significantly in relation to the density compression achieved as compared with shocks of solid crystals, as shown in the example phase diagram in Fig. 1 [8]. The resulting shocked states are then typically in the so-called warm dense matter (WDM) regime. WDM encompasses a broad range of phase space from a fraction of solid density to several times compression, and temperature from a few thousand to millions of degrees Kelvin. Laboratory experiments in general are difficult in this regime, and even as progress is made the resulting accurate data is still quite sparse.

WDM represents matter for which the ions are in a liquid-like state with motion that may be moderately to strongly correlated while at the same time the quantum nature of the electron structure is still an important aspect. Theoretically WDM lies at the confluence of plasma physics, condensed matter physics, and liquid theory [9], as such the regime is inherently challenging because its thermodynamics cannot be framed in terms of small perturbations from ideal, solvable models. This difficulty presents the need for accurate first principles simulations to provided data over the WDM regime. Though computationally expensive, and in fact prohibitive for temperature much above 100,000 K, quantum molecular dynamics (QMD) based on density functional theory can provide constraining EOS data where there is very little or no data at all [10]. In the case of porous materials it has also been of

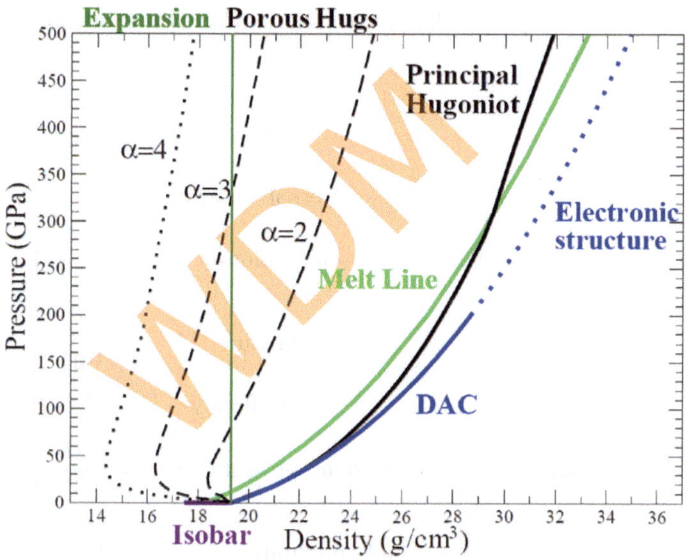

Fig. 1 Phase space and constraining data for a porous material (gold) [8]. Above the ambient solid density, 19.3 g/cm^3, isothermal diamond anvil cell (DAC) data or melt curve data may be available. Porous shocks, indicated by the α labeled curves where increasing α, the ratio of the porous volume to the non-porous volume, indicates increasing porosity, result in warm dense matter (WDM) conditions which is challenging for both theory and experiment

interest, to determine the energy offset due to the porosity (i.e., the surface energy) of the initial conditions for shock experiments, but this has been done using classical potential molecular dynamics as the systems are too large to simulate via QMD [11].

2 Theoretic Background

Before proceeding to the next section on constructing the EOS, we provide here some of the basic thermodynamic relations and considerations needed. To begin we will work primarily with the Helmholtz free energy $F(V, T)$, additionally the Gibbs free energy $G(P, T)$ will be useful particularly when considering phase transitions. When taken with the proper variable dependence (V, volume or P, pressure and T, temperature as we have indicated) both F and G are thermodynamic potential functions, meaning that all other thermodynamic properties may be obtained by differentiation alone [12]. Their relations to entropy S and internal energy E are

$$F = E - TS \tag{1}$$

$$G = E - TS + PV. \tag{2}$$

The EOS will be completely derived from F, and have the following fundamental thermodynamic relations:

$$\text{Entropy} \quad S = -\left(\frac{\partial F}{\partial T}\right)_V \tag{3}$$

$$\text{Internal energy} \quad E = F + TS \tag{4}$$

$$\text{Specific heat at constant volume} \quad C_V = \left(\frac{\partial E}{\partial T}\right)_V \tag{5}$$

$$\text{Pressure} \quad P = -\left(\frac{\partial F}{\partial V}\right)_T \tag{6}$$

$$\text{Enthalpy} \quad H = E + PV \tag{7}$$

$$\text{Specific heat at constant pressure} \quad C_P = \left(\frac{\partial H}{\partial T}\right)_P \tag{8}$$

$$\text{Isothermal bulk modulus} \quad K_T = -V\left(\frac{\partial P}{\partial V}\right)_T \tag{9}$$

$$\text{Volume coefficient of thermal expansion} \quad \alpha = \frac{1}{V}\left(\frac{\partial V}{\partial T}\right)_P \tag{10}$$

$$\text{Grüneisen parameter} \quad \Gamma = V\left(\frac{\partial P}{\partial E}\right)_V . \tag{11}$$

Of these, the last four are most directly applicable for relating experimental results and the theoretical EOS. First C_P and α are measurements of the thermal response at fixed pressure. These measurements are available for most materials at atmospheric pressure and up to at least the melt temperature. Next the isothermal $P(V)$ curve, and hence K_T, is measured directly through static compression experiments such as diamond anvil cell. This is usually at room temperature, but can be performed along lower and elevated isotherms. Lastly the Grüneisen parameter Γ is a very important thermodynamic parameter used to help quantify the relationship between the thermal and elastic properties of a solid. An important definition of Γ is

$$\Gamma = \frac{\alpha K_T V}{C_V}, \tag{12}$$

which shows Γ composed of individual measurable physical properties, each of which varies significantly with temperature, but the ratio of these properties as given above does not vary greatly with temperature, and sometimes not at all [13]. Many approximations for Γ have been developed, the choice and implementation of which will greatly affect the EOS thermal response over the entire density range.

We have discussed the importance of the shock Hugoniot for porous materials being that it may be the only experimental data available. So finally here we list the Rankine–Hugoniot jump conditions which relate the initial state to the final shocked state along the Hugoniot curve [14].

$$\rho_0 U_s = \rho_1 (U_s - u_p) \tag{13}$$

$$P_1 - P_0 = \rho_0 U_s u_p \tag{14}$$

$$E_1 - E_0 = \tfrac{1}{2}(P_1 + P_0)(V_0 - V_1) \tag{15}$$

The initial state is represented by the 0 subscript and the final state is represented by the 1 subscript. The first two equations allow us to relate the shock velocity U_s and the particle velocity u_p, which are usually experimentally observed, to the pressure P and density ρ of the shocked state. The last equation can be used to calculate from the EOS the Hugoniot curve given the initial pressure and density of the experiment.

For the porous case the initial density is some fraction of solid density, which does introduce some complexities. First is that we must determine the proper initial conditions. As a first approximation one chooses E_0 to be the internal energy of the ambient solid, because the particles are at ambient pressure, but takes the volume or density to be at the average porous density. The next issue is that compaction will play a key role until the shock is strong enough to produce an equilibrium, homogeneous state. So we cannot constrain the EOS, without inclusion of a compaction model below this threshold, which must be determined.

3 General Equation of State Construction

The search for the equations of state of materials has a long history. Various branches of physics were developed or originated from the equations of state, while in return more complex formulations of the EOS were due to the development of modern physics [15, 16]. Indeed the study of porous materials is one area where this back and forth continues. Early on developments through the studies of gases lead to the first relations of pressure, volume, and temperature eventually leading to the well-known ideal gas law

$$PV = nRT, \tag{16}$$

here the pressure P, volume V, and the temperature T are related through the number of moles of the gas n, and the ideal gas constant $R \approx 8.314\,\text{J/(mol K)}$. Later van der Walls and others extended the range of applicability of the gas equation of state by including underlying physics. In the first half of the last century high pressure studies led by P.W. Bridgman led to solid state equations of state, and there exists now as for gases a myriad of equations of state. One early form is due to Murnaghan

$$P = \frac{K_0}{K_0'}\left[\left(\frac{V}{V_0}\right)^{-K_0'} - 1\right] \tag{17}$$

which relates the pressure of a solid compressed to volume V from the volume at zero pressure V_0 through the zero pressure bulk modulus K_0 and its pressure derivative K_0'. Equations (16) and (17) may be referred to as incomplete EOSs as they represent only some portion of phase space. Beyond these cases of the gas system and the solid cold curve, however, temperature excitations in solids in the form of phonons, disordered liquids, and the challenging case of warm dense matter must all be described. For a complete and wide-ranging EOS we need to consider all of the regions of phase space as well as phase boundaries such as the melt curve of a material.

Clearly there is no simple relation for the equation of state of a material across the broad ranges of temperature and pressure we wish to represent. Extrapolating the ideal gas law to too low of temperature is inaccurate especially in the denser cases, as is extrapolation of the Murnaghan EOS to compression much higher than 10% or 20% increase from ambient density. There exist varied approaches for generating wide ranging equations of state [1, 13, 15, 17–20] some focusing on greater accuracy in some regions than in others. In some cases great attention is paid to the ionization state of the atoms, this is important for fluids and gases of higher temperature and lower density approaching the plasma state [21]. At lower temperatures and densities dissociation of molecules and chemical thermal equilibrium of species can be accounted for [22]. The approach described in the remainder of this section is to use the physics models and limited range equations of state that best represent particular phases of a material, where the model works best, and stitching together the limiting case and data constrained results. Because of the large ranges of density and temperature, typically from 10^{-6} to 10^4 g/cm^3 and 0 to 10^5 eV, respectively, and the various combinations of theoretical models, the EOS is stored and used in tabular form on a density and temperature grid as opposed to being a set of analytic equations. This then requires accurate interpolation, particularly across phase boundaries and when performing hydrodynamic simulations with the EOS [23].

In order to construct an EOS with the broad range of thermodynamic conditions and encompassing the varied phases, it is best practice to work with one of the thermodynamic potential functions, and we choose to work with the Helmholtz free energy as a function of the volume and temperature $F(V, T)$. Regardless of choice one must in the end be able to calculate and compare with all experimental results.

At this point we make the first modeling choice and also approximation to the EOS. That is, to use the standard and common choice of separating the free energy into three independent components [24, 25]

$$F(V, T) = F_0(V) + F_e(V, T) + F_i(V, T). \tag{18}$$

Here the first contribution F_0 is the so-called cold curve, which is the energy of the system at zero temperature and as a function of the compression only. The second contribution F_e is the contribution from the thermal excitation of the electrons. While the third term F_i is the thermal contribution of the ions to the free energy, and includes the zero-point energy. The immediate approximation is that there is no explicit term for the electron–ion interaction at finite temperatures. Still for the accuracy required for the EOS this form has proven capable of accurately representing many systems, given that the appropriate models and basic input data whether theoretic or experimental are utilized. Below we provide details of each of these components, and also consider the multiphase construction as well as mixtures of materials.

3.1 Cold Curve

The cold curve F_0, or the zero temperature isotherm, is the only contribution that depends on the volume alone and not the temperature. The difficulty, however, is that the cold curve must extend to many times ambient density while also extending down to near zero density. The low density case is important for the study of porous materials, which may have densities from 20–30% to near 100% of ambient density for powders, and even down to just a few percent of ambient density for aerogels. While the very porous shock case heats readily with shock, even at temperatures of tens of thousands of degrees, neglecting the cold curve contribution can result in errors of 40% in the pressure of the shocked state.

The basic EOS approach for this is to separate the cold curve into separate regions as shown in Fig. 2. The middle density or low compression region extends from ambient density, ρ_0 or just below to a compression somewhere around 2–4 times ρ_0. In this region a semiempirical fit is generally used, for example, the Murnaghan EOS of Eq. (17) or other forms developed over the years by Birch et al. [26] (and the references therein). Alternatively a tabulated cold curve could be used over the

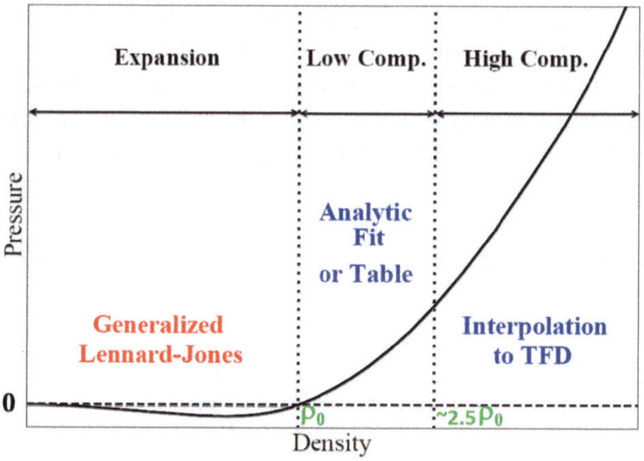

Fig. 2 Example of a cold curve and the various regions of density compression, which are represented by different underlying models. Little data is generally available to constrain the extrapolated expansion and high compression regions

finite density range. In the low compression region, we are typically interested in compression much greater than 10–20% that is accurately captured by Eq. (17) and so we use the finite-strain EOS of Birch [27, 28] where the pressure is given as the following expansion:

$$P_0(\rho) = \left\{9\left[1 + \frac{3}{2}\left(K_0' - 4\right)\eta\right] + \sum_{j=1}^{N_{\text{terms}}} \frac{C_j \eta^{j+1}}{(j+2)!}\right\}\frac{K_0 \eta}{3}\left(\frac{\rho}{\rho_0}\right)^{5/3} \tag{19}$$

and the corresponding free energy contribution is

$$F_0(\rho) = \left\{\frac{9}{2}\left[\eta^{2/3} + \left(K_0' - 4\right)\eta^3\right] + \sum_{j=1}^{N_{\text{terms}}} \frac{C_j \eta^{j+3}}{(j+3)!}\right\}\frac{K_0}{\rho_0} + E_{\text{shift}}. \tag{20}$$

Here ρ is the density with $\eta = 1/2[(\rho/\rho_0)^{2/3} - 1]$, and the input parameters are the bulk modulus K_0 and its pressure derivative K_0' at the reference density ρ_0 as well as the expansion coefficients C_j and the overall energy shift E_{shift}. The advantage of this expansion is accurately representing the cases where shock pressure above 10 Mbar can provide accurate data for density compressions greater than two [29].

In the high compression region the cold curve is given by the Thomas–Fermi–Dirac (TFD) approximation. Basically the TFD approximation is an ion of charge Z in a sphere. The sphere has a volume such that it represents the prescribed material density given the atomic mass of the ion. The electron density is then found using the semiclassical Thomas–Fermi statistical approximation within this sphere such that the total electron number is equal to Z and that the derivative of the density goes to zero at the sphere radius. The Dirac term indicates that the local exchange potential for the electrons is included as well as the direct Coulomb potential from the ion and the electron density itself. This approximation is believed to be accurate at densities above 10 Mbar. This does, though, vary by material and in some cases may need to be at higher pressures and the only verification of that is through shock data or simulations such as QMD. The total cold curve must be continuous and so we interpolate between the middle density region and the high density region as follows:

$$F_0(\rho > \rho_{TF}) = [E_{TF}(\rho) - E_{TF}(\rho_{TF})]y(\rho) + \Delta E, \tag{21}$$

with

$$y(\rho) = 1 + \frac{b_1}{\rho} + \frac{b_2}{\rho^{4/3}}. \tag{22}$$

Here E_{TF} is the zero temperature result of the TFD calculation. The parameters ΔE, b_1, and b_2 are found by matching the energy, pressure, and $dP/d\rho$ at the input upper compression match point ρ_{TF}.

Finally in the expansion region, the Lennard-Jones and related potentials describe interactions between neutral particles which are strongly repulsive at short range and then become attractive due to, for example, van der Walls forces. The transition to the generalized-Lennard-Jones model used at low density is given by Kerley [30]

$$F_0(\rho < \rho_{LJ}) = f_1\rho^{f_2} - f_3\rho^{f_{LJ}} + E_{\rm coh}, \tag{23}$$

where ρ_{LJ} is the lower compression cutoff match point, f_{LJ} is an input exponent, and $E_{\rm coh}$ is the input cohesive energy. The coefficients f_1, f_2, and f_3 are then found by requiring that the energy, pressure, and $dP/d\rho$ are continuous at ρ_{LJ}. This yields the characteristic minimum in the pressure and the limit of zero pressure at zero density that is seen in Fig. 2. The match itself is usually done at the ambient reference density or at just a few percent below so $\rho_{LJ} \approx \rho_0$.

Using this fit approach is good if the parameters in Eq. (20), such as the bulk modulus, are available. This could be through DAC isothermal compression data extrapolated to zero temperature or from first principles calculations such as density functional theory (DFT) or quantum Monte Carlo. However if one only has shock Hugoniot data available one may employ the Mie–Grüneisen EOS. In this approach one assumes that the Grüneisen parameter is a function of the density only. Then integrating Eq. (11) we have

$$P_a - P_b = \rho\Gamma(E_a - E_b), \tag{24}$$

where every term is evaluated at ρ. One is then free to choose a as the zero temperature isotherm and b as the Hugoniot. Finally, since $P_a = \rho^2 \frac{dE_a}{d\rho}$ the cold curve may be determined by solving the differential equation

$$\rho^2 \frac{dE_a}{d\rho} - P_b = \rho\Gamma(E_a - E_b). \tag{25}$$

Of note is that E_b is determined from the Rankine–Hugoniot shock equations

$$E_b - E_i = \frac{1}{2}(P_b + P_i)\left(\frac{1}{\rho_i} - \frac{1}{\rho_b}\right), \tag{26}$$

where the i subscripts indicate the initial conditions. This approach is direct and only requires an input of $\Gamma(\rho)$. However, for porous materials shocks often lead to significant heating, so inclusion of the thermal electron component is also required. In this case one must choose an F_e and an F_i contribution, whether F_i is Mie–Grüneisen or not, and then optimize the cold curve to reproduce the experimental Hugoniot.

3.2 Electron Thermal Contribution

The electron thermal contribution F_e in Eq. (18) arises primarily from the excitations of the electrons into higher energy states. This is quite clear within the framework of Kohn–Sham DFT as the many body electron problem is mapped to a system of non-interacting particles occupying a spectrum of eigenstates. At finite temperature, these states are populated according to Fermi–Dirac statistics and so higher temperature leads to higher internal energy. One could model the electron contribution as a uniform electron gas, but the question would be at what density. For example, aluminum has a valence of three electrons at ambient conditions, but will ionize more electrons as the temperature increases and there is also pressure driven ionization to be considered.

The common approach to approximate the electron thermal contribution is to use the TFD model at finite temperature [31], which is semiclassical but still provides the basic quantum statistical contribution to the free energy. The model is just as described in the previous section except that the electron density is treated at finite temperature instead of zero temperature. In order to arrive at the F_e contribution one subtracts the cold curve TFD energy from the finite temperature free energy at the calculated temperature and the same density. There are no suitable analytic models for arbitrary ion charge Z, temperature, and volume. So this must be tabulated on a grid over the entire range desired for the EOS and interpolated. The short coming in this model is that it does not include the discrete eigenspectrum of the bound states and so no shell structure effects are seen in the energy, but the Thomas–Fermi (TF) approach does provide a smoothing through this. Inclusion of the Dirac exchange is a significant improvement to the TF only approximation that was first employed at finite temperature [32]. Later corrections to the TF model were made by including electron density gradient terms [33], but this has little change for F_e as most of the difference is in the cold curve and not in the thermal excitations.

In order to include shell effects one may solve an average atom in the same way as in the TFD model, but instead of solving for the electron density in the TF approximation one can employ DFT and solve for the single particle states which in turn gives the density. This is the Kohn–Sham average atom (KSAA) approach [34, 35]. This is significantly more complex than TF involving the need to solve for both bound and continuum states, which require different approaches. Again the F_e contribution is found from subtracting the $T = 0$ result.

There is no method to extract this term from experimental results, but as it is defined we can, over some ranges of density and for temperatures up to a few eV, perform accurate DFT calculations. For example, if we perform a crystal calculation at zero temperature, we may perform a subsequent calculation with the ions frozen and the electrons at elevated temperature where the difference in free energy is F_e. Figure 3 shows the comparison of these benchmark results (KS-fcc) with both the TFD and with KSAA approaches described above. The benchmark simulation and the KSAA are in good agreement up to 50,000 K, while at that temperature the TFD is deviating by 30% in the internal energy. However this difference is noticeably

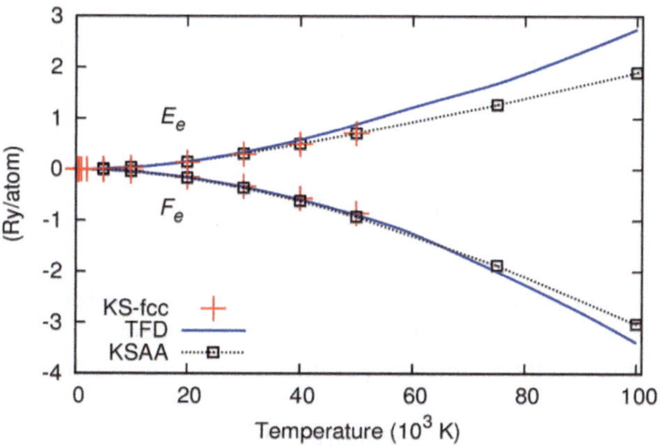

Fig. 3 Calculations for the electron thermal contribution for aluminum at $\rho = 2.7$ g/cc. The KS-fcc results are the benchmark results. Good agreement is shown with KSAA calculations for both the internal E_e (upper curves) and free F_e (lower curves) energies, which means structure is not that important for F_e. Finally TFD is shown in blue and has slightly better agreement for F_e than for E_e

smaller for the free energy. Still it seems that the KSAA is the more ideal approach that can cover a significant portions of the EOS range. The KSAA results are from the Tartarus code [36] and the KS-fcc results were calculated in the Quantum-Espresso [37] plane wave code using projector augmented wave formalism[38]. In both cases the local density approximation for the exchange-correlation [39] was used.

3.3 *Ion Thermal Contribution*

The ion thermal contribution, F_i, is the effect of the motion of the ions due to finite temperature in the material. Often at ambient pressure many thermal properties are measured which can constrain this contribution, including the thermal expansion as well as the specific heat. Diamond anvil cell data at elevated temperatures can also provide information at higher pressures. However, at even higher temperatures and pressures, and in particular for porous materials, the shock Hugoniot provides the only insight into the thermal contributions. The Hugoniot data itself are almost always absent temperature measurement, which makes it difficult to go beyond simple models.

At low temperature, we have a solid and can consider the effect of phonon vibrations [13, 20, 40]. We use for the solid phases ion thermal contributions given by a Debye model, which is derived using the harmonic approximation for the lattice vibrations. In some cases the specific heat above the Debye temperature cannot be well fit with this model and hence anharmonic corrections are needed. But largely the model is extremely effective. The free energy for the Debye model is given by

$$F_i(\rho, T) = \frac{R}{M}\left[\frac{9}{8}\theta + 3T\ln(1 - e^{-\theta/T}) - TD(\theta/T)\right], \tag{27}$$

where

$$D(x) = \frac{3}{x^3}\int_0^x \frac{y^3}{e^y - 1}dy \tag{28}$$

is a Debye function for the energy, R is the ideal gas constant, M is the atomic weight, T is the temperature. Also θ is the Debye temperature which is dependent on the density through the Grüneisen parameter, γ, by the relation

$$\gamma = d\ln\theta/d\ln\rho. \tag{29}$$

It is necessary to provide $\gamma(\rho)$ and additionally θ_{ref}, the value of θ at the reference density to fix the constant of integration, when solving for θ in Eq. (29). The parameter θ_{ref} is the Debye temperature that is typically available from experimental results. It may also be calculated from first principles calculations and phonon analysis. With this F_i may be evaluated everywhere. Of note is that γ is the ion thermal contribution to the total Grüneisen parameter Γ of Eq. (11) for the total EOS, which contains the electron thermal contribution as well.

There are many options for modeling γ, the simplest being constant γ or constant $\rho\gamma$. We impose an analytic model for γ in order to determine θ. The model is given by the following equations:

$$\gamma(\rho \geq \rho_{\text{ref}}) = \gamma_\infty + \frac{\rho_{\text{ref}}}{\rho}(2\gamma_{\text{ref}} - 2\gamma_\infty + \gamma_R') + \left(\frac{\rho_{\text{ref}}}{\rho}\right)^2(\gamma_\infty - \gamma_{\text{ref}} - \gamma_R'), \tag{30}$$

$$\gamma(\rho < \rho_{\text{ref}}) = \gamma_0 + \frac{\rho}{\rho_{\text{ref}}}(2\gamma_{\text{ref}} - 2\gamma_0 - \gamma_L') + \left(\frac{\rho}{\rho_{\text{ref}}}\right)^2(\gamma_0 - \gamma_{\text{ref}} + \gamma_L'). \tag{31}$$

Here γ_0 and γ_∞ are the values at $\rho = 0$ and $\rho = \infty$. γ_L' and γ_R' are the left and right logarithmic density derivatives of γ at ρ_{ref}. That is, $\gamma' = d\gamma/d\ln(\rho)$, from just below and above the reference density. Though the model allows these to be different, they are typically the same value. ρ_{ref} is the reference density and γ_{ref} is the value of γ at that density. The ideal gas limit is $\gamma = 2/3$ which can be used for the limiting case.

In terms of constraint we often have experimentally the thermal expansion and the heat capacity, and theoretically the Dulong–Petit limit for which the heat capacity approaches $3R$. After the materials melt, however, there is little constraint. Eventually at high enough temperature the ionic contribution to the heat capacity must be that of an ideal gas which is $3/2R$. So for the liquid case all models must attempt to extrapolate down from the high temperature limit through the non-ideal case, such as in the Cowan model [41], or interpolate between some best choice of characterization of the liquid near melt and the ideal gas limit, such as in the

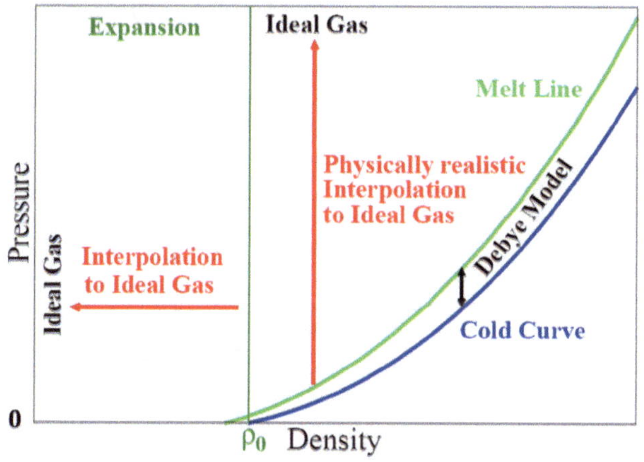

Fig. 4 Modeling of the liquid and warm dense matter regime necessitates interpolation from near melt conditions to the ideal gas limit for both expansion and increased temperature

models of Johnson [42], Kerley [43], and Chisolm and Wallace [44]. This situation is highlighted in Fig. 4. We note here that in the study of highly porous materials the available experimental shock data are mostly in the expansion region at higher temperatures and pressures, which is precisely where we have the least constrained models.

Each of the last three models are commonly used for the liquid regime. The equations of these are a bit tedious to list so we refer the reader to the original papers [42–44]. The approaches are similar but do differ in some aspects. Johnson interpolates the heat capacity from a Debye model with $C_v = 3R$ through the liquid to the high temperature limit. This interpolation has defined temperatures (depending on version) which guide the interpolation, for example, reaching $9R/4$ at temperature five times the melt temperature. Additionally, the model enforces the integral of the specific heat to give the known high temperature entropy. The Kerley model is based on perturbation theory of hard sphere systems. In principle, only the cold curve is needed to derive the thermodynamic properties for the fluid since it contains all the information about the intermolecular forces needed for the model. In practice, Kerley implements four adjustable parameters to include quantum corrections and to permit the user to make empirical adjustments that improve agreement with experiment. In the Chisolm model, the free energy is represented by an interpolation between a low temperature expression and the high temperature free energy value. The low temperature value is based on the liquid vibration-transit theory [45] of Wallace and collaborators in which the liquid has both a Debye-like vibrational energy as well as a diffusive or transit contribution. The interpolation to the ideal limit is a smooth but arbitrary function.

Recent developments in quantum molecular dynamics methods have led to the ability to extract the ion contribution from numeric simulation [46]. The method involves performing a standard QMD simulation for the energy and then

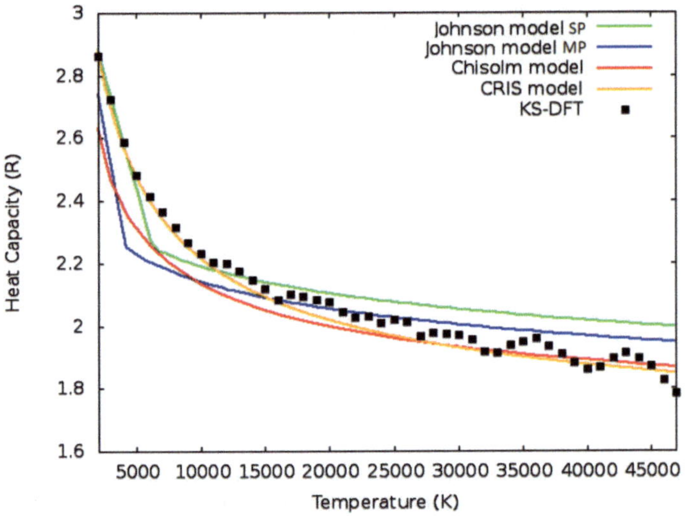

Fig. 5 Results for the constant-volume heat capacity of Al at 2.7 g/cm^3 as a function of temperature. QMD results [46] are denoted by black dots; predictions of the Johnson [47, 48], the Chisolm, and the CRIS models [49] by solid lines. The heat capacity, C_V is given in unit of the ideal gas constant R

recalculating the energy at all the same ion geometries in order to separate the energy components as given in Eq. (18). The specific heat from this approach has been compared to different EOSs for aluminum that have been constructed using the different models. The comparison is shown in Fig. 5. In this case the Kerley (CRIS) model [49] seems to be doing best, while for the Johnson model the transition point is producing an unphysical sharp change. This is seen in two different EOSs that use the Johnson model. The single phase Johnson model (SP) [47] does better at low temperatures just above melt, while the multiphase Johnson model (MP) [48] does better at intermediate and high temperatures. Finally, the Chisolm model was used with the exact same EOS parameters as the Johnson model (MP) but produces a physically smooth curve and agrees with the CRIS model above 20,000 K.

It should be noted that these liquid regime models have in previous EOS work been used to represent the entire EOS including the transition of model form at a calculated melt curve or using a smoothed latent heat. In our EOS work, we use multiphase construction, which is described in the next section. So, we only use these models in the liquid phase, and the melt curve is determined by the Gibbs free energy of that and the solid phases.

3.4 Multiphase Equations of State

We have thus far described the methods for generating the equation of state only for individual phases either solid or liquid. But materials have at least a single solid and fluid phase, and possibly many solid phases. There are two approaches to model this. The first is the older approach, which is to emulate the multiphase material through what is essentially a single phase. Here, one first constructs the cold curve, which though continuous might have switching behavior to emulate a solid–solid phase change. Then for the ion thermal contribution one could use the Johnson, Kerley, or Chisolm models. All of these can implement the Debye model to represent the solid and calculate a Lindemann melt curve, then from melt apply the liquid model interpolation to the ideal regime as detailed in the previous section. One issue here is that this uses the solid cold curve for the liquid regime, and if solid–solid transitions have been emulated in the cold curve, this will propagate through the liquid regime. Another is that solid–solid transitions that occur with temperature increase as opposed to pressure increase cannot be treated, without adding some sort of switching in the ion thermal model.

The second approach is the true multiphase approach. This entails constructing an EOS for each individual phase. Each has its own cold curve, ion and electron modeling. This includes the cold curve for the liquid, which of course is not physical and cannot be experimentally determined, but is still a component of the EOS. With the individual phases constructed, the multiphase EOS is constructed by determining the equilibrium state at all pressures and temperatures, which is the state that has the minimum Gibbs free energy [50]. This can also be done via a different procedure using the Helmholtz free energy [16]. The melt curve and other phase boundaries arise naturally as opposed to using an approximate model. An example of this is shown for CeO_2 in the next section. Finally looking forward, if one has the individual phases in combination with a kinetics model for the material, then hydrodynamics codes can simulate time-dependent phase transitions of shock experiments. The difficulty in the second approach is that in current application the EOS is a table on grid and in simulations interpolating in and across regions of mixed phases can be problematic.

3.5 Mixing of Equations of State

Often there may be the need to have an EOS of a mixture of materials. In many dynamic situations, materials will mix. Depending on the degree of mixture various models may be appropriate. Simulations often assume that a system is decomposed into a set of volumes consisting of known quantities of the materials described by EOSs. However, the description of any volume element is in general expected to be unlike the available EOSs if it contains more than one material. The materials within the volume may be unmixed or immiscible, out of thermodynamic equilibrium, or in an extreme case, homogenized to the molecular level. Porous materials represent another specific case. The voids in the material in the experimental system may be filled with gas, be it air or something else, instead of vacuum. It has been shown producing a mixed EOS, in this case considering nitrogen gas, can provide improved results for shocked powders [51].

There exists only a finite set of EOS, and so an efficient mixing of two EOSs is advantageous for simulation. For this we use Amagat's mixing rule (also known as pressure matching or additive volume mixing), which simply states that the mix components are at the same pressure and temperature. The total volume is then the sum of the component volumes [52]. This is described by the equations

$$P(V,T) = P_i(V_i,T) \tag{32}$$

$$V = \sum_i V_i, \tag{33}$$

where P and V are the pressure and volume of the mixture, P_i and V_i are the pressure and density of component i, and T is the temperature. This rule has recently been studied by comparing with QMD simulations for warm dense systems [53]. This case where mixing has occurred at the molecular level is the extreme where interspecies quantum mechanics becomes most important and classical mixing models are expected to be most frustrated. The results show that a mixing rule based off Amagat's pressure equilibration scheme provides both accurate pressures and energy given accurate pure materials equations of state. Also, in experiment strongly shocked aluminum and nickel mixtures show agreement with this mixing rule as well with another mixing rule, Dalton's law (partial pressures) [54]. While these validations are useful, the mixing rules are not perfect and in less extreme regions where there is more chemistry the accuracy of the rules will diminish.

Other overall mixing schemes include those of Gibbs potential mixing [55] and more accurate enthalpy mixing [56]. Besides complete EOS mixing there is also a variety of mixing rules for calculating Hugoniot curves from pure EOS contributions, including accurate u_p, particle velocity, mixing [57, 58].

4 Powder, Aerogel, and Foam Applications

Having discussed the fundamentals we now consider a few specific examples. First we consider a multiphase EOS for granular CeO_2, then we look at SiO_2 or quartz which is a shock release standard in aerogel form, and finally foam CH, which has application in inertial confinement fusion experiments.

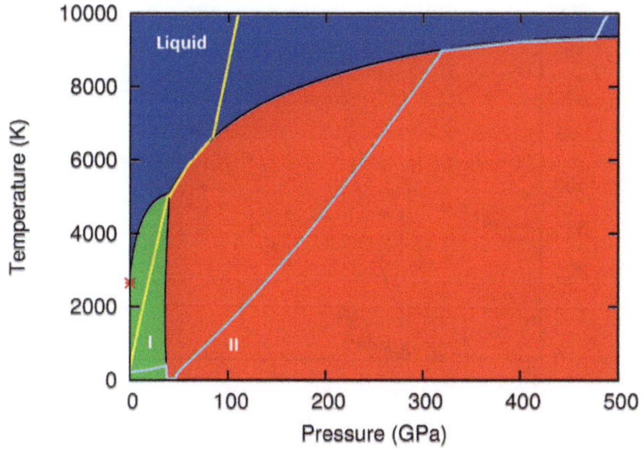

Fig. 6 The phase diagram of CeO_2 exhibits a pressure induced solid–solid transition at 35 GPa and has an ambient melt temperature of about 2600 K. The lower (cyan) curve shows the solid density 7.215 g/cm^3 Hugoniot, while the upper (yellow) curve is the 4.0 g/cm^3 porous Hugoniot

4.1 Ceria Powder—CeO_2

Cerium(IV) oxide, CeO_2, has many technological applications at moderate temperatures and pressures, but also is important as an archetypal f-electron system to study pressure induced phases transitions, as well as temperature induced melting and dissociation. It also falls in the category of being a metal-oxide which is only readily available in granular, powder form. Although through sintering CeO_2 with 96% solid density has been achieved, all current shock data available is from porous densities [11, 59]. However, in this case there has been single crystal DAC measurements as well as thermal expansion and specific heat measurements, which help to identify and constrain the two solid phases. A broad range of QMD calculations have been done also for the liquid regime, extending from 2 to 20 g/cm^3 [11]. Given these data, a multiphase EOS has been constructed following the outline of the previous section. In Fig. 6 the phase diagram is shown as a function of pressure and temperature. Additionally the Hugoniot curves for initial densities 4.0 (upper curve) and 7.215 (lower curve) g/cm^3 are shown.

In this case the experimental shock Hugoniot data was not used to constrain the EOS. This was because the QMD data was completely encompassing of the shock data and therefore the EOS was based on fitting the liquid regime models to the QMD data only and then validating that against the experiments. The results can be seen in Fig. 7. Good agreement is shown for the EOS at both the experimental densities which are at 55% (4.0 g/cm^3) and 28% (2.0 g/cm^3) of the solid density of 7.215 g/cm^3, although the experiment yields slightly higher pressures than those calculated from the EOS for the more porous case. An important point is how significantly the porous material heats when shocked. As noted previously more porous samples will heat more readily with compression. For example, here the

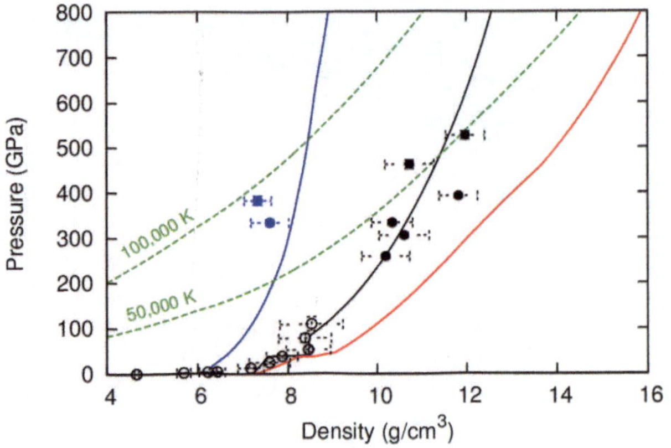

Fig. 7 Experimental and EOS Hugoniots results for CeO_2. The solid lines represent the EOS Hugoniots at initial densities 2.0 g/cm^3 (blue), 4.0 g/cm^3 (black), and 7.215 g/cm^3 (red). The closed symbols are Z-machine experiments and the open symbols are gas gun results [11, 59]. Also the dashed lines show the EOS 50,000 and 100,000 K isotherms

28% sample reaches 100,000 K at pressure less than 500 GPa, while the shocked solid doesn't reach that temperature until over 2400 GPa. This makes having the correct thermal modeling critical for systems when high levels of porosity are to be modeled correctly. Comparing Figs. 6 and 7 the 7.215 g/cm^3 Hugoniot curve can be seen to go through the solid–solid transition at just over 8 g/cm^3 and 35 GPa, and has a mixed phase region from 300 to 400 GPa before being completely melted at 13.5 g/cm^3.

The current EOS was constructed using the formalism of Sect. 3, including the cold curve of Eqs. (19) and (20), the Debye model for the solid phases and the liquid model of Johnson [42] for the ion thermal contribution, and the TFD approximation for the electron thermal contribution. It was also based on the earlier multiphase EOS of Chisolm [60], with the major difference being in the liquid regime, which is improved due to shock Hugoniot and QMD data [11]. The parameters of the EOS are given in Table 1.

4.2 *Silica Aerogel—SiO_2*

The ambient phase of SiO_2, α-quartz, has for some time been an important impedance match standard for shocks above 100 GPa. This is because above shock melt at about 5000 K and 100 GPa quartz becomes a conducting fluid. So the quartz Hugoniot itself has been extremely well studied and is precisely known. However, quartz has also become important as a shock standard in another way, and that is in the aerogel form. An aerogel is a very highly porous material with interconnected structure, similar to a foam but created by replacing the liquid in a gel with gas.

Table 1 Parameters of the CeO_2 EOS

Phase							
	Solid I	Solid II	Liquid		Solid I	Solid II	Liquid
	Cold curve				Liquid		
ρ_0 (g/cm^3)	7.305	8.12	5.18	$\theta_{\rm ref}$ (K)	625	625	–
B (GPa)	230	330	21.9	$\gamma_{\rm ref}$	1.83	1.0	0.768
dB/dP	4.0	4.25	4.9	γ'_R	−1.0	−1.0	−0.095
C_1	–	−295	2125[a]	γ'_L	−1.0	−1.0	−0.095
ρ_{LJ} (g/cm^3)	7.503	7.034	4.83	γ_0	5.0	1.0	0.0
f_{LJ}	1.0	1.5	1.0	γ_∞	0.667	0.667	0.863
$E_{\rm coh}$ (kcal/mol)	55	120	50	$T_{\rm melt}$ (K)	–	–	1561
ρ_{TF} (g/cm^3)	18.04	13.92	13.71	$\rho_{\rm melt}$ (g/cm^3)	–	–	6.5
$E_{\rm shift}$ (MJ/kg)	0.0	0.0	1.10				

[a]Additionally the liquid uses parameters $c_2 = -24{,}206$ and $c_3 = 88{,}876$

In the case of silica the aerogel can be used as impedance match standard when studying the shock release of materials [5].

An EOS was constructed for SiO_2 to encompass both the high density liquid regime relevant to the α-quartz shock as well as to extend down to the aerogel region. The same formal approach was used as for CeO_2, and the parameters are available in Refs. [61, 62]. This was done by performing a large set of quantum molecular dynamics calculations for liquid phase SiO_2 from 0.5 to 11 g/cm^3 and for temperature from 5000 to 1,000,000 K [61]. In this instance orbital-free DFT was used in order to extend the calculations above 100,000 K, where the Kohn–Sham DFT approach is intractable. As in the case of CeO_2 the calculations were then used to constrain the ion thermal and cold curve models used to generate the EOS table.

The shock Hugoniot results of the EOS are compared with experimental results for four different initial densities in Fig. 8. The four initial densities, from highest to lowest and top to bottom in the top panel of Fig. 8, are 4.29 g/cm^3 representing the high density stishovite crystal phase, 2.65 g/cm^3 representing α-quartz, 2.20 g/cm^3 representing fused silica, and 0.200 g/cm^3 representing silica aerogel. In all cases good agreement is found with experiment. In the case of α-quartz the Hugoniot is so precisely known through experiment that it is represent by a dashed line only.

SiO_2 is one of the rare cases where temperature measurements have been done along the Hugoniot. In fact they have been done for all four cases just listed. The EOS is in very good agreement with the temperature experiments for the α-quartz and the fused silica case; also for stishovite the agreement is good, but the EOS is just at the higher limit of the error bars for the experiment [62] as shown in the bottom panel of Fig. 8. For the aerogel at 200 mg/cm^3 (not shown), the experimental results are 10–20% lower temperature than the EOS or quantum molecular dynamics results. The differences have been examined previously [10] and it is believed that the pyrometry technique used may have not been sampling in the equilibrium shocked state. As noted in the introduction, while accurate experimental temperature

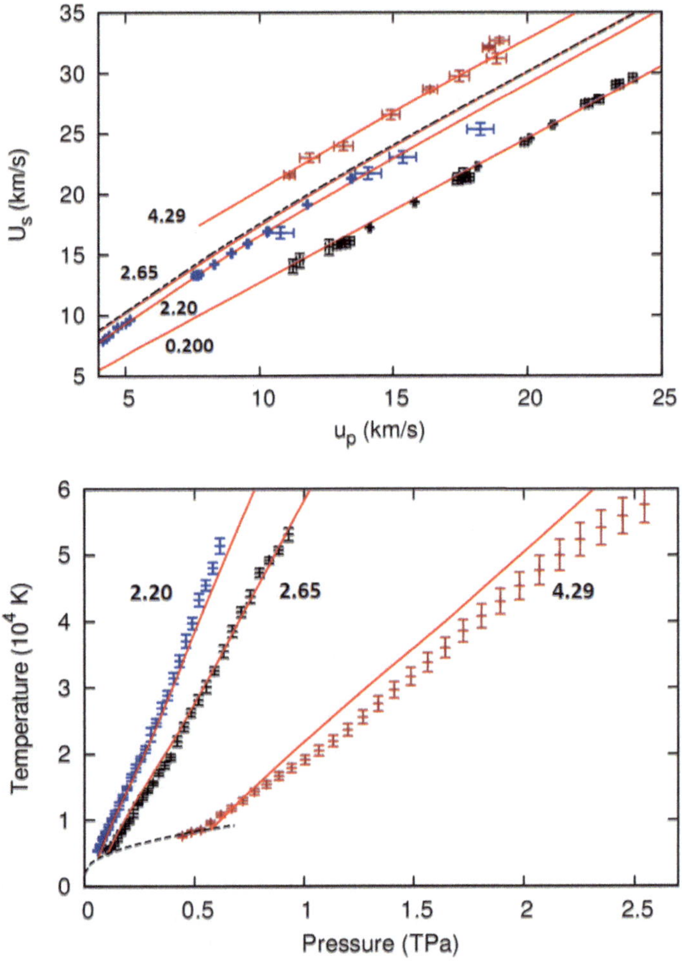

Fig. 8 Top: Shock Hugoniot experiments for SiO_2 compared with the EOS showing very good agreement. The initial densities from top to bottom are 4.29, 2.65, 2.20, and 0.200 g/cm^3. Only the last which is an aerogel is porous. Bottom: Shock experiment temperatures are shown with the EOS results for the 2.20, 2.65, and 4.29 g/cm^3 cases from left to right[10, 62–64]

results on the Hugoniot would be very useful for EOS construction they are quite difficult to obtain, especially in the WDM region.

4.3 Polystyrene Foam—CH

Lastly, attention turns to CH foam. CH itself has been well studied in shock experiments and in particular for its design as the primary ablator for inertial

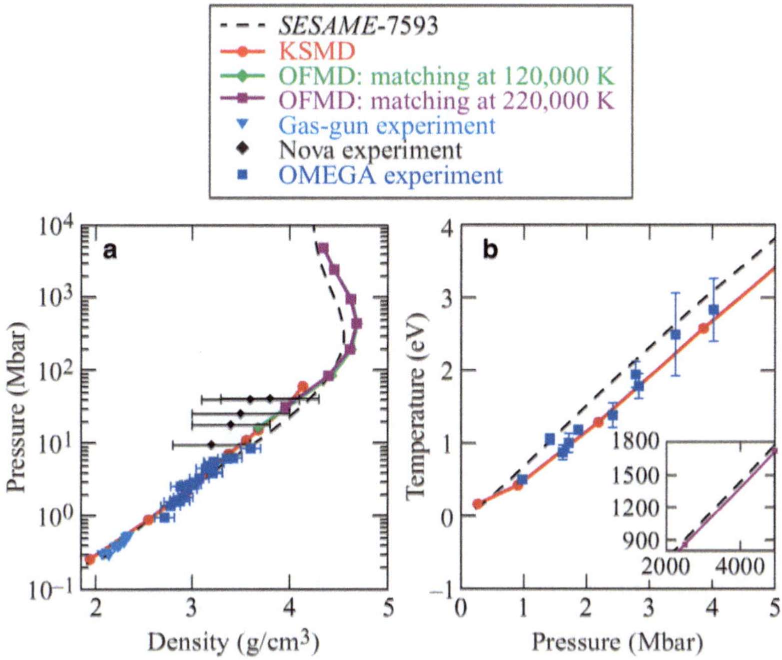

Fig. 9 Pressure (**a**) and temperature (**b**) results for the CH principle Hugoniot at the solid initial density of 1.04 g/cm^3 [65]

confinement fusion (ICF) experiments. In this application the CH undergoes wide variation in pressure and temperature, and an accurate EOS is needed for the simulation of the ICF experiment.

To this end a slightly different approach has been taken, which is to tabulate completely over the region of interest the EOS through quantum molecular dynamics simulations (FPEOS) [65]. This is different from the use of QMD data for CeO_2 and SiO_2, which is used to constrain analytic models, and instead generates the tabular EOS directly. This has been done for CH densities from 0.1 to 100 g/cm^3 and temperatures from 1000 to 4,000,000 K. In order to achieve results for such high temperatures, a compromise must be made, however. That approach is to switch from the standard Kohn–Sham density functional theory, to orbital-free density functional theory which is based on the Thomas–Fermi approximation, and can be applied at high temperatures. In fact, this approximation is correct in the high temperature limit. For the lightest element (hydrogen) these two methods can overlap regions of high accuracy, but even for carbon some interpolation is needed between the high and low temperature sets of calculations.

The results for the ambient solid density ($\rho_0 = 1.04$ g/cm^3) are shown in Fig. 9. The FPEOS results, which are the KSMD and OFMD, compare well with the experiments which have been completed up to very high pressures using laser experiments (Nova, Omega). Also compared is Sesame EOS 7593, which is based

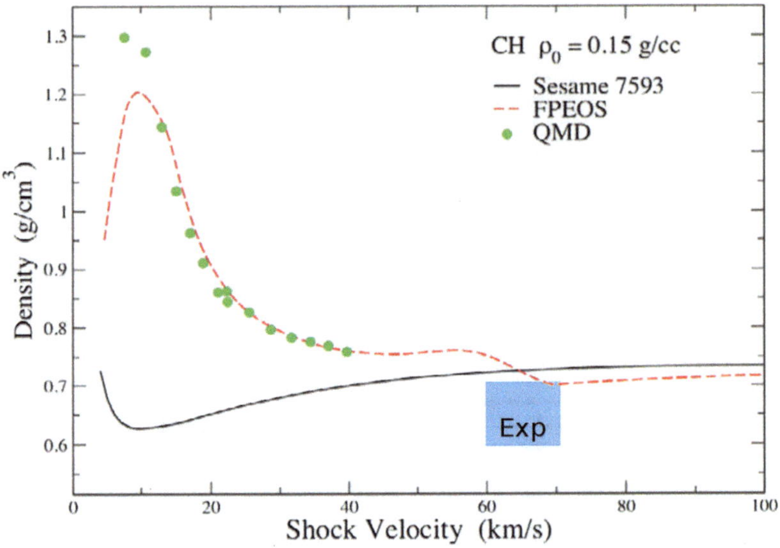

Fig. 10 Density vs. shock velocity results for the CH Hugoniot with initial foam density of 0.15 g/cm^3 [24, 65]. The Exp box indicates the range of experimental results from laser shocked foams of initial density 0.137 g/cm^3 [66]

on the modeling of Sect. 3. It can be seen that the FPEOS calculations predict CH being slightly stiffer in the pressure range of 5–80 Mbar, but softer in higher pressures up to ~5 Gbar. The FPEOS temperature seems to be in better agreement with the experimental results, both slightly lower than Sesame 7593.

With this validation the FPEOS has been applied to ICF simulations, where higher as well as lower density conditions than ρ_0 exist. One would like, though, experimental validation of the EOS at low densities. Recent experiments have looked at strong shocks of CH foams [66]. For example, a 0.137 g/cm^3 foam was shocked by a laser drive and achieved shock velocities of 60–70 km/s. Based on EOS predictions from FPEOS and Sesame 7593, the shocked density was about 0.6–0.7 g/cm^3. Lastly X-ray Thompson scattering diagnostics were used to find temperatures of about 26 eV. This experimental range of results is represented in the blue "Exp" box of Fig. 10. This is not far from the expected results shown by the EOS curves, though the experiment is a little ambiguous as hydrocode simulations indicate there may have been preheating which expanded the foam to a lower initial density. Although the FPEOS and Sesame mostly agree in the solid density case, Fig. 9, it is interesting for the shocked foam that they predict the opposite behavior for the weaker shock cases. The Sesame 7593 curve is in agreement with the earlier shock data [67] it was based on with shock velocities from 3 to 9 km/s (not shown). It is an open question as to whether that older shock data is flawed or the QMD is failing to capture all aspects of the shock physics. Also shown are independent QMD calculations which do agree with the FPEOS result.

5 Concluding Remarks

We have given a general description and methodology for generating equations of state. Though these methods involve mostly simple and well-known models, they can be highly effective. Difficulties encountered are often within the warm dense matter regime, where the models are in a region of interpolation. This is also the region where porous shock experiments provide data, but such experiments are very difficult. Further, at low pressure the coupling of compaction, material strength, and EOS for high porosity remains a very challenging area of research in terms of development and calibration of those material models. Given this, quantum molecular dynamics has proven to be a critical tool for providing constraining data for equations of state at extreme conditions.

Quantum molecular dynamics first appeared in the early 1990s and it has been in the last 10–20 years that the approach has proliferated with the increased access to high performance computing. However, extreme conditions remain a challenge due to the increased computational cost with temperature for the method. As such calculations at solid densities and below remain intractable above temperatures of about 10 eV. Even then a single density–temperature grid point may run on 1000+ cores on a high performance computer for a week or more. This means the expense of the calculations may limit how and when quantum molecular dynamics calculations are applied to EOS generation.

As we first noted, the EOS of any porous material will be identical to the EOS of the equivalent non-porous material. The standard models that are used to generate an EOS can effectively represent the dynamic loading of highly porous materials; however, that data is often not strongly constrained. In the future with more accurate experiments we can assess the theoretic calculations and EOS model forms that are currently used and continue their development to provide greater accuracy EOS.

Acknowledgements The author gratefully acknowledges informative and fruitful collaborations with Scott Crockett, Carl Greeff, Eric Chisolm, and Anthony Fredenburg. The author also thanks Tracy J. Vogler for reading of the manuscript and discussion. This work was performed under the auspices of the United States Department of Energy under contract DE-AC52-06NA25396.

References

1. Crockett S (2009) Fundamentals of SESAME equations of state. Los Alamos National Laboratory Tech. Reps. LA-UR 08-03679, LA-UR 08-03860, LA-UR 08-03681, LA-UR 09-00450, LA-UR 09-00451
2. Boettger JC (2013) Theoretical equations of state for porous/granular materials. Los Alamos National Laboratory Tech. Rep. LA-UR-13-24560
3. McQueen RG, Marsh SP, Taylor JW, Fritz JN, Carter WJ (1970) In: Kinslow R (ed) High-velocity impact phenomena. Academic, New York
4. Fredenburg DA, Koller DD, Rigg PA, Scharff RJ (2013) High-fidelity Hugoniot analysis of porous materials. Rev Sci Instrum 84:013903

5. Knudson MD, Lemke RW (2013) Shock response of low-density silica aerogel in the multi-Mbar regime. J Appl Phys 114:053510
6. Boettger JC (2010) Data requirements for a sesame EOS. Los Alamos National Laboratory Tech. Rep. LA-UR-10-08072
7. Basic Research Needs for High Energy Density Laboratory Physics, Report on the Workshop on High Energy Density Laboratory Physics Research Needs Nov. 15–18 2009 (2010) U.S. Department of Energy, Office of Science and National Nuclear Security Administration
8. Boettger JC, Honnell KG, Peterson JH, Greeff C, Crockett S (2012) Tabular equation of state for gold. In: AIP conference proceedings, vol 1426, p 812
9. Graziani F, Desjarlais MP, Redmer R, Trickey SB (2014) Frontiers and challenges in warm dense matter. Lecture notes in computational science and engineering, vol 96. Springer, Berlin
10. Falk K, McCoy CA, Fryer CL, Greeff CW, Hungerford AL, Montgomery DS, Schmidt DW, Sheppard DG, Williams JR, Boehly TR, Benage JF (2014) Temperature measurements of shocked silica aerogel foam. Phys Rev E 90:033107
11. Weck PF, Cochrane KR, Root S, Lane JMD, Shulenburger L, Carpenter JH, Sjostrom T, Mattsson TR, Vogler TJ (2018) Shock compression of strongly correlated oxides: a liquid-regime equation of state for cerium(IV) oxide. Phys Rev B 97:125106
12. Zemansky MW, Dittman RH (1997) Heat and thermodynamics, 7th edn. McGraw-Hill, New York
13. Anderson OL (1995) Equations of state for geophysics and ceramic science. Oxford University Press, Oxford
14. Zel'dovich YB, Raizer YP (2002) Physics of shock waves and high-temperature hydrodynamic phenomena. Dover, Mineola
15. Eliezer S, Ghatak A, Hora H (2002) Fundamentals of equation of state. World Scientific, Singapore
16. Wallace DC (2002) Statistical physics of crystals and liquids. World Scientific, Singapore
17. Godwal BK, Sikka SK, Chidambaram R (1983) Equation of state theories of condensed matter up to about 10 TPa. Phys Rep 102:121
18. Sengers JV, Kayser RF, Peters CJ, White HJ (2000) Equations of state for fluids and fluid mixtures. Experimental thermodynamics, vol 5. Elsevier, Amsterdam
19. Gathers GR (1994) Selected topics in shock wave physics and equation of state modeling. World Scientific, Singapore
20. Zharkov VN, Kalinin VA (1971) Equations of state for solids at high pressures and temperatures. Springer, New York
21. Brush SG (1967) In: Rouse CA (ed) Progress in high temperature physics and chemistry, vol 1. Pergamon Press, Oxford, p 1
22. Hakel P, Kilcrease DP (2004) Atomic processes in plasmas. In: Cohen J, Mazevet S, Kilcrease D (eds) AIP conference proceedings. AIP, New York, p 168
23. Pimentel DA (2016) EOSPAC user' manual: version 6.3. Los Alamos National Laboratory Tech. Rep. LA-UR-14-29289
24. Lyon SP, Johnson JD (1992) SESAME: The Los Alamos National Laboratory Equation of State Database. Los Alamos National Laboratory Tech. Rep. LA-UR-92-3407
25. Kerley GI (1991) User's manual for PANDA II: a computer code for calculating equations of state. Sandia Report SAND88-2291
26. Holzapfel WB (1998) Equations of state for solids under strong compression. Int J High Pressure Res 16:81
27. Birch F (1977) Isotherms of the rare gas solids. J Phys Chem Solids 38:175
28. Jeanloz R (1981) Finite-strain equation of state for high-pressure phases. Geophys Res Lett 8:1219
29. Knudson MD, Desjarlais MP (2009) Shock compression of Quartz to 1.6 TPa: redefining a pressure standard. Phys Rev Lett 103:225501
30. Kerley GI (1981) User's manual for PANDA: a computer code for calculating equations of state. Los Alamos National Laboratory Tech. Rep. LA-8833-M

31. Cowan RD, Ashkin J (1957) Extension of the Thomas-Fermi-Dirac statistical theory of the atom to finite temperatures. Phys Rev 105:144
32. Feynman RP, Metropolis N, Teller E (1949) Equations of state of elements based on the generalized Fermi-Thomas theory. Phys Rev 75:1561
33. Perrot F (1979) Gradient correction to the statistical electronic free energy at nonzero temperatures: application to equation-of-state calculations. Phys Rev A 20:586
34. Liberman DA (1979) Self-consistent field model for condensed matter. Phys Rev B 20:4981
35. Wilson B, Sonnad V, Sterne P, Isaacs W (2006) Purgatorio – a new implementation of the Inferno algorithm. J Quant Spectrosc Radiat Transf 99:658
36. Starrett CE (2015) A Green's function quantum average atom model. High Energy Density Phys 16:18
37. Giannozzi P, Baroni S, Bonini N, Calandra M, Car R, Cavazzoni C, Ceresoli D, Chiarotti GL, Cococcioni M, Dabo I, Dal Corso A, Fabris S, Fratesi G, de Gironcoli S, Gebauer R, Gerstmann U, Gougoussis C, Kokalj A, Lazzeri M, Martin-Samos L, Marzari N, Mauri F, Mazzarello R, Paolini S, Pasquarello A, Paulatto L, Sbraccia C, Scandolo S, Sclauzero G, Seitsonen AP, Smogunov A, Umari P, Wentzcovitch RM (2009) QUANTUM ESPRESSO: a modular and open-source software project for quantum simulations of materials. J Phys Condens Matter 21:395502
38. Blöchl PE (1994) Projector augmented-wave method. Phys Rev B 50:17953
39. Perdew JP, Zunger A (1981) Self-interaction correction to density-functional approximations for many-electron systems. Phys Rev B 23:5048
40. Wallace DC (1972) Thermodynamics of crystals. Wiley, New York
41. Cranfill CW, More R (1978) IONEOS: a fast, analytic, ion equation-of-state routine. Los Alamos National Laboratory Tech. Rep. LA-7313-MS
42. Johnson JD (1991) A generic model for the ionic contribution to the equation of state. High Pressure Res 6:277
43. Kerley GI (1980) Perturbation theory and the thermodynamic properties of fluids. II. The CRIS model. J Chem Phys 73:478
44. Chisolm E, Wallace DC (2004) Extending the CCW EOS II: extending the nuclear contribution to high temperatures. Los Alamos National Laboratory Tech. Rep. LA-UR-04-3948
45. Wallace DC (1997) Statistical mechanics of monatomic liquids. Phys Rev E 56:4179
46. Burnett SC, Sheppard DG, Honnell KG, Sjostrom T (2018) Sesame-style decomposition of KS-DFT molecular dynamics for direct interrogation of nuclear models. In: AIP conference proceedings
47. Crockett SD (2004) Analysis of Sesame 3720: a new aluminum equation of state. Los Alamos National Laboratory Tech. Rep. LA-UR-04-6442
48. Sjostrom T, Crockett S, Rudin S (2016) Multiphase aluminum equations of state via density functional theory. Phys Rev B 94:144101
49. Cox GA, Christie MA (2015) Fitting of a multiphase equation of state with swarm intelligence. J Phys Condens Matter 27:405201
50. Chisolm E, Greeff C, George D (2005) Constructing explicit multiphase equations of state with OpenSesame. Los Alamos National Laboratory Tech. Rep. LA-UR-05-9413
51. Mulford R, Swift DC (2009) Polystyrene foam product EOS as a function of porosity and fill gas. Los Alamos National Laboratory Tech. Rep. LA-UR-09-05358
52. Abdallah J (1984) User's manual for Grizzly. Los Alamos National Laboratory Tech. Rep. LA-10244-M
53. Magyar RJ, Root S, Mattsson TR (2014) Equations of state for mixtures: results from density-functional (DFT) simulations compared to high accuracy validation experiments on Z. J Phys Conf Ser 500:162004
54. Bradley PA, Loomis EN, Merritt EC, Guzik JA, Denne PH, Clark TT (2018) Experimental validation of thermodynamic mixture rules at extreme pressures and densities. Phys Plasmas 25:012710
55. Duval GE, Taylor SM (1971) Shock parameters in a two component mixture. J Compos Mater 5:130

56. Nayak B, Menon SVG (2017) Generalized enthalpy based equation of state for multi-component mixtures. AIP Adv 7:045113
57. Fredenburg DA (2010) Shock compaction and impact response of thermite powder mixtures. Thesis, Georgia Institute of Technology
58. Petel OE, Jett/'e FX (2010) Comparison of methods for calculating the shock Hugoniot of mixtures. Shock Waves 20:73
59. Fredenburg DA, Koller DD, Coe JD, Kiyanda CB (2014) The influence of morphology on the low- and high-strain-rate compaction response of CeO2 powders. J Appl Phys 115:123511
60. Chisolm E (2014) A multiphase equation of state for cerium (IV) oxide. J Phys Conf Ser 500:032004
61. Sjostrom T, Crockett S (2015) Orbital-free extension to Kohn-Sham density functional theory equation of state calculations: application to silicon dioxide. Phys Rev B 92:115104
62. Sjostrom T, Crockett S (2017) A liquid regime equation of state for silicon dioxide. AIP Conf Proc 1793:050010
63. Millot M, Dubrovinskaia N, Cernok A, Blaha S, Dubrovinsky L, Braun DG, Celliers PM, Collins GW, Eggert JH, Jeanloz R (2015) Shock compression of stishovite and melting of silica at planetary interior conditions. Science 23:418
64. Hicks DG, Boehly TR, Eggert JH, Miller JE, Celliers PM, Collins GW (2006) Dissociation of liquid silica at high pressures and temperatures. Phys Rev Lett 97:025502
65. Hu SX, Collins LA, Goncharov VN, Kress JD, McCrory RL, Skupsky S (2015) First-principles equation of state of polystyrene and its effect on inertial confinement fusion implosions. Phys Rev E 92:043104
66. Falk K, Fryer CL, Gamboa EJ, Greeff CW, Johns HM, Schmidt DW, Smid M, Benage JF, Montgomery DS (2017) X-ray Thomson scattering measurement of temperature in warm dense carbon. Plasma Phys Control Fusion 59:014050
67. Marsh SP (1980) LASL shock Hugoniot data. University of California Press, Berkely

Low-Pressure Dynamic Compression Response of Porous Materials

D. Anthony Fredenburg and Tracy J. Vogler

1 Introduction

The removal of porosity at high rates of strain causes unique phenomena to occur during the dynamic loading of porous materials with respect to their solid counterparts. Most notably, the removal of porosity can generate very high temperatures that can result in the onset of melting or other solid–solid phase transitions to occur at impact conditions much lower than those required for initially solid materials [1, 2]. Under these strong impact conditions, the material response is largely controlled by the equation of state, details of which are discussed in the chapter "Equation of State Modeling for Porous Materials" by Sjostrom. At more modest impact conditions where stresses in the porous material are of similar order as the strength of the underlying solid, the processes involved in the removal of porosity become critical, and both intrinsic and extrinsic properties of the starting material can have an effect on the dynamic response.

The mechanisms involved with the removal of porosity depend on the material, stress level, and strain rate. Nesterenko [3] divided the loading experienced by porous materials into quasistatic and dynamic regimes. In the quasistatic regime changes in the underlying structure of the material are relatively small for metals, while for higher strength brittle materials the structure may undergo significant fracture and fragmentation. Further, under quasistatic conditions the time scales of loading are long enough that particle rearrangement can occur, resulting in the

D. A. Fredenburg
Los Alamos National Laboratory, Los Alamos, NM, USA
e-mail: dafreden@lanl.gov

T. J. Vogler (✉)
Sandia National Laboratories, Livermore, CA, USA
e-mail: tjvogle@sandia.gov

T. J. Vogler, D. A. Fredenburg (eds.), *Shock Phenomena in Granular and Porous Materials*, Shock Wave and High Pressure Phenomena,
https://doi.org/10.1007/978-3-030-23002-9_2

filling of voids with relatively little deformation of the particles. In the dynamic regime, the time scale of loading is much shorter such that both low and high strength materials undergo "substantial morphological" changes from their initial state. Because particle rearrangement does not occur, the removal of porosity must be accommodated solely by deformation and/or fracture of the underlying material. Under these conditions features such as material jets, vortices, and localized melting will occur; these are the characteristic features of porous materials shocked in the dynamic regime. A common feature of shock-loaded porous materials is nonuniformity of the material. At pressures where porosity persists, contact points between grains can experience much higher stresses than the bulk, while at higher pressures jetting and related phenomena can cause small regions of much hotter material.

Developing an improved understanding and ability to model the dynamic response of porous materials shocked to moderate stresses is critical to many applications. For example, explosive and impact loading is a means by which initially porous materials can be consolidated to nearly fully dense bodies while preserving specific features of the original microstructure not possible by more conventional high-temperature, high-pressure sintering methods [4]. For mixtures of materials specifically designed to undergo exothermic reactions under shock loading (e.g., explosives, thermites), the extent to which energy is consumed and distributed among the component materials during compaction is key to determining the onset of the reaction threshold (see chapter "Applications of Reactive Materials in Munitions" by Peiris and Bolden-Frazier). Further, even under the extreme impact conditions imparted by very high energy inputs such as shaped charges, buried explosives, and hypervelocity planetary impacts (see chapter "Planetary Impact Processes in Porous Materials" by Collins et al.), the evolution of wave propagation over finite distances dictates that regions exist some distance away from the primary impact zone that will experience moderate dynamic stresses and, thus, must be characterized if a full system modeling approach is required.

The measurement and interpretation of experimental results and the subsequent calibration and validation of models at moderate dynamic stresses presents a unique set of challenges. Specifically, at these stresses the measured experimental response includes contributions from the removal of porosity, the deformation and/or fracture of the constituent material, as well as the equation of state. In addition, heterogeneity in the underlying material brings the standard assumptions of equilibrium behind the shock front into question. It is with these and other challenges in mind that the present chapter is framed. Presented first is an overview of the experimental techniques, diagnostics, and resultant uncertainties that are typical when studying the dynamic material response at moderate stresses. Next, the principle approaches for modeling the behavior of porous materials at the continuum and mesoscale levels are given. Having covered the fundamentals of experimental and modeling techniques, the chapter continues with a discussion of several aspects of the dynamic response of porous materials that are observed specifically in the moderate stress regime. To conclude, the authors provide their thoughts regarding several topics that remain outstanding, such that progress in these areas could significantly advance the state of the art in our understanding and ability to interpret and model the response of initially porous materials under moderate shock loading.

2 Experimental Techniques Applied to Compaction

A number of different dynamic loading platforms and diagnostic configurations have been utilized over the years to measure the response of porous materials at moderate stresses. While laser and pulsed power platforms have been added to more traditional gas guns and explosive drives, the diagnostics used to measure the shock consolidation response have also evolved. This section focuses on modern experimental techniques for obtaining the shock Hugoniot of porous materials in the compaction regime. First, details of the canonical experiment used to measure the shock Hugoniot response are given. Next, details relevant to preparing and accurately characterizing competent, loose, and pressed porous bodies are presented. This is followed by a discussion of the instrumentation used to measure characteristics of the shock state. Uncertainties associated with the experimental and diagnostic techniques are presented next. Finally, this section concludes with a brief look at alternate experimental techniques for characterizing the dynamic response of porous materials under more complex loading conditions.

2.1 The Canonical Experiment

Although specific details vary, most planar shock experiments on porous materials take the form of the canonical experiment sketched in Fig. 1. Input is provided by a flyer plate launched by a gas gun or by other means such as an explosive or magnetic drive. The flyer impacts the target at a known velocity, driving a shock wave directly into the sample material or first into a driver layer, as shown in Fig. 1. In most instances the flyer plate and driver plate (if used) are composed of a known material whose shock response is well characterized. The experiment is further diagnosed with time of arrival measurements (gauges, pins, velocity interferometry, etc.) on the planes corresponding to the impact and rear surfaces of the sample to measure the transit time of the shock through the material. In addition, a backing window is often used on both competent and loose samples to facilitate velocity interferometry and to help contain the material during the experiment. For highly competent porous samples a reverse ballistic impact configuration may be used, where the target material is mounted into the projectile and impacted into a stationary window or witness material. However, given the relatively high accelerations imparted to projectiles during launch, care must be taken in reverse ballistic configurations to

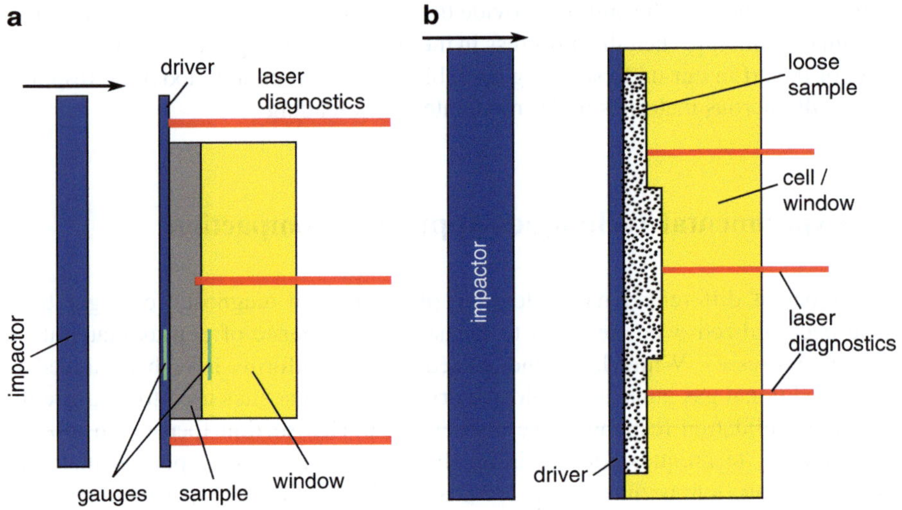

Fig. 1 Schematic of the canonical experiment for (**a**) competent porous solids and (**b**) loose or tap density powder samples

ensure that projectile launch does not crush or otherwise modify the initial state of the sample material.

Target designs for loose or pressed granular samples must incorporate a cell or other containment vessel to ensure the sample dimensions can be measured and maintained from target loading to execution of the experiment. An example of a containment geometry designed specifically for loose or tap density granular samples is illustrated in Fig. 1b, where the granular material is loaded into the two-step target through single or multiple fill-holes. Studies involving both two- [5] and five-step [6] targets have been performed. In the design shown in Fig. 1b, the cell serves as both a containment vessel and a transparent window for velocity interferometry. Alternate designs consisting of samples at a single thickness (one-step) are also used and are well suited for granular samples that require some amount of pre-compression. In these instances the containment vessel, or cell, can be integral to or separate from the window material. Both the single- and multi-step target configurations each come with their own set of advantages and limitations, such that a combination of these two designs is likely required to fully characterize the range of initial conditions achievable for a given material. In either design, the primary in situ measurement is time of arrival at the impact and rear surfaces, similar to the measurements performed for competent target materials.

2.2 Samples

Similar to high-precision dynamic studies on solid bodies, initial characterization and sample preparation are critical for obtaining the highest fidelity measurements on porous materials. To this end, the most important characteristics to be considered are accurate metrology, homogeneity within a given sample, and consistency between samples for multiple measurements along the Hugoniot. For competent porous bodies where target samples can be fabricated specifically or removed from a larger body, such as foams, additively manufactured trusses, and other porous solids, relatively straightforward machining methods can be used to obtain the desired sample size and shape. Furthermore, the ability to move, rotate, and inspect all surfaces of a competent body facilitates full metrological characterization. For the canonical experiment discussed in Sect. 2.1, accurate characterization of the sample dimension in the direction of shock propagation is the most important, where tolerances similar to solids can be achieved for competent porous bodies.

Loose or pressed granular materials present additional challenges in initial sample preparation and characterization. For a shock experiment, the powder is typically contained in a cell made of plastic or metal, such as PMMA or aluminum. Metrology is first performed on the cell components, such that the empty cavity volume is accurately characterized prior to powder filling. Filling the fixed volume with a known mass of powder can be done by simply pouring in the powdered material or by using a combination of pouring and agitation/tapping for a more uniform sample. Such an approach is generally acceptable for particles that are 10 s–100 s of microns in size and have relatively simple geometries. As particle sizes are reduced to less than 1 μm, electrostatic forces between the particles can cause clustering and agglomeration between the particles, leading to a decrease in the flowability and sample uniformity. Spherical particles tend to have improved flowability during agitation/tapping, but block-like particles have also been found to work well [5–8]. Packing fractions for tapped samples are typically in the range of 20–50%, but higher values can be achieved using specific particle size distributions, e.g., bi-modal or tri-modal [9]. Unfortunately, while there are a number of techniques for testing the flowability of powders for industrial scales, there is not an equivalent approach for quantitatively characterizing similar processes for the small amounts of powder typically associated with shock experiments.

An alternate approach to pouring and tapping is to press the powder to a higher density, allowing for a much broader range of achievable densities. Depending upon the powder morphology and strength, initial densities from just above the pour density to those approaching the crystalline density can be achieved. The upper limit on densities achieved by this method is typically governed by the size of the available press or the need to avoid deforming the cell hardware in which the powder is pressed. In some target designs, it may be necessary to press into a cell containing a transparent window for velocity interferometry (discussed in the next section). In these instances only high strength window materials, such as single crystal sapphire or quartz, can be used during pressing. Other common window

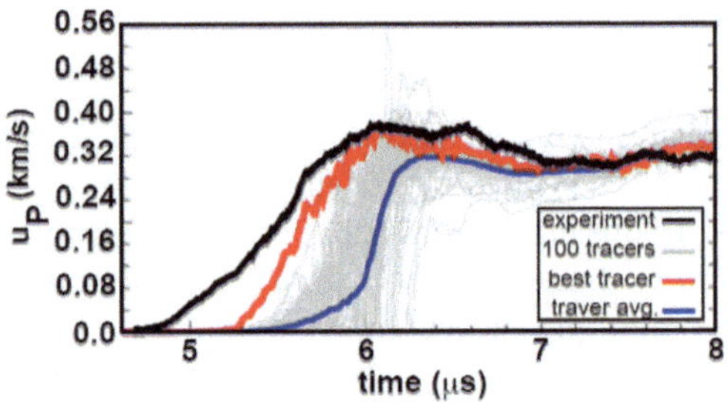

Fig. 2 Propagated material velocity profile measured using VISAR at the window interface for a sand sample impacted at 0.413 mm/μs, compared against simulated wave profiles from 100 tracer particles. The "best tracer" is the one that provides the best agreement with the experimental data. Image adapted from [13]

materials, such as lithium fluoride (LiF) and PMMA, cannot be subjected to the high stresses during pressing as LiF is prone to fracture under loading and PMMA will deform plastically. If these latter two windows are to be used in a target configuration requiring pressing, they must be inserted into the target after pressing is complete. Following completion of the pressing process, an accurate measurement of the sample volume must be obtained. If the compact density is high enough, metrology can be performed directly on the pressed surface and target [10]. However, for lower initial density samples or for relatively ductile materials it may be necessary to use a buffer material of known thickness between the powder surface and the tip of the metrology probe to ensure that the probe itself does not modify the target surface.

Sample homogeneity is also important for accurate determination of the porous compaction Hugoniot. If heterogeneities in the target sample are measured using computed tomography or other through-sample imaging techniques, they must be accounted for in any analysis of the experiment, or mitigated against by ensuring experimental diagnostics are located away from the heterogeneities. Further consideration should also be given to the spatial dimensions of the underlying pore or solid component with respect to those of the target and diagnostic, such that the target geometry should be large with respect to underlying sample features, and the features should be small in comparison with any diagnostic probe. Regarding target size, mesoscale simulations investigating the propagation of shock waves for different particle configurations have shown evolution in wave propagation characteristics near the impact face as well as nonuniformity in the shock front of order several particles thick [11, 12]. As such, sample sizes should be maximized in the direction of shock propagation to minimize the effects of heterogeneities in the underlying features and for steady shock waves to develop. Furthermore, particle-level simulations on dry sand have shown that significant deviations from the measured bulk wave profile can exist if the spot size of the diagnostic probe is comparable to or smaller than the underlying features of the porous material [13]. This effect is shown in Fig. 2, where simulated diagnostic probes in the form of tracer particles shown by the light gray lines measure a wide variety of propagated wave profiles, while the tracer average response possesses a shape characteristic of the bulk experimental measurement, but significantly offset in time. If sample or diagnostic probe sizes are such that the underlying microstructural features are expected to influence the experimental measurements, direct numerical simulation approaches should be used to better interpret any measured result.

At a spatial scale larger than that of an individual grain or pore, homogeneity of the sample density within and across the target is also an important consideration. Sample preparation techniques discussed above can only provide measurements of the bulk geometric density, and, depending on the characteristics of the porous material or loading process, significant heterogeneities in local densities can exist. For example, agitation of a loose granular material with a bi-modal particle size distribution can lead to the well-known Brazil nut effect, in which larger particles accumulate at the top and smaller particles accumulate at the bottom, leaving an inhomogeneous void distribution across the sample thickness [14]. This can be especially pronounced for a mixture of two distinct particle species of differing densities. More complicated non-uniformities can also arise in loose and pressed samples such as faults, voids, and tunnels. In instances where limited diagnostic probes are used to measure the shock velocity through a porous sample, the presence of these features can cause difficulties in interpreting the measurements, as local density heterogeneities near the diagnostic probe may result in transit times not representative of the bulk. In an effort to characterize features of the initial microstructure and understand how their presence affects shockwave propagation, scientists have begun using radiography and computed tomography to image the initial sample heterogeneities and link those heterogeneities to features of the measured wave profiles. However, this is not yet a routine aspect of experiments, in part because the cell containing the powder can sometimes negatively interfere with x-ray penetration.

Aspect ratios of the porous targets are also important. While these considerations are not unique to porous samples, the relatively low wave speeds often found in porous samples requires that particular care must be taken when designing experiments. For experiments where the porous material is surrounded by a cell, window, or baseplate of higher impedance, the sample diameter-to-thickness ratio must be sufficiently large that edge waves from the surrounding cell or Rayleigh waves at the rear powder/window interface do not pollute the shock diagnostic prior to the primary measurement. As a rule of thumb, a diameter-to-thickness ratio of 10:1 is desirable; however, even this may not be sufficient for highly distended samples. If some information is known about the shock response of the material a priori, computational simulations can be used to better understand the limiting aspect ratio for a particular experimental configuration.

The final sample consideration is atmosphere. To the best of our knowledge, there have been no detailed studies on the role of atmosphere in the low-pressure shock response of porous materials. Elliot and Staudhammer [15] examined the role of the gas in the porous sample on consolidation of stainless steel powders and found that increasing it from 1 or 100 atmospheres (10^5 to 10^7 Pa) inhibited consolidation of the part somewhat. Indeed, in shock consolidation one typically evacuates the sample, but there is no general consensus on the evacuation of the target material in an experiment intended to measure material response. For a material with open porosity, the sample can be evacuated simultaneously with the target chamber by means of a vent or other opening. For closed porosity samples or those whose pores are filled with a specific gas or liquid, the sample must be robustly sealed to avoid spurious interactions from the surrounding evacuated target chamber.

2.3 Instrumentation

Investigations focusing on the shock compaction region of initially porous materials typically occur at impact velocities up to 1–2 km/s and stresses less than several 10s of GPa. As such, some of the instrumentation required to diagnose these experiments have unique considerations with respect to similar experiments conducted at higher impact velocities and stresses. In the canonical experiment (see Fig. 1), impact velocities less than ~1 km/s can be measured quite accurately using a series of shorting pins or laser interrupts. As impact velocity is increased, uncertainties associated with these methods increase, such that optical velocimetry measurements of the accelerated projectile offer more accurate measurements of the impact velocity.

Characterization of the impactor tilt in porous experiments is typically obtained using shorting pins to measure the angle of the impacting projectile directly or optical velocimetry to measure the difference in jump-off times at the rear surface of the drive plate, as shown in Fig. 1a. Assuming planarity of the impactor, which can usually be readily obtained by the use of relatively thick impactors at the modest velocities associated with compaction, measurements of the impactor tilt allow for arrival times of the shock entering the target to be determined at the spatial locations corresponding to the propagated velocimetry probes, a key component for obtaining accurate shock velocities. For a more in-depth discussion of the methods used to characterize impact velocity, tilt, and bow for porous materials, see Fredenburg et al. [10].

Following impact, the Hugoniot state in a one-dimensional loading experiment is most commonly determined by measuring the shock velocity, U_S, or and longitudinal stress, σ. Combined with an inferred material velocity, u_P, from impedance matching [16] and a measured initial porous density, ρ_{00}, the conservation of momentum across a shock discontinuity, $\sigma = \rho_{00} U_S u_P$ (for a material initially at rest), allows one to solve for the remaining unknown. The full thermodynamic state of the material is further determined by applying the conservation equations for mass and energy. Therefore, accuracy for a calculated Hugoniot state depends largely on how well one can characterize the initial density, material velocity, and shock velocity or stress for a given experiment. Modern investigations for determining the shock velocity and stress in porous bodies largely employ optical velocimetry and gauge techniques, respectively. Of the gauge techniques, manganin [17] and polyvinylidene fluoride [18] gauges are the most prevalent, while the VISAR [19] and heterodyne/photon Doppler velocimetry (PDV) [20] methods are the most common velocimetry techniques.

A schematic illustrating the gauge and velocimetry techniques, along with representative time-resolved outputs for each is given in Fig. 3. One can see that the two techniques differ in the spatial location of their probes. In gauged experiments the primary diagnostic is the impact gauge, which must further be contained within a protective layer, or buffer, typically of similar shock impedance to the gauge itself. The protective layer shields the gauge from the high temperatures associated with void collapse and the heterogeneous nature of the compaction shock (see Eakins and Thadhani [21]) that may otherwise cause degradation or failure of the gauge. The impact gauge package configuration permits direct measurements of the shock stress in the gauge prior to impacting the sample, σ_{gi}, as well as the stress in the sample as it is reflected back into the impact gauge, σ_{sH}. Placement of a second gauge behind the sample further allows for a direct measurement of the stress associated with the second shock or release, σ_{sr}. Shock transit times through the porous target, Δt_S, can also be measured using the combined impact/rear gauge package method; however, calculation of shock velocities using this technique must take into account the thickness of the gauge package and protective layers. With gauge package and buffer thicknesses on the order of several millimeters, shock velocities in agreement with impedance matched values have been demonstrated for silica powders of varying initial density after accounting for the gauge package thickness [22], while for much thinner gauge packages (10s of microns) agreement has been demonstrated without correction for a system of nano-sized iron powders by [23].

In contrast to gauged experiments, those diagnosed using optical velocimetry directly measure the shock transit time through the sample. Depending on the geometry of the target, velocimetry measurements at the impact plane can provide full characterization of the input shockwave profile, measurements of shock arrival time only, or some combination of the two. For example, if the containment shown in Fig. 3 is composed of an optically transparent material, the velocity history at the impactor/containment interface can provide a direct measurement of the input shockwave profile. However, in target configurations where the containment (see Fig. 3) is not optically transparent, velocimetry at the impact plane is typically a free surface velocity measurement of the impactor, which may or may not contain information other than shock jump-off times. Evaluating the velocity history entering the porous material is important at the low-to-moderate stresses relevant to compaction because many of the standards used to construct the experimental target fixtures exhibit non-steady wave characteristics in this regime. For example, powder targets pressed in situ require an impactor with moderate strength to support the loads applied during the pressing operation. Thus, impactors in this configuration may be composed of metals such as aluminum or copper which can exhibit two wave structures under shock loading due to elastic–plastic transitions [24], or iron containing steels which can exhibit similar wave structures due to phase

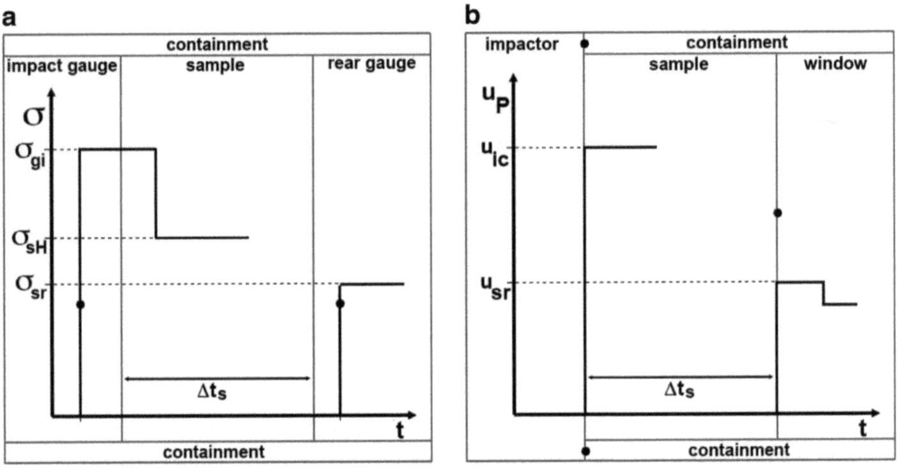

Fig. 3 Schematic illustrating the (gray) experimental configurations for (**a**) gauge and (**b**) optical velocimetry diagnosed experiments. Representative wave profiles taken at the spatial position indicated with • are overlain to provide perspective on differences in timing between the two techniques. In this configuration impact occurs from left to right

transformations [25]. Therefore, care must be taken in the experimental design to ensure that a planar steady wave is imparted into the target material when applying the Rankine–Hugoniot jump conditions.

Velocimetry measured at the rear, or propagated, surface of the target sample provides a measure of the shock transit time as well as insight into the physical mechanisms of compaction. For example, the amount of temporal dispersion in a propagated velocity history can be linked with the dissipative nature of the material being shocked, as shown in Fig. 4a. In this comparison sand and CeO_2 powders at similar initial percentages of their theoretical maximum densities (% TMD) are shock loaded and produce very different propagated wave profiles. For sand, the shock rise time spans several hundred nanoseconds with significant curvature at both low and high material velocities, while that for CeO_2 exhibits a sharp rise over several nanoseconds followed by curvature as the velocity approaches its maximum. For the two experiments in Fig. 4a, differences in the theoretical density, particle size, and Hugoniot stress (sand/CeO_2; ρ_0 = 2.6/7.2 g/cm^3, Φ = 200/7 µm, σ_H = 0.8/1.8 GPa) make it difficult to link specific features of the initial and shocked states to the dispersion measured in the wave profiles. However, regardless of its origin, this dispersion must be accounted for when performing an analysis of the Hugoniot state and its corresponding uncertainties. See Sect. 4.2 for further discussion of the rise time for shocks in porous materials.

When using velocimetry to measure propagated wave profiles, some researchers [5, 6, 26] have used buffers between the measurement plane and the granular sample. This is done to protect the reflective surface, particularly important for VISAR measurements, and to "average out" the velocities so that a single velocity

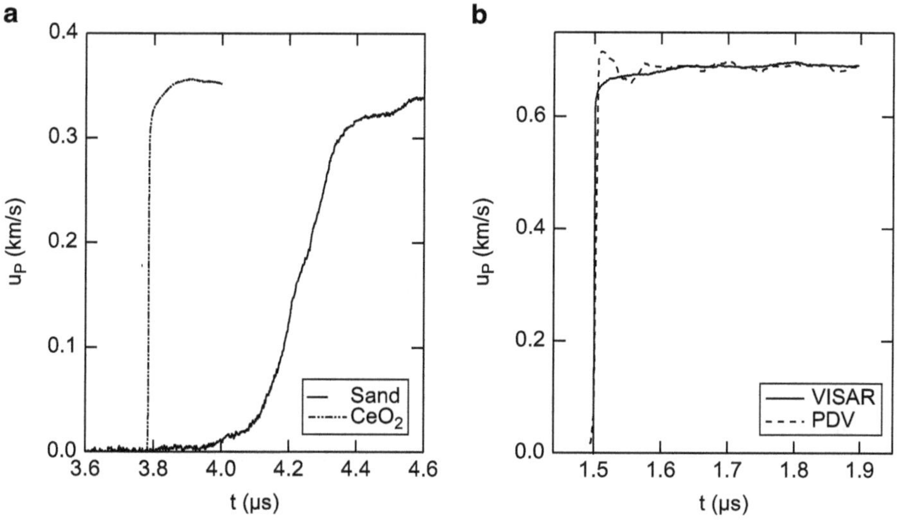

Fig. 4 (**a**) VISAR transmitted wave profiles for ρ_{00} = 59% TMD sand and ρ_{00} = 55% TMD CeO_2 shocked to ρ = 92% and 79% TMD, respectively, illustrating the variation in shock rise times. (**b**) Transmitted wave profiles measured from a PDV line-out at the powder rear surface and a VISAR at the buffer/window interface on CeO_2 powder; traces are shifted in time for comparison. Data for sand from Brown et al. [5] and from Fredenburg et al. [26] for CeO_2

history is obtained. An extreme case of this is that of Winter and McShane (this volume), who used a 6 mm stainless steel witness plate as a buffer to obtain PDV measurements of additively manufactured truss structures with pores of order millimeters in size. In a study on CeO_2 powders, Fredenburg et al. [10] measured propagated wave profiles after traveling through an 8 μm Al foil in contact with the powder using PDV and after traveling through the foil and a 0.5 mm PMMA buffer using VISAR. Their results are shown in Fig. 4b, where both profiles are observed to display similar characteristics in the initial rise and late time velocity plateau regions. However, in the transition region into the shock plateau, the two methods give different responses, with PDV at the powder/foil interface exhibiting an oscillatory behavior characteristic of its ability to measure multiple velocity fields, and VISAR at the buffer/window interface exhibiting a smooth transition characteristic of PMMA [27].

With sample heterogeneity (material + void) underlying all experiments on porous materials, experimental efforts in recent years have been undertaken to better understand this heterogeneity using spatially resolved velocimetry as well as advanced X-ray sources. Trott et al. [8] used line-VISAR to study the spatially resolved shock response of pressed sugar, examining heterogeneities in transmitted wave profiles with respect to impact velocity, sample thickness, and particle size distribution. Using similar methods, Baer and Trott [28] investigated the response of ordered stackings of tin spheres. In more recent investigations, synchrotron X-ray sources are being used to probe the heterogeneous nature of shock propagation in porous materials in situ (see chapter "X-Ray Phase Contrast Imaging of Granular Systems" by Jensen et al.). Experimental techniques such as these, when coupled with complementary mesoscale modeling techniques, offer great promise toward increasing our physical understanding of the particle-level mechanisms active during compaction and further interpretation of the bulk measurements that have been common practice for the past several decades.

2.4 *Uncertainty and Error Analysis*

In the previous section, much of the discussion of instrumentation was focused on velocimetry techniques for measuring the transit time of a shock through a known initial thickness sample. While knowledge of the shock transit time is useful, the true value of a shock compaction measurement is in being able to determine how the material responds in the stress-density plane, i.e., how much porosity is removed and densification of the bulk occurs for a given applied stress. Further, a good measurement also strives to have low uncertainties in the calculated quantities of stress and density ($\rho = 1/V$), such that this information may be used to accurately calibrate and/or validate a continuum compaction model (see the next section for more discussion of compaction models).

A transit time measurement can be converted to a point on the stress-density plane using the conservation equations, and it is in these relations where uncertainties are captured. For stress:

$$\sigma = \rho_0 U_S u_P \tag{1}$$

such that uncertainties in the initial density, shock velocity, and material velocity contribute proportionally to the total uncertainty in stress. Therefore, a skilled experimentalist will work to design a target fixture and diagnostic suite that reduces the individual uncertainties in ρ_0, U_S, and u_P to keep the uncertainty in stress low. However, the functional relationship for density

$$\rho = \rho_0 \frac{U_S}{U_S - u_P} \tag{2}$$

introduces larger the fact that values for the shock and material velocities can be quite similar in the low stress compaction regime. While a requisite for the development of a shock wave is that $U_S > u_P$, if U_S is only slightly larger than u_P, then the denominator of Eq. (2) is small, such that even small uncertainties in either the shock or material velocities can result in relatively substantial uncertainties in density. This is a feature unique to initially porous materials at low pressures, and it imposes real limitations in accurately characterizing the shock compressed density.

While the conservation relations serve as the underlying basis for all uncertainty analyses, several specific methods have been applied to porous materials and are discussed further here. A key work on the analysis of uncertainties for shock experiments performed on solid materials is that of Mitchell and Nellis [29]. They considered both random uncorrelated uncertainties and systematic uncertainties to estimate the total uncertainties in both the measured and calculated Hugoniot state values. This approach was applied to porous granular materials by Fredenburg et al. [10], who noted that the assumption of sample homogeneity in the initial density state is likely not as strictly enforced for porous materials as it is for solids. As such, the authors acknowledged the presence of inhomogeneities in granular samples and suggested tomography as a potential means to characterize sample heterogeneities prior to an experiment. However, in the absence of tomography they advocate that multiple measurements of the shock velocity for a given experiment may be able to account for some of the sample heterogeneities in the analysis of experiments.

Recently, Root et al. [30] utilized a forward Monte Carlo approach for calculating uncertainties in the high-pressure shocked states of hydrocarbon foams when impedance matching with a known standard is utilized. In this work, the authors too note the presence of sample heterogeneities, but in this case are focused on local heterogeneities at the scale of the grain/pore. For the foams investigated, which had pore sizes of order 1 μm, they argue that if the transit time of the shock across a pore is less than the temporal resolution of the diagnostic (in this case VISAR), then the shock can be treated as an equilibrium state and the Rankine–Hugoniot jump conditions can be applied. While the Monte Carlo approach has not yet been utilized in published reports on the low-pressure shock response of porous materials, the authors are currently working to extend this approach to that regime.

While both analysis methods are useful in ascribing uncertainties to a single experimental data point, an alternate approach to capturing the uncertainty in the compaction response for a given initial density state is to perform a statistical number of experiments at the same initial conditions of density and impact velocity and evaluate the scatter observed between the points. While the authors know of no studies that have performed of order ten or more repetitions of a single experiment, more limited studies have found that the spread between the Hugoniot states of multiple experiments performed under nominally identical conditions is similar to the uncertainty bounds assigned to any given data point. An example of this behavior is shown in Fig. 5, which shows the spread in the calculated Hugoniot volumes for experiments on sand and copper under nominally similar impact conditions. Inspection of these data sets reinforce the observation that underlying heterogeneities at one or more length scales in porous materials are the principal driver for uncertainties in the Hugoniot state. For a more in-depth study of the role heterogeneities may play in the shock compression response of initially porous materials, one may also look at alternate loading techniques.

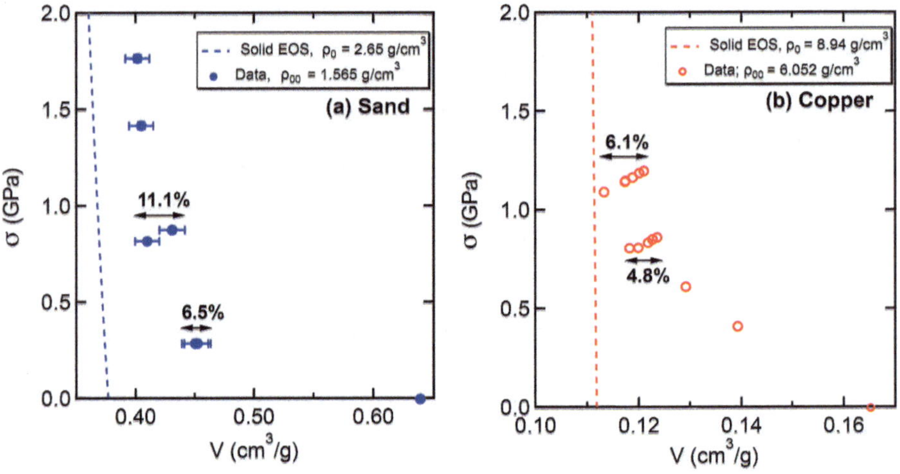

Fig. 5 Compaction data for (**a**) sand [5] illustrating uncertainties associated with individual data points and (**b**) copper [31] illustrating the spread in Hugoniot states for similar initial condition experiments. Percentages are with respect to the ambient solid specific volume

2.5 Non-planar Loading and Alternate Experimental Platforms

The canonical experiment discussed in Sect. 2.1 is used to measure the response of porous materials under the idealized conditions of one-dimensional planar shock loading. As such, application of the conservation equations to these types of experiments yields only the components of stress and strain in the direction of shock propagation. Given the complex nature of shock waves in real-world applications (see chapters "Planetary Impact Processes in Porous Materials" by Collins et al., "Recent Insights into Penetration of Sand and Similar Granular Materials" by Omidvar et al., and "Applications of Reactive Materials in Munitions" by Peiris and Bolden-Frazier), the shock compaction response under multi-dimensional loading must be understood and characterized for the development of accurate models. Over the years, several different approaches have been developed for this purpose, though none have been used extensively.

Of specific interest to the relatively low stress regimes associated with compaction is the ability to measure the deviatoric (shearing) stress response, which allows for characterization of the strength (deviatoric) response under dynamic loading. Tang and Aidun [32] have reviewed the experimental and theoretical aspects of the combined pressure and shear technique, where measurement of the longitudinal and transverse waves is used to study the constitutive properties of solids. These methods have been applied by Sairam and Clifton [33] to granular

Al_2O_3, by Vogler et al. [34] to sand and granular WC, and by LaJeunesse [35] to sand. The experiments consisted of a thin layer of the granular sample sandwiched between two plates that are intended to remain elastic. Impact is planar but at an angle θ (typically 20–30°) to the direction of projectile motion, which generates both longitudinal and shear waves in the target. The amplitude of the shear transmitted through the sample is taken as a measure of its strength. These studies have shown that the strength of the porous material increases approximately linearly with applied pressure, at least over the range studied. The approach is limited to thin samples, which can be difficult to prepare or non-representative for granular materials. Also, the approach has only been applied to granular materials up to about 3 GPa because of the desire to remain in the elastic regime of the confining plates.

Recently, an approach to examine the evolution of non-planar features in a propagating shock wave has been utilized in an attempt to probe the strength of a shocked porous material [36]. In continuum and mesoscale simulations, increased material strength is found to delay the decay of the perturbation. Thus, it should be possible to fit a strength model for the material of interest to perturbation decay data for varying conditions (impact velocity/pressure, volume fraction, etc.). Only limited results with this technique have been reported to date, but additional work on the technique is ongoing.

While all of the experimental approaches discussed above have been planar or quasiplanar, cylindrical geometries have been used extensively as a consolidation technique [37–40] and to study materials science issues [41, 42]. However, to the best of our knowledge there has only been a single reported use of the cylindrical configuration as a means to study the low-pressure dynamic response of a porous material [43]. That study utilized current pulses through conductive coils to radially compress a copper tube filled with sand. PDV diagnostics on the exterior provided quantitative (but incomplete) information on the response of the sand. Additional development of this approach is needed before it will be suitable for characterizing porous material response.

3 Computational Techniques Applied to Compaction

Computational approaches applied to simulating dynamic consolidation processes can typically be divided into two classes. The first models the porous material as a continuum, which requires homogenization of the solid and void components into a single simulated material. The second captures the material at the mesoscale, which includes discreet modeling of both the solid and void components. Both of these approaches offer unique and complementary capabilities as discussed below.

3.1 Continuum Modeling

Using the experimental methods described above, researchers have measured the shock compaction response for a wide variety of materials ranging from relatively soft and ductile metals to hard and brittle ceramics. These data can be further utilized to calibrate or validate continuum-level compaction models used in simulations. Over the years, several different modeling approaches have been developed to capture the continuum-level shock response associated with the removal of porosity. In the first class of models the compaction response is assumed to be only a function of the pressure (mean stress). In the second, the full stress tensor is included in the model. An example of the later approach is given in the chapter "Continuum Modeling of Partially Saturated Soils" by Banerjee and Brannon of this book, so only the former approach is covered here.

Hydrostatic compaction models are used widely across the shock physics community due to their relative ease of implementation. Of these models, the most prevalent is the P-α model, originated by Herrmann [44] and expanded upon by Carroll and Holt [45]. In this framework alpha is a measure of the distention, and is defined by $\alpha(P) = V(P)/V_S(P)$, where $V(P)$ is the specific volume of the porous material at pressure P and $V_S(P)$ is the specific volume of the fully solidified material at the same pressure P. In this analysis, shock energies for the solid and porous materials are assumed equivalent at a given pressure and temperature. Therefore, α captures the extent to which the volume of the porous material differs from that of the solid at a given pressure and energy. When $\alpha \rightarrow 1$, the porous response coincides with that of the initially porous, but now fully solidified material.

Depending on the type of material and range of initial porosity being captured, the specific formulation for $\alpha(P)$ can vary. In much of the early work on metals, see, for example, Herrmann [44], Butcher and Karnes [46], and Boade [31], $\alpha(P)$ was separated into elastic and plastic components, where the extent of the elastic region was found to decrease with increasing initial porosity. Due to the relatively small change in volume associated with the low-pressure elastic region, in some instances $\alpha(P)$ can be sufficiently captured by treating it as a constant, equal to its initial value at zero pressure α_0 [47]. In the region of plastic deformation, large volume changes occur due to the removal of porosity, and a discreet functional form must be applied for $\alpha(P)$. A polynomial in terms of P was first suggested by Herrmann [44]:

$$\alpha(P) = \alpha_0 + \alpha_1 P + \alpha_2 P^2 + \alpha_3 P^3 \tag{3}$$

and was found to adequately capture the compaction response of 17–90% initial density porous iron. For other materials, application of Eq. (3) has been unable to sufficiently capture experimental data, see, for example, Borg et al. [22] and Fredenburg and Thadhani [48], so alternate representations of $\alpha(P)$ have been developed. Other functionals for $\alpha(P)$, such as the exponential [49]:

$$\alpha(P) = 1 + (\alpha_E - 1)e^{-\hat{a}(P-P_E)} \tag{4}$$

and power law [5]:

$$\alpha(P) = \left(\frac{P_S}{P}\right)^{1/n} \tag{5}$$

forms have been proposed to provide alternate relations for empirically fitting experimental compaction data. In Eqs. (4) and (5) the subscript E corresponds to values of α and P at the transition from the elastic to plastic region and $\hat{a}$, P_S, and n are empirically derived fitting parameters.

In complement to the P-α model, the P-λ model has also been developed to capture the evolution of porosity under dynamic loading [50, 51]. In this formulation, λ is the progress variable for porosity evolution and takes the form:

$$\lambda(P) = 1 - e^{-(P_l/Y_l)^n}, \tag{6}$$

where Y_l and n are empirical fitting parameters with the former having units of pressure. In this model, the specific volume of the initially porous material is defined as mixture of the material in its equilibrium equation of state response and in a non-equilibrium elastic response, evolved through $\lambda(P)$. This formulation has been used successfully to capture the low-pressure compaction response in a wide range of both homogeneous and heterogeneous materials [5, 48, 52]. An example from Fredenburg and Thadhani [48] showing the variation in P-α and P-λ model fits with compaction data from a powder mixture of Ti + Si is shown in Fig. 6. Here, the

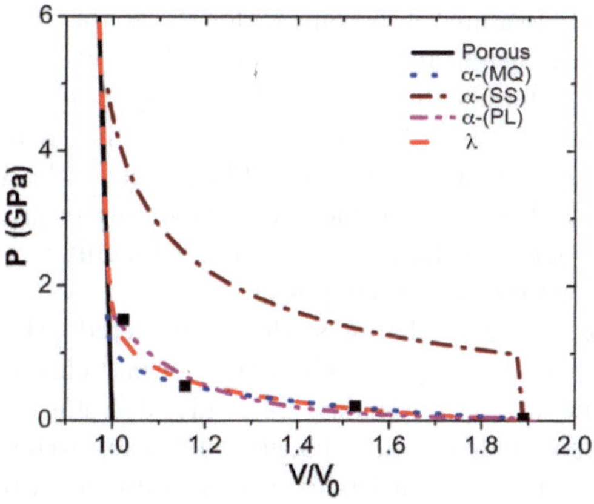

Fig. 6 Shock compaction data for a Ti + Si powder mixture plotted against four distinct continuum model formulations, where α and λ in the legend indicate that a P-α or P-λ model was used in the fit. Details of the functional forms are: (MQ) = $\alpha(P) = 1 + (\alpha_E - 1)\left(\frac{P_S - P}{P_S - P_E}\right)^N$; (SS) = $\alpha(P) = \frac{1}{1 - e^{-3P/2Y}}$; (PL) = Eq. (5), and λ = Eq. (6). Image adapted from [48]

three model forms that fit the data well have empirically derived fitting parameters, while the fourth fit, which shows the largest disagreement with the data, does not include any free fitting parameters. Rather, this fit relies entirely on the static yield strength of the mixture to predict the compaction response.

In more recent years, Collins et al. [53] extended the $\epsilon - \alpha$ [54] model to include data for highly porous initial states. In this model α is again the progress variable for compaction; however, here distention is evolved using knowledge of the volume strain, $\alpha(\epsilon)$ rather than the pressure as it is for the P-α model. By advancing distention using volume strain, the $\epsilon - \alpha$ model allows for increased computational efficiencies in certain hydrocode implementations that would otherwise require simultaneous solution of both the pressure and distention through iterative means. Similar to the models above, the removal of porosity is separated into elastic and plastic regions, and $\alpha \rightarrow 1$ when all porosity is removed. While general in its form, application of this model has been largely focused on simulating high-velocity impacts of planetary bodies, see, for example, Bland et al. [55] and Davison et al. [56].

Each of the three models discussed above are applied similarly in that they require an existing data set to determine optimal values for their empirically derived fitting parameters. Further, all have limited utility in predicting the compaction response for a material and/or initial condition outside of the range for which the model was calibrated. As such, development of a predictive capability for compaction modeling continues to be an active area of research.

3.2 Mesoscale Modeling

Modeling the shock loading of porous materials at the mesoscale is typically undertaken to provide insight into the phenomenology of material behavior at the grain scale. Porous materials are well suited to this type of modeling because they inherently possess intermediate length scale features, that of the grains and pores, which can play an important role in material behavior. In the 1997 predecessor to this volume, Benson [57] reviewed mesoscale modeling in great detail. Since that time, mesoscale modeling techniques have matured significantly, driven primarily by a tremendous growth in computing power.

While early mesoscale modeling studies considered primarily planar shock loading configurations involving a small number of particles, increased computing power has led to much larger domains. Currently, it is not uncommon for planar shock simulations to include domains that are comparable to the sample thicknesses used in shock experiments, i.e., a few to several millimeters. In these simulations, rigid or periodic boundary conditions are often imposed on the lateral faces to simulate the uniaxial strain loading of planar shock experiments. As simulation domains have grown, it has also become possible to consider loading configurations other than planar shocks, such as penetration [58, 59], cylindrical loading [60], and perturbation decay [36].

Current computational capabilities have also made possible the study of mesoscale calculations in three-dimensions (3-D) (cf. [8, 60–62]) though two-dimensional (2-D) calculations are still routinely undertaken due to their much lower computational cost. Borg and Vogler [61] examined the differences in compaction between 2-D and 3-D mesoscale simulations and found that the behavior of WC powder could be accurately simulated with both 2-D and 3-D models. However, different treatments of interparticle interactions were required to achieve agreement between the two techniques, and differences in the temperature and energy distributions between the two cases were noted, suggesting that in investigations where energetics or reactive materials are of primary concern, it may be best to model these systems in 3-D. Thus, while some problems (such as penetration, see chapter "Recent Insights into Penetration of Sand and Similar Granular Materials" by Omidvar, Bless, and Iskander) are inherently 3-D, it may be possible to model other shock phenomena relatively well in 2-D.

Benson [57] also laid out an approach for generating random distributions of circles in a 2-D domain for generating initial microstructures, which has been used and adapted by others over the subsequent years [11, 63]. Similar spherical approaches have also been used in more recent 3-D calculations (cf. [61]). Regarding the generation of synthetic initial microstructures composed of spherical particles, Borg and Vogler [63] examined a number of different aspects of this approach, and emphasized that random arrangements of particles were needed to reduce systematic biases in computational results.

Compaction studies of other 2-D shapes [12, 63] have generally shown that the behavior of inert materials is relatively insensitive to particle shape. However, computational investigations focused on reactive mixtures have shown that shock initiation can be strongly influenced by particle morphology [21]. In studies of explosive powders and their simulants, Baer [64] and Trott et al. [8] used a particle insertion approach to generate realistic arrangements of cubes, while other researchers [65, 66] have imported real experimental microstructures from cross-sections of porous samples to generate complex 2-D domains, primarily for reactive mixtures.

In most mesoscale modeling studies of granular systems, relatively simple constitutive models are used to describe the solid particles. For metals and polymers, material behavior is commonly captured using an equation of state and metal plasticity models to capture the hydrostatic and deviatoric response under shock loading. It is also common practice to model brittle materials in the same fashion, even though fracturing of the constituent particles most likely occurs. This simplification is undertaken largely due to the difficulty in treating fracture in mesoscale simulations, particularly since fracture can quickly drive particle size below the resolution limit of the initial simulation. Despite this shortcoming, there has been some success at matching observed bulk behaviors with mesoscale models that employ these simplified techniques [11, 62].

The execution of mesoscale simulations focused on the shock loading of granular materials can be conducted using several different frameworks. Of these frameworks, the Eulerian, Lagrangian, and molecular dynamics approaches are discussed briefly here. In the Eulerian approach, which is composed of a fixed computational grid through which material moves, hydrocodes such as CTH [67], iSALE [54], and others have been used to study the shock and impact response of granular and porous systems. Benson [57] indicates that the first such study using the Eulerian framework was performed by Williamson [68], with numerous other studies to follow [11, 57, 58, 60–63, 69, 70]. This framework is particularly well suited to the consolidation of porous systems because of its ability to easily handle large material deformations. However, this approach is less robust for simulations where fracture and interfacial interactions are an important component of the system being investigated. Despite these limitations, to date, most mesoscale studies of shock-loaded porous materials have been undertaken using Eulerian methods.

In addition to the Eulerian framework, studies of granular systems have also been undertaken using Lagrangian finite element method (FEM) approaches [59, 71, 72], where the mesh elements move with the material when deformation occurs. This structure has restricted can cause mesh entanglement. In instances where large deformations are expected to occur, Lagrangian methods can be evolved without mesh entanglement using ALE (arbitrary Lagrangian Eulerian) techniques, where remapping of highly strained elements occurs and allows for material to flow through the elements as they do in Eulerian simulations. This remapping is typically performed locally, near the regions of highest deformation, and requires accurate material advection models that are not usually developed for Lagrangian-only frameworks. In addition to facing challenges with respect to large deformations, Lagrangian frameworks are also not currently well suited for granular fracture, though general advancements in cohesive zone algorithms are allowing for fracture to be modeled in Lagrangian codes more readily [73].

Lastly, classical molecular dynamics has found use in simulating the shock response of porous materials because it has the distinct advantage of explicitly treating all aspects of material behavior, including fracture and plastic deformation, at least to the accuracy of the interatomic potential used. While some nanofoams [74] and metal organic frameworks (MOFs) [75] have been accurately represented in MD simulations, it is impossible to represent the experimental microstructures of most porous materials with today's computational capabilities. Instead, simulations must be conducted at length and time scales that are much smaller than those of real shock experiments [76, 77]. However, there are some indications that even simulations at such a disparate scale can still reproduce some aspects of the bulk response. While computational capabilities continue to improve and simulations of larger domains will soon become possible, it will likely still be many years before simulations with length scales of millimeters and time scales of microseconds can be performed using classical molecular dynamics. For a more thorough discussion of this technique, see the chapter "Shock Compression of Porous Materials and Foams Using Classical Molecular Dynamics" by Lane in this volume.

4 Phenomenology

Having covered details of the experimental methods by which Hugoniot data in the compaction region is obtained (Sect. 2), and the theoretical methods by which this data is captured in computational models (Sect. 3 and the chapter "Continuum Modeling of Partially Saturated Soils" from Banerjee and Brannon), the focus now shifts toward some of the unique behaviors observed during the compaction of initially porous materials. It should be noted that in most of the instances discussed, it is required to have rather large data sets for individual material systems or across different material systems to make these comparisons. As such, the authors acknowledge the important role that comprehensive studies on porous materials play in advancing our understanding of these unique phenomenologies.

4.1 Shock Precursors

In competent porous materials, separation between an elastic precursor and the bulk shock wave is commonly observed at low to modest stresses, in a manner similar to fully dense solids. In these systems the precursor amplitude, referred to at the Hugoniot elastic limit (HEL), has been observed to decrease with increasing porosity. For example, Butcher and Karnes [46] found a linear log–log dependence of the yield point with initial density in sintered iron specimens with $\rho_{00} >$ 60% TMD, while Brar et al. [78] showed a linear reduction in the HEL with initial porosity in boron carbide sintered to densities greater than 83% TMD. Additively manufactured truss structures studied by Winter and McShane (see chapter "Additively Manufactured Cellular Materials" in this book) also exhibited elastic precursors under impact loading, though their study only focused on a single initial fractional density of 64% TMD. In these and other investigations on competent porous bodies, the presence and magnitude of the elastic precursor tends to be more likely at low-to-moderate levels of initial porosity.

In systems composed of loose or pressed granular materials, elastic precursors in transmitted stress or velocity profiles are not typically observed due to the dispersive nature of the initial configuration. However, Neal et al. [79] observed substantial precursors in shock-loaded mono-disperse spherical glass beads at low stresses, which became overdriven and disappeared as shock stresses were increased, as shown in Fig. 7. The authors note that the precursor was likely not wholly elastic because it produced substantial densification; rather, it was likely some combination of plastic deformation and particle rearrangement. Lajeunesse et al. [62] observed a less pronounced precursor in coarse-grained sand. In pressed granular sugar, Trott et al. [8] observed precursor waves using spatially resolved velocimetry techniques, and found the temporal duration of the precursor diminished as impact velocity and applied stress increased. Despite different sample preparation methods, initial densities for these studies were about 65% of the theoretical maximum density.

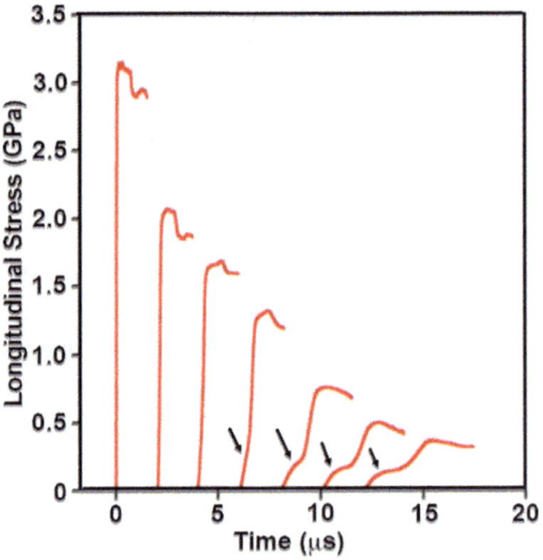

Fig. 7 Evolution of transmitted stress wave profile in spherical glass beads with increasing input stress. Arrows indicate the approximate location of the transition from the initial precursor state to the primary plastic front. Image adapted from Neal et al. [79]

Other studies investigating the dynamic response of granular WC [6], CeO_2 [26], and sand [5] at approximately 55% TMD have not shown the presence of elastic waves. Taken together with the observations on yielding in competent porous bodies, precursor behavior (whether purely elastic or some combination of other deformation mechanisms) is more prevalent at low impact stresses in systems with low initial levels of porosity.

4.2 *Shock Rise Time*

Heterogeneities at length scales of the grains/pores and also those at larger length scales associated with the bulk give rise to unique wave propagation characteristics in shock-loaded porous materials. Evidence of this behavior is most prevalent in the compaction regime, where the applied dynamic loads are not sufficient to overdrive the inter- and intra-granular strength (cf. Nesterenko [3], Sheffield et al. [7], Tong and Ravichandran [80]). Consequently, unique scaling behavior has been observed for initially porous materials that is in stark contrast to solids. In nominally homogeneous solid materials (and some two-phase solids), the strain rate in the shock front scales with stress to the fourth power, $\dot{\epsilon} \propto \sigma^4$ [81]. This so-called fourth power law is widely observed and has become part of the canon for the field of shock physics. However, the introduction of heterogeneity alters this scaling and, in the case of layered materials, has been shown to produce a second power scaling, $\dot{\epsilon} \propto \sigma^2$ [82]. More recently, experimental wave profiles were examined for porous tungsten carbide and sand and were found to display a first-power scaling, $\dot{\epsilon} \propto \sigma$ [5, 6].

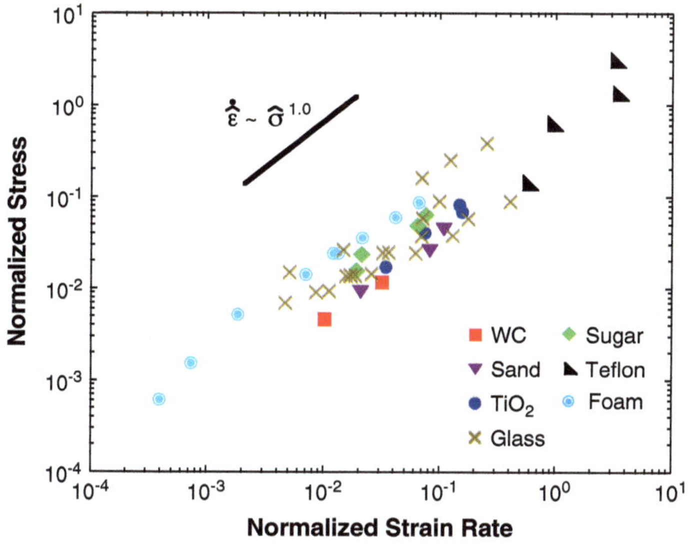

Fig. 8 Normalized stress versus strain rate for shock experiments on a wide range of porous materials showing a first-power scaling relationship. Image adapted from [83]

Vogler et al. [83] examined the three power law scalings and were able to identify non-dimensional groups for the second and first power law scalings that allowed for the response of a wide variety of different materials to be collapsed onto a single curve. The results from this analysis are given in Fig. 8, where it is observed that first power law scaling is exhibited by a very broad range of initially porous material types, from porous polymeric foams to high strength granular ceramics such as tungsten carbide and titanium dioxide. Somewhat surprisingly, the best scaling is obtained without inclusion of a strength or hardness for the material.

In addition, Vogler et al. [83] proposed a simple conceptual model to explain the first-power scaling. The model considers the delay in wave propagation through a series of aligned particles caused by a missing particle. This gap must be closed through motion of the material across the gap, slowing the wave in that region. By relating that delay to a strain rate, they were able to obtain values for the scaling exponent close to unity for reasonable values of the compaction response. Thus, they propose that the mechanism of mass propagation as a means to close voids in the porous material is the essential element in the first-power scaling observed in porous materials. First-power scalings observed experimentally for porous materials have also been found in mesoscale computations based on Eulerian hydrocodes [11, 61], though some effect of the method used to estimate the strain rate has been found. Simulations utilizing a particle-based Lagrangian code have also been found to display similar scaling [83].

4.3 Porosity Enhanced Densification

Another interesting phenomenology associated with porous materials is manifested in the interplay between the thermal and mechanical resistances to densification. For materials subjected to strong shocks (well in excess of the stresses required for complete densification), the conservation of mass, momentum, and energy across a shock front predict decreasing levels of densification for increasing levels of initial porosity at an equivalent applied stress. This results in a "family" of porous Hugoniots with increasing displacement from the solid as initial porosity is increased [84], a behavior governed by the increasing thermal energy associated with shock loading higher porosity materials. As initial shock stresses are decreased, the Hugoniot response transitions from being governed by the thermal (and electronic) response of the continuum to one dominated by the mechanical response of the grains and pores.

By definition, a porous material is composed of at least two components, a solid component with relatively high strength and a gas/void component with minimal resistance to shear. Under compression these two materials behave quite differently, such that compression of the gas/void component is favored over that of the solid constituent. Therefore, at relatively high initial porosities there is little resistance to densification, such that large increases in shock compressed densities can be achieved with low stresses applied to the composite material. Conversely, higher stresses are required to achieve significant densification for materials with low amounts of initial porosity. This is likely one of the contributing factors to the observation of Brown et al. [5], who noted that the P-α model formulation developed by Herrmann [44] which contains an elastic response and a plastic response given by Eq. (3) was unable to fit the data for 59% TMD porous sand. Rather, better agreement was obtained using compaction models that contained exponential and power-law functions, and allowed for large initial increases in shock compressed density at low applied loads.

Further evidence of a transition in the compaction response from that which offers little resistance to densification at high initial porosities, i.e., a purely plastic response, to a more rigid response that must be captured by distinct elastic and plastic regions as initial porosity is reduced is found in the shock response of tungsten. Figure 9 gives the dynamic response for initially porous tungsten at several initial distentions ranging from 28% to 80% TMD along with representative P-α model fits to each data set using Eq. (4). For the lower initial densities of tungsten corresponding to 28% and 55% TMD, the compaction response can be captured well using only the plastic response of Eq. (4). However, as initial density increases the elastic response must be included to achieve sufficient agreement between the data and the model. The data for tungsten, along with that of sand, suggests there is a maximum initial packing fraction at which the compaction response transitions from perfectly plastic to elastic plus plastic for both ductile and brittle materials, and that this transition occurs somewhere between 59% and 66% TMD. It should be noted that the maximum packing fraction for a random arrangement of spheres

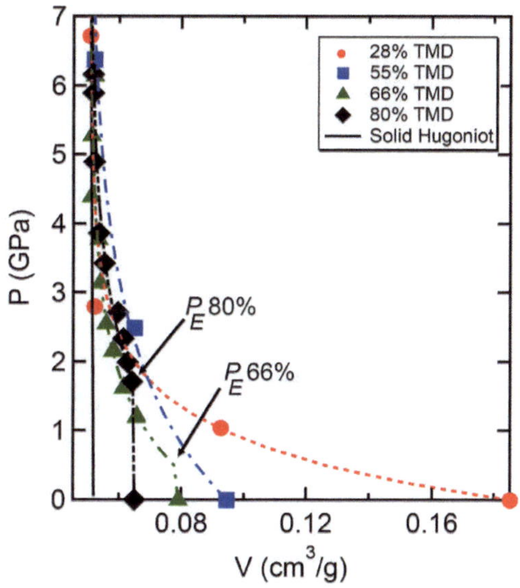

Fig. 9 Compaction data for porous tungsten plotted with P-α model fits from Eq. (4) illustrating the transition from plastic to elastic plus plastic model response with decreasing initial porosity. Data for tungsten is taken from Bakanova et al. [85], Boade [86], and Dandekar and Lamothe [87], with reported percentages of the theoretical maximum density (% TMD) calculated using $\rho_0 = 19.257$ g/cm^3 from the solid equation of state, SESAME 93540 [88]

is approximately 64% TMD, such that this value might serve as a practical upper limit on the transition for initially porous granular materials. However, further experimental and theoretical work is needed to verify this assumption.

4.4 Morphology Effects

Mesoscale material characteristics such as the size and shape of the solid and void/gas components are important characteristic of porous materials in that they can be measured experimentally prior to the dynamic event. If these intrinsic properties can be further related to particular compaction phenomena, then it may be possible to predict certain features of the compaction response by having only limited information about the material initial state. For example, it has been observed in many granular systems that particle morphology and size distribution influences initial packing fractions, see Sect. 2.2. These properties can affect the preparation of granular samples for dynamic experiments, and may influence shockwave characteristics in those materials, as discussed in Sect. 4.2. At present, the role of intrinsic property variations on the bulk dynamic response of a porous compact is not fully understood.

A recent investigation by Fredenburg et al. [26] studied the static and dynamic response of three distinct morphology CeO_2 powders, with initial particle sizes differing by a factor of 20. Under dynamic loading the consolidation responses of the three morphologies, following static compression to the same initial density state, were nearly indistinguishable within the reported uncertainties of the experiments, as shown in Fig. 10. This suggests that it is the initial porosity and the properties

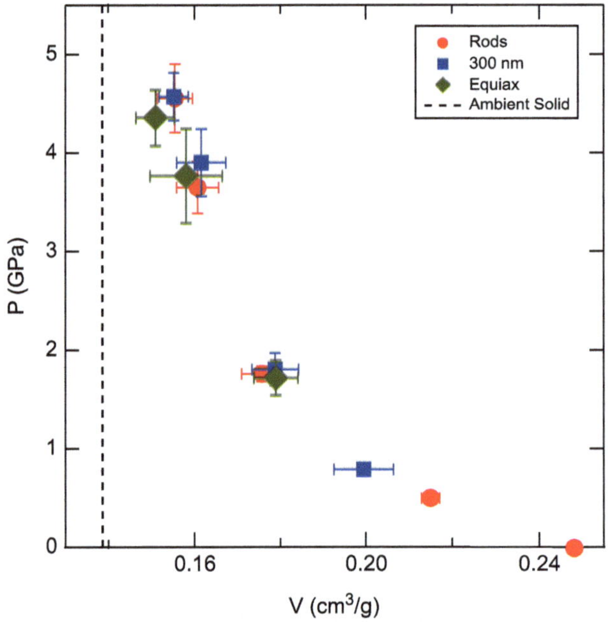

Fig. 10 Dynamic compaction response of three distinct morphologies of CeO_2 powders pressed to an initial porous density of $\rho_{00} = 4.03$ g/cm^3, or 56% TMD, illustrating nominally similar dynamic responses, regardless of morphology

of the bulk material that have the greatest influence on the compaction response, and not the characteristics of the particle morphology. Support for this is offered by Grady [89], who proposed that the relatively low work of fracture leads to a turbulent-like separation of length scales under shock loading. In the case of CeO_2, this implies that the initial particle characteristics have little influence on the response because the particles can so readily fracture into smaller particles. However, Grady's argument does suggest that materials with vastly different particle sizes (orders of magnitude) can behave dissimilarly, but the variation in particle (or void) characteristics required to observe this difference is unknown and may be impractical to examine experimentally.

In contrast to the above argument that the dynamic compaction response is not influenced strongly by initial morphology, studies on granular and porous explosives as well as intermetallic and thermite mixtures have shown that initial characteristics of the particles and pores can have a significant effect on the energetic and reactive response. For example, coarse granular energetic materials such as HMX, TATB, PETN have been found to react more readily than those with fine particles, an observation that is typically attributed to the formation of hot spots under shock loading, which may originate from the closure of voids or the fracture and comminution of grains under dynamic loading. For intermetallic and thermite mixtures, reactivity is favored by initial configurations that promote intimate mixing between the two constituents during shock loading. These types of systems have been covered extensively in the literature, and the interested reader is referred to Davison et al. [90], Baer [64], and Eakins and Thadhani [21] for further details on their consolidation and energetic responses.

5 Outstanding Issues and Directions for Future Work

While significant progress has been made in the field of shock compaction in the past 20 years, there are still a number of outstanding issues related to the experimental characterization of porous materials and targets, the continuum and mesoscale modeling of these systems, and theoretical considerations of their behavior. In this final section, the authors bring attention to several of these issues and look forward to future research that can help move these areas forward.

5.1 Is the Compaction Response a Shock Hugoniot?

The analysis of shock experiments (and even simulations, in many cases) typically utilizes the Rankine–Hugoniot jump conditions to calculate the stress/density/energy state behind the shock front, implicitly assuming that the compaction state reached lies on the porous Hugoniot corresponding to that initial density. Even for cases where the shock wave has been shown to have reached a steady state [6], it is not entirely clear that the use of the jump conditions is appropriate. Such issues have been discussed for other types of materials by Krehl [91]. For example, the material passing through a shock front is treated as having instantaneously gone from its initial to final states by following a so-called Rayleigh line that connects the two states by a straight line in stress-volume space. However, in a porous material, or, indeed, any heterogeneous material, different material elements can have very different histories to the degree that no single material point may actually follow the Rayleigh line.

Another issue arises from the relatively long time scales needed to obtain thermal equilibrium when compared to those required for mechanical equilibrium. With this in mind, it is unlikely that thermodynamic equilibrium is achieved during the time it takes for the shock to transit the sample in the canonical compaction experiment, thus calling into question straightforward application of the jump conditions for initially porous materials. Despite these concerns, it seems likely that porous materials averaged over some suitable volume may indeed satisfy the Rankine–Hugoniot conditions. However, the limits of these approaches have not yet been defined, warranting further theoretical and numerical studies.

5.2 How Are Dynamic and Static Responses Related?

While the focus of this chapter is on the shock response of porous materials, static experiments can also be used to characterize the compression response of initially porous materials. These static experiments are attractive from the standpoint of their relatively low cost with respect to dynamic experiments, allowing for the potential of more in-depth studies to be performed. While static studies are informative, there are a number of reasons to expect that the dynamic and static responses for a single material may be different. First, intrinsic rate dependence of the underlying material could play a role in the consolidation response in a manner similar to strength, where strain-rate dependent models may be required to capture the constitutive response, see, for example, Meyers [92]. More broadly, material deformation under shock loading is confined to much smaller regions, roughly the width of the shock front, as compared to deformation under quasistatic loads, which are applied to the bulk of the material. This distribution in stress, from local to global, is expected to significantly alter the densification response.

Unfortunately, relationships linking the static and dynamic compaction responses for porous materials are poorly understood, and experimental results can often contradict one another. In some cases [6, 93], the shock response was stiffer than the static response for all load levels. Zaretsky et al. [94], on the other hand, found the two corresponded at the lowest stress levels measured but diverged for higher stresses. In contrast, studies of sand by Brown et al. [5] found the dynamic and static compression response to be nearly equivalent, while [95] found the static response of aluminum foam was stiffer than the shock response. These differences likely stem, in part, from the difficulties associated with accurately executing and analyzing both static and dynamic compaction experiments. It is the opinion of the authors that the ability to link densification responses under static and dynamic conditions has the potential to significantly advance compaction theory toward the predictive, and should thus be given further attention.

5.3 How Should Heterogeneity Be Handled During Hugoniot Analysis?

Porous materials, especially foams and granular forms, will almost invariably have significant heterogeneity present on the scale of a few or many times the characteristic length of the grain or pore. Depending upon the scale and extent of heterogeneity, it has the potential to significantly affect the analysis and interpretation of experimental measurements that treat the material as a homogeneous continuum. Thus, it is incumbent upon experimentalists to minimize the heterogeneities that arise during sample preparation or to be sure to include the heterogeneity in any analysis of experiments. In systems with high degrees of heterogeneities, where the heterogeneities themselves as a component of a porous system are important aspects of the system being studied, it may be necessary to characterize the initial samples through tomographic or radiographic techniques prior to performing dynamic experiments. For some material and target systems, this type of pre-shot characterization may be straightforward, but for others it may present significant challenges. However, if the role of heterogeneities is to be fully understood during the process of shock consolidation, then future work must be focused on the characterization and analysis of these features.

5.4 How Should Shock Compaction Be Modeled at the Continuum?

The continuum models for compaction discussed in Sect. 3.1 assume the compaction process is governed entirely by the hydrostatic pressure (mean stress) applied to the material. In contrast, the so-called full stress models, such as those discussed by Banerjee and Brannon in this volume, consider the effect of the entire stress tensor on the material response. Since such full stress models generally require the calibration of a relatively large number of material constants compared to those based solely on pressure, the choice of implementing one model type over the other for modeling and simulation purposes may be material or application dependent. For materials with relatively low inherent strength, the hydrostatic approach may be sufficient for first, even second order accuracies of simulations under a broad range of loading configurations. However, for higher strength materials, calibration of continuum compaction models may require the inclusion of strength, whether it be a model of the full stress form or some other more simplified form.

5.5 What Is the Role of Mesoscale Modeling?

In Sect. 3.2, some aspects of mesoscale modeling of porous materials were discussed. However, none of the approaches considered, Eulerian, Lagrangian, or molecular dynamics, are individually capable of modeling all of the mechanisms (fracture, large inelastic deformations, chemical reactions, etc.) and length and time scales (millimeters and microseconds or larger) that are important to compaction. Despite these limitations, mesoscale modeling approaches have proven useful for providing insights into the mechanical response of grains and pores under shock loading, as well as predicting specific aspects of the continuum response for a variety of porous materials classes. Improvements to the physics of mesoscale models are likely to occur in the relatively near-term future, as investigations into particle-level interactions under shock loading at advanced light sources such as those discussed in the chapter "X-Ray Phase Contrast Imaging of Granular Systems" by Jensen et al. become more common. As these experimental tools mature and become more widely used, they will likely have a profound impact on the fidelity of mesoscale modeling techniques.

6 Conclusion

In this chapter, the authors' intent was to present the physics of shock compaction in a manner that is both foundational and forward-looking. In the first few sections, the foundations of experimental, diagnostics, and modeling approaches were covered, providing a snapshot in time of current capabilities. Some of the unique features and observations of porous material responses to shock loading were then covered, and some of the major outstanding questions in the field were presented. From this mix of perspectives, it is hoped that the reader, and especially those that may be just beginning their careers in the field of shock compression science, finds motivation to push the limits of our current understanding of shock compaction physics for porous materials. While it is most definitely true that significant advancements have been made in this field over the years, significant advancements remain to be realized, and the authors look forward to the new discoveries that lie ahead.

Acknowledgements Sandia National Laboratories is a multimission laboratory managed and operated by National Technology and Engineering Solutions of Sandia, LLC, a wholly owned subsidiary of Honeywell International Inc., for the U.S. Department of Energy's National Nuclear Security Administration under contract DE-NA0003525. Los Alamos National Laboratory is managed by Triad National Security, LLC for the U.S. Department of Energy's NNSA under contract number 89233218CNA000001.

References

1. Grady D, Fenton G, Vogler T (2014) Equation of state for shock compression of distended solids. J Phys. Conf Ser 500(11):152007
2. Fredenburg DA, Koller DD (2014) Interpreting the shock response of porous oxide systems. J Phys Conf Ser 500(11):112025
3. Nesterenko VF (2001) Dynamics of heterogeneous materials. Springer, New York
4. Fredenburg DA, Thadhani NN, Vogler TJ (2010) Shock consolidation of nanocrystalline 6061-T6 aluminum powders. Mater Sci Eng A 527:3349–3357
5. Brown JL, Vogler TJ, Grady DE, Reinhart WD, Chhabildas LC, Thornhill TF (2007) Dynamic compaction of sand. In: Elert M, Furnish MD, Chau R, Holmes N, Nguyen J (eds) Shock compression of condensed matter - 2007. AIP, New York, pp 1363–1366
6. Vogler TJ, Lee MY, Grady DE (2007) Static and dynamic compaction of ceramic powders. Int J Solids Struct 44:636–658
7. Sheffield SA, Gustavsen RL, Anderson MU (1997) Shock loading of porous high explosives. In: Davison L, Horie Y, Shahinpoor M (eds) High-pressure shock compression of solids IV: response of highly porous solids to shock loading. Springer, New York, pp 23–61
8. Trott WM, Baer MR, Castañeda JN, Chhabildas LC, Asay JR (2007) Investigation of the mesoscopic scale response of low-density pressings of granular sugar under impact. J Appl Phys 101:024917
9. Yang R, Zou R, Yu A (2003) Effect of material properties on the packing of fine particles. J Appl Phys 94:3025–3034
10. Fredenburg DA, Koller DD, Rigg PA, Scharff RJ (2013) High-fidelity Hugoniot analysis of porous materials. Rev Sci Instrum 84:013903

11. Borg JP, Vogler TJ (2008) Mesoscale calculations of the dynamic behavior of a granular ceramic. Int J Solids Struct 45:1676–1696
12. Benson DJ (1995) The calculation of the shock velocity - particle velocity relationship for a copper powder by direct numerical simulation. Wave Motion 21:85–99
13. Schumaker MG, Kennedy G, Thadhani NN, Hankin M, Stewart ST, Borg JP (2017) Stress and temperature distributions of individual particles in a shock wave propagating through dry and wet sand mixtures. AIP Conf Proc 1793:120016
14. Jaeger HM (2005) Sand, jams and jets. Phys World 18:34–39
15. Elliot NE, Staudhammer KP (1992) Effect of internal gas pressure on the shock consolidation of 304 stainless steel powders. Marcel Dekker, New York, pp 371–381
16. McQueen RG, Marsh SP, Taylor JW, Fritz JN, Carter WJ (1970) The equation of state of solids from shock wave studies. In: Kinslow R (ed) High-velocity impact phenomena. Academic, New York, pp 293–417
17. Yiannakopoulos G (1990) A review of manganin gauge technology for measurements in the gigapascal range. Tech. Rep. MRL-TR-90-5, Materials Research Labs Ascot Vale, Australia
18. Bauer F (1982) Behavior of ferroelectric ceramics and PVF2 polymers under shock loading. In: Nellis WJ, Seaman L, Graham RA (eds) Shock waves in condensed matter - 1981. AIP, New York, pp 251–266
19. Barker LM, Hollenbach RE (1972) Laser interferometry for measuring high velocities for any reflecting surface. J Appl Phys 43:4669–4675
20. Strand OT, Goosman DR, Martinez C, Whitworth TL, Kuhlow WW (2006) Compact system for high-speed velocimetry using heterodyne techniques. Rev Sci Instrum 77:083108
21. Eakins DE, Thadhani NN (2009) Shock compression of reactive powder mixtures. Int Mater Rev 54:181–213
22. Borg JP, Chapman DJ, Tsembelis K, Proud WG, Cogar JR (2005) Dynamic compaction of porous silica powder. J Appl Phys 98:073509
23. Dai C, Eakins DE, Thadhani NN (2007) On the applicability of analytical models to predict Hugoniot of nano-sized powder compacts. In: Elert M, Furnish MD, Chau R, Holmes N, Nguyen J (eds) Shock compression of condensed matter - 2007. AIP, New York, pp 35–38
24. Razorenov SV, Kanel GI, Baumung K, Bluhm HJ (2002) elastic limit and spall strength of aluminum and copper single crystals over a wide range of strain rates and temperatures. In: shock compression of condensed matter - 2001. AIP, New York, pp 503–506
25. Barker LM, Hollenbach RE (1974) Shock wave study of the alpha to epsilon phase transition in iron. J Appl Phys 45:4872
26. Fredenburg DA, Koller DD, Coe JD, Kiyanda CB (2014) The influence of morphology on the low- and high-strain-rate compaction response of CeO_2 powders. J Appl Phys 115:123511
27. Barker LM, Hollenbach RE (1970) Shock-wave studies of PMMA, fused silica, and sapphire. J Appl Phys 41:4208–4226
28. Baer MR, Trott WM (2004) Mesoscale studies of shock loaded tin sphere lattices. In: Furnish MD, Gupta YM, Forbes JW (eds) Shock compression of condensed matter. American Institute of Physics, New York, pp 517–520
29. Mitchell AC, Nellis WJ (1981) Shock compression of aluminum, copper, and tantalum. J Appl Phys 52:3363–3374
30. Root S, Haill TA, Lane JMD, Thompson AP, Grest GS, Schroen DG, Mattsson TR (2013) Shock compression of hydrocarbon foam to 200 GPa: experiments, atomistic simulations, and mesoscale hydrodynamic modeling. J Appl Phys 114:103502
31. Boade RR (1968) Compression of porous copper by shock waves. J Appl Phys 39:5693–5702
32. Tang Z, Aidun JB (2009) Combined compression and shear plane waves. In: Horie Y (ed) Shock wave science and technology reference library, vol 3. Springer, Berlin, pp 109–167
33. Sairam S, Clifton RJ (1994) Pressure-shear impact investigation of dynamic fragmentation and flow of ceramics. In: Gilat A (ed) Mechanical testing of ceramics and ceramic composites. AMD, vol 197. ASME, New York, pp 23–40
34. Vogler TJ, Alexander CS, Thornhill TF, Reinhart WD (2011) Pressure-shear experiments on granular materials. Report SAND2011-6700, Sandia National Laboratories

35. LaJeunesse JW (2018) Dynamic behavior of granular earth materials subjected to pressure-shear loading. Ph.D. thesis, Marquette University, Department of Mechanical Engineering
36. Vogler TJ (2015) Shock wave perturbation decay in granular materials. J Dyn Behav Mater 1:370–387
37. Carton EP, Verbeek HJ, Stuivinga M, Schoonman J (1997) Dynamic compaction of powders by an oblique detonation wave in the cylindrical configuration. J Appl Phys 81:3038–3045
38. Thadhani NN (1988) Shock compression of powders. Adv Mater Manuf Process 3:493–549
39. Jin ZQ, Thadhani NN, McGill M, Ding Y, Wang ZL, Chen M, Zeng H, Chakka VM, Liu JP (2005) Explosive shock processing of $Pr_2Fe_{14}B/\alpha$-Fe exchange-coupled nanocomposite bulk magnets. J Mater Res 20:599–609
40. Sethi G, Myers NS, German RM (2008) An overview of dynamic compaction in powder metallurgy. Int Mater Rev 53:219–234
41. Nesterenko VF, Meyers MA, Chen HC (1996) Shear localization in high-strain-rate deformation of granular alumina. Acta Mater 44:2017–2026
42. Shih CJ, Meyers MA, Nesterenko VF (1998) High-strain-rate deformation of granular silicon carbide. Acta Mater 46:4037–4065
43. Fenton G, Caipen T, Daehn G, Vogler T, Grady D (2009) Shock-less high rate compaction of porous brittle materials. In: Elert ML, Buttler WT, Furnish MD, Anderson WW, Proud WG (eds) Shock compression of condensed matter - 2009. AIP, New York, pp 1337–1340
44. Herrmann W (1969) Constitutive equations for the dynamic compaction of ductile porous materials. J Appl Phys 40:2490–2499
45. Carroll M, Holt AC (1972) Suggested modification of the p-alpha model for porous materials. J Appl Phys 43:759–761
46. Butcher BM, Karnes CH (1969) Dynamic compaction of porous iron. J Appl Phys 40:2967–2976
47. Carroll M, Holt AC (1974) Shock-wave compaction of porous aluminum. J Appl Phys 45:3864–3875
48. Fredenburg DA, Thadhani NN (2013) On the applicability of the p-alpha and p-lambda models to describe the dynamic compaction response of highly heterogeneous powder mixtures. J Appl Phys 113:043507
49. Boade RR (1970) Principal Hugoniot, second-shock Hugoniot, and release behavior of pressed copper powder. J Appl Phys 41:4542–4551
50. Grady DE, Winfree NA, Kerley GI, Wilson LT, Kuhns LD (2000) Computational modeling and wave propagation in media with inelastic deforming microstructure. J Phys IV 10:15–20
51. Grady DE, Winfree NA (2001) A computational model for polyurethane foam. In: Staudhammer KP, Murr LE, Meyers MA (eds) Fundamental issues and applications of shock-wave and high-strain-rate phenomena. Elsevier, New York, pp 485–491
52. Fenton G, Grady D, Vogler T (2015) Shock compression modeling of distended mixtures. J Dyn Behav Mater 1:103
53. Collins GS, Melosh HJ, Wunneman K (2011) Improvements to the ϵ-α porous compaction model for simulating impacts into high-porosity solar system objects. Int J Impact Eng 38:434–439
54. Wunnemann K, Collins GS, Melosh HJ (2006) A strain-based porosity model for use in hydrocode simulations of impacts and implications for transient crater growth in porous targets. Icarus 180:514–527
55. Bland PA, Collins GS, Davison TM, Abreu NM, Ciesla FJ, Muxworthy AR, Moore J (2014) Pressure-temperature evolution of primordial solar systems during impact-induced compaction. Nat Commun 5:5451
56. Davison TM, Collins GS, Bland PA (2016) Mesoscale modeling of impact compaction of primitive solar system models. Astrophys J 821:68
57. Benson DJ (1997) The numerical simulation of the dynamic compaction of powders. In: Davison L, Horie Y, Shahinpoor M (eds) High-pressure shock compression of solids IV: response of highly porous solids to shock loading. Springer, New York, pp 233–255

58. Borg JP, Vogler TJ (2008) Mesoscale simulations of a dart penetrating sand. Int J Impact Eng 35:1435–1440
59. Dwivedi SK, Teeter RD, Felice CW, Gupta YM (2008) Two dimensional mesoscale simulations of projectile instability during penetration in dry sand. J Appl Phys 104:083502
60. Homel MA, Herbold EB (2017) On mesoscale methods to enhance full-stress continuum modeling of porous compaction. In: Chau R, Germann T, Oleynik I, Peiris S, Ravelo R, Sewell T (eds) Shock compression of condensed matter - 2015, vol 1793. AIP, New York, p 080010
61. Borg JP, Vogler TJ (2013) Rapid compaction of granular materials: characterizing two and three-dimensional mesoscale simulations. Shock Waves 23:153–176
62. Lajeunesse JW, Hankin M, Kennedy GB, Spaulding DK, Schumaker MG, Neel CH, Borg JP, Stewart ST, Thadhani NN (2017) Dynamic response of dry and water-saturated sand systems. J Appl Phys 112:015901
63. Borg JP, Vogler TJ (2009) Aspects of simulating the dynamic compaction of a granular ceramic. Model Simul Mater Sci Eng 17:045003
64. Baer MR (2007) Mesoscale modeling of shocks in heterogeneous reactive materials. In: Horie Y (ed) Shock wave science and technology reference library. Springer, Heidelberg, pp 321–356
65. Benson DJ, Conley P (1999) Eulerian finite-element simulations of experimentally acquired HMX microstructures. Model Simul Mater Sci Eng 7:333–354
66. Eakins D, Thadhani NN (2007) Discrete particle simulation of shock wave propagation in a binary Ni+Al powder mixture. J Appl Phys 101:043508
67. McGlaun JM, Thompson SL, Elrick MG (1990) CTH: a three-dimensional shock wave physics code. Int J Impact Eng 10:351–360
68. Williamson RL (1990) Parametric studies of dynamic powder consolidation using a particle-level numerical model. J Appl Phys 68:1287–1296
69. Derrick JG, LaJeunesse JW, Davison TM, Borg JP, Collins GS (2018) Mesoscale simulations of shock compaction of a granular ceramic: effects of mesostructure and mixed-cell strength treatment. Model Simul Mater Sci Eng 26:035009
70. Borg JP, Maines WR, Chhabildas LC (2014) Equation of state and isentropic release of aluminum foam and polyvinylidene fluoride systems. J Appl Phys 115:213515
71. Dwivedi SK, Pei L, Teeter RD (2015) Two-dimensional mesoscale simulations of shock response of dry sand. J Appl Phys 117:085902
72. Xi CQ, Li QM (2017) Meso-scale mechanism of compaction shock propagation in cellular materials. Int J Impact Eng 109:321–334
73. Labanda NA, Giusti SM, Luccioni BM (2018) Meso-scale fracture simulation using an augmented Lagrangian approach. Int J Damage Mech 27:138–175
74. Gunkelmann N, Rosandi Y, Ruestes CJ, Bringa EM, Urbassek HM (2016) Compaction and plasticity in nanofoams induced by shock waves: a molecular dynamics study. Comput Mater Sci 119:27–32
75. Banlusan K, Strachan A (2016) Shockwave energy dissipation in metal-organic framework MOF-5. J Phys Chem C 120:12463–12471
76. Cheruka MJ, Germann TC, Kober EM, Strachan A (2014) Shock loading of granular Ni/Al composites. Part I: mechanics of loading. J Phys Chem C 118:26377–26386
77. Lane JMD, Thompson AP, Vogler TJ (2014) Enhanced densification under shock compression in porous silicon. Phys Rev B 90:134311
78. Brar NS, Rosenberg Z, Bless SJ (1992) Applying Steinberg's model to the Hugoniot elastic limit of porous boron carbide specimens. In: Schmidt SC, Dick RD, Forbes JW, Tasker DG (eds) Shock compression of condensed matter. North-Holland, Amsterdam, pp 467–470
79. Neal WD, Chapman DJ, Proud WG (2012) Shock-wave stability in quasi-mono-disperse granular materials. Eur Phys J Appl Phys 57:31001
80. Tong W, Ravichandran G (1997) Recent developments in modeling shock compression of porous materials. In: Davison L, Horie Y, Shahinpoor M (eds) High-pressure shock compression of solids IV: response of highly porous solids to shock loading. Springer, New York, pp 177–203

81. Swegle JW, Grady DE (1985) Shock viscosity and the prediction of shock wave rise times. J Appl Phys 58:692–701
82. Zhuang S, Ravichandran G, Grady DE (2003) An experimental investigation of shock wave propagation in periodically layered composites. J Mech Phys Solids 51:245–265
83. Vogler TJ, Borg JP, Grady DE (2012) On the scaling of steady structured waves in heterogeneous materials. J Appl Phys 112:123507
84. Zel'dovich YaB, Raizer YuP (2002) Physics of shock waves and high-temperature hydrodynamic phenomena. Dover, Mineola, NY
85. Bakanova AA, Dudoladov IP, Sutulov YN (1974) Shock compressibility of porous tungsten, molybdenum, copper, and aluminum in the low pressure domain. J Appl Mech Tech Phys 15:241–245
86. Boade RR (1969) Dynamic compression of porous tungsten. J Appl Phys 40:3781–3785
87. Dandekar DP, Lamothe RM (1977) Behavior of porous tungsten under shock compression at room temperature. J Appl Phys 48:2871–2897
88. Crockett S (2017) Sesame equation of state 93540. SESAME Database
89. Grady DE (2010) Length scales and size distributions in dynamic fragmentation. Int J Fract 163:85–99
90. Davison L, Horie Y, Shahinpoor M (eds) (1997) High-pressure shock compression of solids IV: response of highly porous solids to shock loading. Springer, New York
91. Krehl POK (2015) The classical Rankine-Hugoniot jump conditions, an important cornerstone of modern shock wave physics: ideal assumptions vs. reality. Eur Phys J H 40:159–204
92. Meyers MA (1994) Dynamic behavior of materials. Wiley, New York
93. Lagunov VA, Stepanov VA (1963) Measurements of the dynamic compressibility of sand. Zhurnal Prikladnoi Mekhaniki i Tehknicheskoi Fiziki 4:88–96
94. Zaretsky E, Asaf Z, Ran E, Aizik F (2012) Impact response of high density flexible polyurethane foam. Int J Impact Eng 39:1–7
95. Barnes AT, Ravi-Chandar K, Kyriakides S, Gaitanaros S (2014) Dynamic crushing of aluminum foams: part I - experiments. Int J Solids Struct 51:1631–1645

Continuum Modeling of Partially Saturated Soils

Biswajit Banerjee and Rebecca Brannon

1 Introduction

Engineering applications of continuum constitutive models of soils at high rates of loading can be classified into two categories. In the first category are problems that require the computation of the penetration of soils by impactors and the spread and momenta of ejecta from such events [1, 2]. The second important category includes problems that involve explosions in soils and the effect of imparted momentum on nearby objects and structures [3–5]. Both types of problem involve severe shock compression followed by tensile expansion and failure of the soil. Few models are able to predict the response of soils for the entire range of stress states and strain rates that these problems involve.

If the soil contains pore liquids, the problem becomes even more complex and rigorous solutions at multiple coupled length and time scales become necessary for accurate predictions. To date, such coupled approaches have proved to be numerically intractable [6, 7] necessitating the development of continuum models that can reduce the computational demand. Such continuum models for high strain-rate applications have typically been purely empirical (see, e.g., [8, 9]), involving either the interpolation of tabulated data or simplified fits to experimental data. In this chapter we describe a slightly more physics-based approach that exploits developments in quasistatic modeling of soils containing pore fluids.

Numerous phenomenological models have been proposed in the scientific literature for predicting stresses and deformations in partially saturated granular and

B. Banerjee
Parresia Research Limited, Auckland, New Zealand

R. Brannon (✉)
University of Utah, Salt Lake City, UT, USA
e-mail: rebecca.brannon@utah.edu

T. J. Vogler, D. A. Fredenburg (eds.), *Shock Phenomena in Granular and Porous Materials*, Shock Wave and High Pressure Phenomena,
https://doi.org/10.1007/978-3-030-23002-9_3

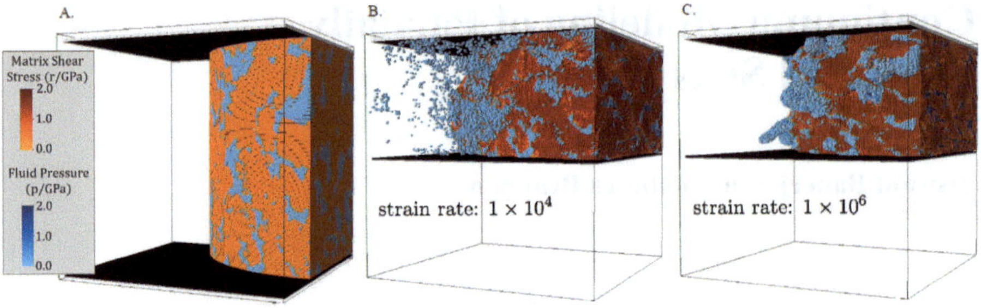

Fig. 1 Simulations of compression of a cylinder of porous rock (orange) fully saturated with a pore fluid (blue) help to justify neglecting fluid–solid relative motion at sufficiently high rates (effectively treating open porosity as if it were closed from inertial head loss) [44]

porous media [10, 11].[1] These models are typically expressed in terms of either total or "effective" stresses (the stress experienced by the solid skeleton in the absence of pore fluids) [16, 27, 41]. Most studies deal with quasistatic or long-wavelength conditions for which the choice of stress measure in the momentum equations has been explored extensively [10, 12, 25, 29, 42, 43]. It is less clear which stress measure is most appropriate for dynamics, mainly because the definition of total stress is ambiguous in the presence of relative motions of the pore fluids, although (as illustrated in Fig. 1) this issue is less concerning in extremely high-rate applications that suppress relative motion of fluid and solid constituents and therefore eliminate the need to simultaneously solve separate momentum equations for each constituent. Investment in multiscale modeling (cf. [6]) seems necessary to identify appropriate forms for macroscale constitutive relations, regardless of whether conditions are quasistatic or dynamic.

The first few sections of this chapter provide equations governing fully or partially saturated sand. The theory is necessarily idealized in order to accommodate

[1]Other examples can be found in the following publications and the references cited in them. In [12] the validity of the effective stress model for quasistatic loading of soils with high pore pressure (6 MPa) is reconfirmed. A complete thermodynamic description of multiphase soils and connections between microscopic and macroscopic quantities can be found in [13–17]. An alternative to the "pressure equilibrium" assumption in mixtures and a rigorous definition of fluid pressure are explored in [18]. Alternative approaches based on effective medium theory can be found in [19–21]. In [22, 23], the "zero air-pressure" assumption is used to extend the small deformation theory of partially saturated soils to quasistatic large deformations. In more recent papers, the importance of suction is acknowledged but the water and air phases are typically assumed to be incompressible [24–26]. The mixture theory developed by Hassanizadeh was extended to finite strains and dynamics in [27–30]. Phenomenological models have also been developed at a regular pace [31–33]. For instance, models for the interaction between damage and fluid flow in porous rocks can be found in [34–36]. A cap evolution model for partially saturated soils is described in [37]. Configurational mechanics-based theories and models have also been developed, e.g., [38]. Scant attention has been paid to the deformability of the fluid phases but a few works do address that issue, e.g., [39, 40].

practical constraints (such as deployability into a massively parallel host code), so it must be understood that it includes only features that seem to be absolutely crucial for high-rate loading of granular media containing fluids. Model parameterization procedures are provided along with evidence of continued gaps in both data and model features.

As a practical matter, such models must be tailored to specific applications. Our focus on explosive loading, for example, justifies neglecting heat flow (but not dissipation), permeability (but not fluid compressibility), creep (but not rate-induced strengthening), etc. To legitimately use quasistatic data to help parameterize a high-rate model, these low-rate effects in the data must be "stripped out" or otherwise accounted for.

The constitutive model accommodates arbitrarily large bulk rotation (and rotation rates) via Lie derivatives, which means that the constitutive model is evaluated in the material frame, with the updated state transformed to the spatial frame by the host code. For simplicity, efficiency, and practical model calibration, the constitutive model is limited to applications having approximately stationary *reference* stretch directions, which is a very good approximation for many engineering applications, including buried explosives. In particular, for an initially isotropic medium, a model that uses a multiplicative decomposition of the deformation gradient is expected to reduce to a form having additive velocity gradients and therefore additive logarithmic strain rates whenever reference stretch directions are stationary, allowing a multiplicative model to be directly compared with one that uses an additive decomposition of strain rates [45] or additive velocity gradients [46] from the outset.

Any claim of superiority of a multiplicative decomposition requires showing that it agrees with data better than its otherwise equivalent additive version in cases of non-rotating reference stretch **directions** (not stretch magnitudes, as is commonly done). It is crucial to use actual data for such an assessment to avoid any temptation to inspect trends under unrealistically large material distortions (e.g., beyond any reasonable applicability of the constitutive model, such elasticity beyond the elastic limit). An assumption of stationary reference stretch directions is equivalent to having negligible $\mathbf{Z} = \text{skew}\mathbf{L} - \dot{\mathbf{R}} \cdot \mathbf{R}^{-1}$, where $\mathbf{L}$ is the spatial velocity gradient, and $\mathbf{R}$ is the polar rotation tensor. Physically, $\mathbf{Z}$ is the part of vorticity that is *not* coming from overall material rotation, and its norm $\|Z\|$ is an indicator of the rate of rotation of reference stretch directions. This "material vorticity," as we call it, can be proved to equal $\mathbf{Z} = \mathbf{VD} - \mathbf{DV}$, in which $\mathbf{V}$ is the left stretch from the polar decomposition, and $\mathbf{D}$ is the symmetric part of the velocity gradient [47]. If $\|Z\|$ is negligible during the times when accurate constitutive predictions are needed, then an algorithmically simple additive-strain-rate model is well justified, and the logarithmic strain rate itself is furthermore accurately approximated by the sym$\mathbf{L}$. To gain a sense of how good this approximation is even in conditions that *do* involve significant rotation of the reference stretch directions, the middle part of Fig. 2 shows a set of parallelograms that have been deformed in familiar conditions of simple shear. The lowest (red) parallelogram is the spatial configuration, while the blue parallelogram is the same deformation with overall rotation removed to

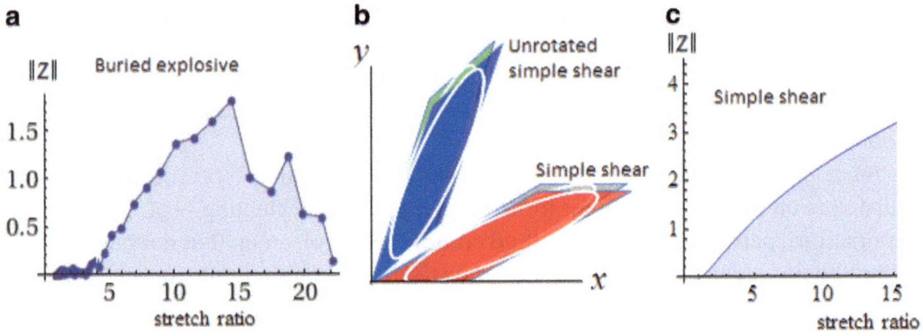

Fig. 2 Rationale for exploiting an additive strain-rate decomposition in buried-explosive simulations: (**a**) A $\|Z\|$ intensity (which quantifies the part of vorticity not attributable to overall material rotation) is seen to be negligible through a material stretch ratio of $\sim$5 in the most severely deformed zone of a buried-explosive simulation, well past the degree of stretch for complete loss of cohesion and hence, because the large deformations occur with negligible reference stretch rotation, the motion is within a range for which additive and multiplicative models can be made to agree with each other. (**b**) Visualization of simple shear at a stretch ratio $\sim$5 (=eccentricity of the inscribed ellipse). The red and blue are, respectively, spatial and unrotated configurations; the background images correspond to an approximation of Hencky strain rate by the symmetric part of the velocity gradient, which is clearly a good approximation in simple shear (and even better in buried-explosive deformations to the same stretch since the error is negligible when $\|Z\|$ is small). (**c**) A plot of $\|Z\|$ in simple shear provides a basis of comparison for much smaller values of $\|Z\|$ reached in the buried-explosive simulation at a comparable loading rate

produce a pure reference stretch. The principal directions of the inscribed ellipse are the stretch directions, shown at the moment when the deformed state has a stretch ratio (eccentricity of the ellipse) of $\sim$5. Knowing that the reference stretch directions are initially at $\pm 45°$, this simple-shear deformation has a visible but still only slight counter-clockwise rotation of the reference stretch directions. The background (green and gray) configurations show the effective deformations solved by a constitutive model that approximates the reference Hencky strain rate by the symmetric part of the unrotated symmetric part of the velocity gradient. Even in this case of rotating reference stretch directions, the error in deformed shape is visibly small enough that one would not expect the associated kinematics-related errors in the constitutive predictions to be even marginally significant in comparison to other modeling errors. Moreover, the simple shear case in Fig. 2b and c is more severe than what is observed in buried-explosive simulations, where the $\|Z\|$ indicator in Fig. 2a is clearly far smaller than what is observed in simple shear at a comparable strain rate and stretch. In conclusion, since buried-explosive simulations involve negligible rotation of *reference* principal stretch directions and since experimental results are invariably lacking for conditions of rotating principal directions (needed to properly validate a multiplicative model), it follows that there is no motivation to suffer the extra expense of a multiplicative model in comparison to efficient additive (reference) strain rate models.

Continuum constitutive models for granular media often suffer challenges of a solid, fluid, and gas. In a buried-explosive simulation, for example, the Jacobian might vary non-monotonically by orders of magnitude (from very stiff highly compressed states during the initial blast, to gas-like free expansion during ejecta propagation, finishing with recompression upon impacting an obstacle). A Lagrangian finite-element code cannot handle the large distortions, while an Eulerian approach typically corrupts the constitutive model's evolving internal state variables. To alleviate these problems, many simulations in this chapter use the open-source computational framework Uintah [48], which was selected because of its well-supported option to solve the momentum equations via the material point method (MPM) [49].

Homogenized models for granular media are of dubious value unless their implementation accounts for the fact that the aggregates (sand grains) are not sufficiently small to reach the continuum limit. As aleatory heterogeneity cannot be neglected, any local and deterministic theory (like the one described here) ought to be invoked within a host code using nonlocality and/or scale-dependent property variability to benefit both predictiveness and convergence. Despite its crucial importance, this topic is discussed in detail elsewhere in the literature [50, 51]. The scope of this chapter is, for brevity, focused exclusively on key features needed in the local component of the constitutive model with the understanding that these other considerations of scale effects and uncertainty must ultimately be incorporated as part of the wrapper to such a model. Additionally, as more "physics-based" models emerge that attempt to directly represent the effect of sub-continuum phenomena such as force chains developed between sand grains, a mathematical framework is emerging in parallel to handle the evaluation of literally million or trillion-term summations that would need to be evaluated to infer bulk behavior from those theories. Such methods, for example, have been applied to account for the evolution of literally hundreds of thousands of microcracks in ceramics [52], but have not yet been applied to granular media.

This chapter is limited to the following topics, selected as being either crucial in granular modeling efforts or ripe for future research:

1. Governing equations for an idealized (and open-source) constitutive model for partially or fully saturated soil subjected to high-rate loading.
2. Model parameterization procedures, with examples of fitted data.
3. Model behavior under varying porosity and saturation and validation against split-Hopkinson pressure bar experimental data.
4. Numerical challenges in both in the constitutive model and its host code.

2 The ARENA Model

This chapter describes a constitutive model called ARENA.[2] By limiting its scope to features that seem to be crucial for applications involving high-rate loading of partially saturated soils, ARENA is reasonably robust in buried-explosive simulations. The constitutive relations include a nonlinear bulk modulus model and a nonlinear Drucker–Prager limit surface with a compression cap that is fitted to crush-curve data. Building on Homel et al. [54] (which itself was specialized from the multipurpose KAYENTA geomechanics model [55]), pore pressure in ARENA is modeled as an isotropic contribution to backstress with a better-developed theory for partial saturation.[3]

Herein, details are omitted for any governing equations that are already well documented in preceding work. For example, the ARENA model adopts the same Duvaut-Lions viscoplasticity and damage theories that are documented in [60], and any host code using ARENA can be written to accommodate statistical variability and scale effects as documented in [51]. Features available in the more complicated KAYENTA model *(not* ARENA*)* include third-invariant dependence, kinematic hardening, advanced multipurpose thermodynamic equations of state, etc., all of which are well documented elsewhere and could easily be restored into the ARENA model at a considerable cost in code maintenance and robustness for relatively little gain in comparison to the model features retained as essential for soil-blast simulations (where, for example, approximately adiabatic—not isentropic—conditions provide simplifying constraints to a general EOS formulation). ARENA is superior to KAYENTA in its support for fully and partially saturated states as well as its model for adjusting parameters based on initial sample preparation (e.g., compaction through shaking), so these features are a primary focus of this chapter.

The ARENA model assumes a three-phase porous medium with constituents $\alpha = \{s, w, a\}$, where s is the solid skeleton, w is water, and a is air. Each material point is considered to be the center of mass of a representative volume element (RVE) region (ω) of volume dv and mass dm.[4] The region ($\omega^\alpha \subset \omega$) occupied by phase α at each material point has a volume dv^α and mass dm^α such that

$$dv = dv^s + dv^w + dv^a \quad \text{and} \quad dm = dm^s + dm^w + dm^a. \tag{1}$$

The *volume fraction* (f^α) of phase α at each point is defined as

[2] ARENA (available under open source [48]) has been verified against several analytical tests [53] and validated against data for a variety of sands at different moisture and initial states.

[3] Not part of KAYENTA, these governing equations are based on work of Uzuoka and Borja [56] (see also [27–30, 56–58]) which is, in turn, a simplification of the averaging theory proposed by Hassanizadeh and Gray [13, 14, 16, 17, 59].

[4] This assumption can significantly increase computational overhead if a highly expanded domain impacts an obstacle after a blast event. The problem arises because an RVE does not exist on this distended scale, and might be alleviated via enriched basis functions [61].

$$f^\alpha = \frac{dv^\alpha}{dv} \quad \text{so that} \quad \sum_\alpha f^\alpha = 1. \tag{2}$$

The volume fraction of the solid skeleton is

$$f^s = 1 - \phi, \quad \phi = \frac{dv^w + dv^a}{dv}, \tag{3}$$

where the ϕ is the *porosity*. The volume fractions of water and air are

$$f^w = \phi S_w, \; f^a = \phi S_a = \phi(1 - S_w), \quad S_w = \frac{dv^w}{dv^w + dv^a}, \tag{4}$$

where S_w is the *saturation*. The Biot parameter is a measure of the influence of porosity on the elastic bulk modulus:

$$B := 1 - \frac{\widetilde{K}^s(\widetilde{p}^s)}{K_s(p^s)}, \tag{5}$$

where K_s is the bulk modulus of the solid phase, p^s is the net pressure in the solid phase, $\widetilde{K}^s$ is the bulk modulus of the porous solid (i.e., the tangent slope of pressure vs. logarithmic strain of the composite material with all but the solid phase and void removed), and $\widetilde{p}^s$ is the pressure in the porous solid. Physical arguments leading to specific formulas used to compute ARENA's Biot parameter (which are, unfortunately, too detailed to be accommodated in this small overview chapter) can be found in a ResearchGate open-access technical report [62].

The governing equations are simplified with other (arguably tenuous) assumptions (such as pressure equilibrium between phases) and adiabatic conditions (justifying a mechanical model without ignoring dissipation or thermodynamic consistency). ARENA does not account directly for temperature dependence of material properties (which can certainly be done as described in KAYENTA documentation [55]), and consequently ARENA is designed for events involving one primary loading interval (e.g., initial shock compression) followed by a secondary unloading to a significantly distended state (free flight of disaggregated particles), and only qualitatively reasonable support for recompaction upon impact with an obstacle.

2.1 Constitutive Model

In rate form, the constitutive model is[5]

---

[5]This model is applied in the "unrotated" configuration to satisfy frame indifference. The applications of interest are assumed to have negligible rotation of *reference* stretch directions, thus

$$\begin{aligned}\dot{\sigma} &= \dot{\sigma}_{\text{eff}} + \dot{\alpha} = \dot{\sigma}_{\text{eff}} - \dot{B}\overline{p^w}\boldsymbol{I} - B\dot{\overline{p^w}}\boldsymbol{I} \\ &= g_1(\sigma, \eta, d^s) - g_2(\sigma, \eta, d^s)\overline{p^w} - g_3(\sigma, \eta, d^s)B,\end{aligned} \tag{6}$$

where g_1, g_2, and g_3 are assumed constitutive relations that can be expressed as

$$\begin{aligned} g_1(\sigma, \eta, d^s) &= \dot{\sigma}_{\text{eff}} = \mathbf{C}^{\text{pe}}_{\text{eff}} : d^s \\ g_2(\sigma, \eta, d^s) &= \dot{B} = -\frac{1}{K_s}\dot{\tilde{K}}_s + \frac{K_d}{K_s^2}\dot{K}_s = -\frac{1}{K_s\mathcal{B}}\left[\frac{\partial K_d}{\partial \overline{p^w}} - \frac{K_d}{K_s}\frac{\partial K_s}{\partial \overline{p^w}}\right]\text{tr}(d^s) \\ g_3(\sigma, \eta, d^s) &= \dot{\overline{p^w}} = \frac{1}{\mathcal{B}}\text{tr}(d^s). \end{aligned} \tag{7}$$

In the above equations, $\mathbf{C}^{\text{pe}}_{\text{eff}}(\sigma, \eta)$ is an effective elastic–plastic tangent modulus, $K_d(\sigma, \eta)$ and $K_s(\sigma, \eta)$ are the bulk moduli that contribute to the Biot parameter, and $\mathcal{B}(\sigma, \eta)$ is defined in a fairly complicated formula given in ARENA's fully detailed manual [62].

High-rate engineering applications of high-rate events often involve small enough rotation of reference principal stretch directions to easily justify an assumption that

$$d^s = d^{\text{e}} + d^{\text{p}} = \dot{\varepsilon}^{\text{e}} + \dot{\varepsilon}^{\text{p}}, \tag{8}$$

where ε^{e} is an elastic strain and ε^{p} is a plastic strain.[6] In addition to plastic strain, ARENA tracks the following internal variables (η):

$$\eta = \{\overline{p^w_{\text{e}}}, \overline{p^w_{\text{p}}}, B^{\text{e}}, B^{\text{p}}, \phi^{\text{e}}, \phi^{\text{p}}, S^{\text{e}}_w, S^{\text{p}}_w, \overline{X}^{\text{e}}, \overline{X}^{\text{p}}\}, \tag{9}$$

where $\overline{p^w_{\text{e}}}$ is the part of the pore pressure that goes to zero after a load that leads to plastic deformation is removed, $\overline{p^w_{\text{p}}} = \overline{p^w} - \overline{p^w_{\text{e}}}$ is the unrecoverable pore pressure which remains nonzero after removal of the load, B^{e}, B^{p} are the Biot parameter

making the symmetric part of the velocity gradient, d, a very good approximation to the rate of Hencky strain and hence (in this approximation) conjugate to Cauchy stress σ.

[6]As this theory is described in the context of an additive decomposition of strain rates, it carries with it an implicit potential limitation that reference stretch **directions** remain approximately stationary or that, by the time such rotations become large, the stress in the material is negligible due to disaggregation. These assumptions are quite reasonable in high-rate buried-explosive applications for which the model was designed. In this context, the unrotated symmetric part of the velocity gradient d equals the rate of reference Hencky strain $\dot{\varepsilon}$ and is conjugate to the unrotated Cauchy stress σ. Because multiplicative decompositions of the deformation gradient should (for initially isotropic media) become additive in these conditions, any claim of their superiority must be backed with (1) demonstrated equivalence to simpler additive models if stretch directions are stationary and (2) compellingly better agreement with validation data when stretch directions rotate.

values associated with elastic and inelastic deformation, ϕ^{e} is the recoverable porosity when the load is removed, ϕ^{p} is the unrecoverable or "unloaded" porosity, S_w^{e} is the recoverable saturation, S_w^{p} is the unrecoverable saturation, and $\overline{X}^{\mathrm{e}}$, $\overline{X}^{\mathrm{p}}$ are the hydrostatic compressive strengths associated with elastic and inelastic deformations, respectively.

2.1.1 Elasticity Model

The ARENA models allows for the possibility that during elastic deformation the elastic behavior of the soil can be coupled to the plastic behavior. This seems to be an essential feature to include because, for example, crushing out pore space must affect not only strength but also stiffness. Recognizing that stress depends not only on elastic strain but also on evolving internal variables, the rate equation for elastic behavior can be written as (see [60])

$$\dot{\boldsymbol{\sigma}} = \mathbf{C}^{\mathrm{e}} : \boldsymbol{d}^{\mathrm{e}} - \dot{\lambda}\boldsymbol{Z}, \tag{10}$$

where the purely elastic stiffness tensor, $\mathbf{C}^{\mathrm{e}}(\boldsymbol{\sigma}, \overline{p_{\mathrm{e}}^{w}}, B^{\mathrm{e}}, \phi^{\mathrm{e}}, S_w^{\mathrm{e}}, \overline{X}^{\mathrm{e}})$, is defined as

$$\mathbf{C}^{\mathrm{e}} = \frac{\partial \boldsymbol{\sigma}}{\partial \boldsymbol{\varepsilon}^{\mathrm{e}}} + \frac{\partial \boldsymbol{\sigma}}{\partial \overline{p_{\mathrm{e}}^{w}}} \otimes \frac{\partial \overline{p_{\mathrm{e}}^{w}}}{\partial \boldsymbol{\varepsilon}^{\mathrm{e}}} + \frac{\partial \boldsymbol{\sigma}}{\partial B^{\mathrm{e}}} \otimes \frac{\partial B^{\mathrm{e}}}{\partial \boldsymbol{\varepsilon}^{\mathrm{e}}} + \frac{\partial \boldsymbol{\sigma}}{\partial \phi^{\mathrm{e}}} \otimes \frac{\partial \phi^{\mathrm{e}}}{\partial \boldsymbol{\varepsilon}^{\mathrm{e}}} + \frac{\partial \boldsymbol{\sigma}}{\partial S_w^{\mathrm{e}}} \otimes \frac{\partial S_w^{\mathrm{e}}}{\partial \boldsymbol{\varepsilon}^{\mathrm{e}}} + \frac{\partial \boldsymbol{\sigma}}{\partial \overline{X}^{\mathrm{e}}} \otimes \frac{\partial \overline{X}^{\mathrm{e}}}{\partial \boldsymbol{\varepsilon}^{\mathrm{e}}}, \tag{11}$$

where $\boldsymbol{\varepsilon}^{\mathrm{e}}$ is the elastic part of the strain that is energy conjugate to the *unrotated* Cauchy stress (see [63] for possible strain measures). In the plastic coupling term in (10), $\dot{\lambda}$ is the plastic flow rate parameter (equal to the magnitude of the plastic strain rate when using normalized yield normal and flow tensors as advocated in [64]), and $\boldsymbol{Z}(\boldsymbol{\sigma}, \overline{p_{\mathrm{p}}^{w}}, B^{\mathrm{p}}, \phi^{\mathrm{p}}, S_w^{\mathrm{p}}, \overline{X}^{\mathrm{p}}; \boldsymbol{\varepsilon}^{\mathrm{p}})$ is a rank-2 elastic–plastic coupling tensor given by

$$\boldsymbol{Z} = h_p \frac{\partial \boldsymbol{\sigma}}{\partial \overline{p_{\mathrm{p}}^{w}}} + h_B \frac{\partial \boldsymbol{\sigma}}{\partial B^{\mathrm{p}}} + h_\phi \frac{\partial \boldsymbol{\sigma}}{\partial \phi^{\mathrm{p}}} + h_{S_w} \frac{\partial \boldsymbol{\sigma}}{\partial S_w^{\mathrm{p}}} + h_X \frac{\partial \boldsymbol{\sigma}}{\partial \overline{X}^{\mathrm{p}}}, \tag{12}$$

where

$$\dot{\overline{p_{\mathrm{p}}^{w}}} = \dot{\lambda} h_p, \quad \dot{B}^{\mathrm{p}} = \dot{\lambda} h_B, \quad \dot{\phi}^{\mathrm{p}} = \dot{\lambda} h_\phi, \quad \dot{S}_w^{\mathrm{p}} = \dot{\lambda} h_{S_w}, \quad \dot{\overline{X}}^{\mathrm{p}} = \dot{\lambda} h_X. \tag{13}$$

The "hardening" functions h_p, h_B, h_ϕ, h_{S_w}, and h_X require additional constitutive assumptions. For example, as explained in [60], a theoretical form for h_ϕ can be derived by approximating the solid matrix to be plastically incompressible. As explained in [55], the h_X function can be determined directly from the so-called crush curve for pressure vs. volumetric strain in hydrostatic loading.[7] During purely elastic loading, the coupling term is zero because $\dot{\lambda}$ is zero.

To simplify our calculations and the parameter calibration process, we assume that

$$\frac{\partial \boldsymbol{\sigma}}{\partial \boldsymbol{\varepsilon}^e} = \mathbf{C}^{\mathrm{e}}_{\sigma} = \left(\widehat{K} - \tfrac{2}{3}\widehat{G}\right) \boldsymbol{I} \otimes \boldsymbol{I} + 2\widehat{G}\,\mathbf{I}^{(s)}, \tag{14}$$

where $\widehat{K}$ and $\widehat{G}$ are, respectively, the tangent bulk and shear moduli of the mixture (both of which are generally functions of $\boldsymbol{\sigma}$, $\overline{p^w}$, B, ϕ, and S_w), $\boldsymbol{I}$ is rank-2 identity tensor, and $\mathbf{I}^{(s)}$ is the minor-symmetric part of the rank-4 identity tensor. The bulk and shear moduli of the mixture are determined using the dry moduli as described below. As detailed in [65], apparent pressure dependence of the shear modulus would be better justified by using a hyperelastic formulation (cf. [29]).

Bulk Modulus—Dry Soil

Data for dry sands can be fit reasonably well to the form[8]

$$\frac{\bar{p}_{\mathrm{eff}}}{K_s(\bar{p}_{\mathrm{eff}})} = b_0\, \overline{\varepsilon^e_v} + \frac{b_1 (\overline{\varepsilon^e_v})^{b_4}}{b_2 (\overline{\varepsilon^e_v})^{b_4} + b_3}. \tag{15}$$

Here, $b_0 > 0$, $b_1 > 0$, $b_2 > 0$, $b_3 > 0$, and $b_4 > 1$ are fitting parameters, $\overline{\varepsilon^e_v}$ is the volumetric elastic strain in the matrix, K_s is the bulk modulus of the solid grain material, and

$$\bar{p}_{\mathrm{eff}} := -\tfrac{1}{3}\mathrm{tr}(\boldsymbol{\sigma}_{\mathrm{eff}}) = -\tfrac{1}{3}\mathrm{tr}(\boldsymbol{\sigma} - \boldsymbol{\alpha}). \tag{16}$$

The bulk modulus of the solid grains is assumed to be given by

$$K_s(\overline{p^s}) = K_{s0} + n_s\,(\overline{p^s} - \overline{p^s_0}), \tag{17}$$

where K_{s0} and n_s are material properties, and $\overline{p^s_0}$ is a reference pressure.

The tangent bulk modulus of the dry soil is defined as

[7] For high-rate applications, a crush curve that is measured in quasistatic conditions must be converted to a form that removes low-rate creep (and other effects from heat transfer, fluid seepage, etc.) that would not occur in dynamic loading. The constitutive model must be furthermore supplemented with viscoplasticity parameters needed to predict apparent strengthening (beyond hardening) that does pertain to slow loading.

[8] This particular model has been developed by the authors to fit experimental data on Colorado Mason sand. Existing bulk modulus models, e.g., the KAYENTA model, were found to be inadequate for sand. Even if the numerical model is revised to use tabular data for this type of function, this approximate form can potentially serve as an interpolation function that would be more accurate (filling unknown gaps in data better) than a piecewise-linear fit—especially for low-data situations and for extrapolation beyond available data.

$$K_d(\bar{p}_{\text{eff}}) := \frac{d\bar{p}_{\text{eff}}}{d\overline{\varepsilon_v^e}}. \tag{18}$$

Then, using (15),

$$K_d(\bar{p}_{\text{eff}}) = \frac{[K_s(\bar{p}_{\text{eff}})]^2}{[K_s(\bar{p}_{\text{eff}}) - n_s\bar{p}_{\text{eff}}]}\left[b_0 + \frac{b_1 b_3 b_4 (\overline{\varepsilon_v^e})^{b_4-1}}{\left[b_2(\overline{\varepsilon_v^e})^{b_4} + b_3\right]^2}\right]. \tag{19}$$

To express (19) in closed form in terms of $\bar{p}$ we have to eliminate $\bar{\varepsilon}_v^e$. But a closed form expression for the volumetric elastic strain cannot be derived from the pressure model. So we find an approximate form of (15) by assuming $b_0 \to 0$, which is valid at moderate to large strains. Then, from (15) with $b_0 = 0$, we have

$$\overline{\varepsilon_v^e} \approx \left[\frac{b_3\bar{p}_{\text{eff}}}{b_1 K_s(\bar{p}_{\text{eff}}) - b_2\bar{p}_{\text{eff}}}\right]^{1/b_4}. \tag{20}$$

Shear Modulus—Dry Soil

A non-constant shear modulus may be needed to fit experimental data and to prevent negative values of Poisson's ratio in some simulations. In those situations, a variable Poisson's ratio (ν) is defined as

$$\nu(\bar{p}_{\text{eff}}) = \nu_1 + \nu_2 \exp\left[-\frac{K_d(\bar{p}_{\text{eff}})}{K_s(\bar{p}_{\text{eff}})}\right], \tag{21}$$

where ν_1 and ν_2 are the material parameters. The shear modulus is computed using the Poisson's ratio and the dry bulk modulus:

$$G_d(\bar{p}_{\text{eff}}) = \frac{3K_d(\bar{p}_{\text{eff}})(1-2\nu)}{2(1+\nu)}. \tag{22}$$

Elastic Moduli—Partially Saturated Soil

Saturated and dry soils are assumed to have the same shear modulus[9]:

$$\widehat{G}(\bar{p}_{\text{eff}}, \overline{p^w}, \overline{\varepsilon_v^p}, \phi, S_w) = G_d(\bar{p}_{\text{eff}}). \tag{23}$$

The tangent bulk modulus, $\widehat{K}$, of the partially saturated soil is computed using a variant of the Biot–Gassmann model for fully saturated rocks that is valid for long-wavelength displacements [67–69]. As detailed in [62], the resulting expression for the tangent bulk modulus of a partially saturated medium is given by

[9]Experiments at quasistatic strain rates indicate that the saturation can affect the shear modulus [66]. However, these effects are important only at low confining pressures. We are not aware of any analogous experimental studies for high-rate loading.

$$\widehat{K}(\bar{p}_{\text{eff}}, \overline{p^w}, \overline{\varepsilon_v^p}, \phi, S_w) = K_d(\bar{p}_{\text{eff}}) + \frac{B^2(\bar{p}_{\text{eff}})}{\dfrac{B(\bar{p}_{\text{eff}})}{K_s(\bar{p}_{\text{eff}})} + \phi\left(\dfrac{1}{K_f(\overline{p^w})} - \dfrac{1}{K_s(\bar{p}_{\text{eff}})}\right)}, \tag{24}$$

where K_d is the bulk modulus of the drained soil, K_f is the bulk modulus of the pore fluid, and K_s is the bulk modulus of the solid grains. For numerical tractability, upcoming sample calculations adopt simple idealized models. Namely, as detailed in [62], a Murnaghan EOS is used for the drained soil and fluid, while air is modeled as an ideal gas (selected not for realism but for tractability and to achieve appropriate trends in limiting cases with the expectation that other components of the model have far greater influence during the initial explosive loading phase).

2.1.2 Porosity, Saturation, Volumetric Strain

Having separate equations of state for the constituents (solid, fluid, and gas) requires theories for partitioning the total volumetric strain into strains for each phase. Details for the fairly standard assumptions invoked to do this are provided in the full ARENA theory manual [62]. A summary is given below.

Recall that the saturation is defined as

$$S_w = \frac{v^w}{v^a + v^w} = 1 - S_a \quad \Longrightarrow \quad \frac{v^a}{v^w} = \frac{1 - S_w}{S_w}, \tag{25}$$

where v^α is the volume occupied by phase α in the pore volume. Also, the porosity is defined as

$$\phi = \frac{v^a + v^w}{v^s + v^a + v^w} \quad \Longrightarrow \quad 1 - \phi = \frac{v^s}{v^s + v^a + v^w}. \tag{26}$$

The volumetric strain in each phase is defined as

$$\varepsilon_v^\alpha = \ln\left(\frac{v^\alpha}{v_0^\alpha}\right) \quad \text{where} \quad \alpha = \{s, w, a\}, \tag{27}$$

where v_0^α is the initial volume of phase α. The total volumetric strain of the mixture is

$$\exp(\varepsilon_v) = (1 - S_0)\phi_0 \exp(\varepsilon_v^a) + S_0\phi_0 \exp(\varepsilon_v^w) + (1 - \phi_0)\exp(\varepsilon_v^s), \tag{28}$$

where S_0 is the initial saturation and ϕ_0 is the initial porosity. With the assumption that the total volumetric strain can be additively decomposed into elastic and plastic parts, we have

$$\varepsilon_v = \varepsilon_v^{\text{e}} + \varepsilon_v^{\text{p}}. \tag{29}$$

Therefore,

$$\exp(\varepsilon_v^{\mathrm{p}}) = (1 - S_0)\phi_0 \frac{\exp(\varepsilon_v^a)}{\exp(\varepsilon_v^{\mathrm{e}})} + S_0\phi_0 \frac{\exp(\varepsilon_v^w)}{\exp(\varepsilon_v^{\mathrm{e}})} + (1 - \phi_0)\frac{\exp(\varepsilon_v^s)}{\exp(\varepsilon_v^{\mathrm{e}})}. \tag{30}$$

The equation for the evolution of saturation (assuming $\overline{p^w} = \overline{p^a}$) is

$$S_w(\overline{p^w}) = \frac{\mathcal{C}}{1 - S_0 + \mathcal{C}}, \quad \mathcal{C} := S_0 \exp\left[\overline{\varepsilon_v^a}(\overline{p^w}) - \overline{\varepsilon_v^w}(\overline{p^w})\right]. \tag{31}$$

Consider a soil that has undergone compressive loading with plastic volumetric deformation. Phase continuity implies that, after unloading, the saturation is different from the initial value of saturation even when the hydraulic conductivity of the material is zero. In that case, Eq. (31) implies that the residual saturation is a function of nonzero strains in the water and air. In turn, because air and water are assumed to be elastic, the saturation is a function of a residual pressure. This residual pressure can be identified with a "plastic" pore pressure which is a function of the plastic volumetric strain. The pore pressure computed in the ARENA model is this "plastic" pore pressure.

Recall that the porosity (ϕ) can be defined using

$$1 - \phi = \frac{v^s}{v} \quad \implies \quad \frac{1-\phi}{1-\phi_0} = \frac{v^s}{v_0^s}\frac{v_0}{v} = \exp(\overline{\varepsilon_v} - \overline{\varepsilon_v^s}) = \frac{\exp(\varepsilon_v^s)}{\exp(\varepsilon_v)}. \tag{32}$$

If we know the total volumetric strain and the volumetric strains in the fluids, we can use Eq. (28) to write the above as

$$1 - \phi = \frac{\exp(\varepsilon_v) - (1 - S_0)\phi_0 \exp(\varepsilon_v^a) - S_0\phi_0 \exp(\varepsilon_v^w)}{\exp(\varepsilon_v)}. \tag{33}$$

When the pore water and air pressure are equal to each other, and also equal to the intrinsic pressure in the solid grains), the porosity equation (33) can be expressed as

$$\phi(\overline{p^w}) = (1 - S_0)\phi_0 \exp\left[\overline{\varepsilon_v}(\overline{p^w}) - \overline{\varepsilon_v^a}(\overline{p^w})\right] + S_0\phi_0 \exp\left[\overline{\varepsilon_v}(\overline{p^w}) - \overline{\varepsilon_v^w}(\overline{p^w})\right]. \tag{34}$$

The primary cause of the residual pressure in a soil that has undergone volumetric plastic deformation is the irreversible change in porosity. This "unloaded" porosity is the quantity that is used in the ARENA model.

2.1.3 Rate-Dependent Plasticity

Strain-rate dependence is modeled using the classical Perzyna/Duvaut-Lions approach, where rate-dependent plasticity effects are applied after a rate-independent solution has been found [60]. The viscoplastic rate of deformation tensor in Duvaut-Lions form is

$$\boldsymbol{d}^{\text{vp}} = \begin{cases} (\mathbf{C}^{\text{e}})^{-1} : \left(\dfrac{\boldsymbol{\sigma} - \mathcal{P}\boldsymbol{\sigma}}{\tau}\right) & \text{if } f(\boldsymbol{\sigma}, \boldsymbol{\eta}) > 0 \\ 0 & \text{otherwise,} \end{cases} \tag{35}$$

where $\mathbf{C}^{\text{e}}$ is the elastic stiffness tensor, $\boldsymbol{\sigma}$ is the stress (which can be outside the rate-independent yield surface), $\boldsymbol{\eta}$ are internal variables, τ is a viscoplastic relaxation time, and $\mathcal{P}\boldsymbol{\sigma} = \boldsymbol{\sigma}_{qs}$ is the rate-independent (quasistatic) stress. It can be shown that [62] the rate-dependent (dynamic) stress state can be computed as

$$\boldsymbol{\sigma}^{n+1} = \boldsymbol{\sigma}_{qs}^{n+1} + \left[\left(\boldsymbol{\sigma}_{n+1}^{\text{trial}} - \boldsymbol{\sigma}^{n}\right) - \left(\boldsymbol{\sigma}_{qs}^{n+1} - \boldsymbol{\sigma}_{qs}^{n}\right)\right] R_H + \left(\boldsymbol{\sigma}^{n} - \boldsymbol{\sigma}_{qs}^{n}\right) r_h, \tag{36}$$

where the superscripts n and $n+1$ indicate states at times t_n and $t_{n+1} = t_n + \Delta t$, respectively, and

$$R_H := \frac{1 - \exp\left(-\Delta t/\tau^n\right)}{\Delta t/\tau^n} \quad \text{and} \quad r_h := \exp\left(-\Delta t/\tau^n\right). \tag{37}$$

The relaxation parameter τ is modeled as

$$\tau = T_1\, \dot{\varepsilon}^{-T_2}, \quad \dot{\varepsilon} := \|\boldsymbol{d}\|, \tag{38}$$

where T_1, T_2 are material parameters, and $\boldsymbol{d}$ is the symmetric part of the spatial velocity gradient. Next, we describe the rate-independent plasticity model used to compute $\boldsymbol{\sigma}_{qs}$. For simplicity, we assume that the internal state variables are not rate-dependent.

2.1.4 Rate-Independent Plasticity

For rate-independent plastic deformations (or as a submodel that is later revised to include rate effects) we use the canonical phenomenological model that consists of a yield function, a flow rule and the associated consistency condition, and models of internal variable evolution. The model is supplemented with a collapsing limit surface when damage is included.

ARENA Yield Function/Limit Surface

If the volumetric and deviatoric components of the total stress are

$$\bar{p} := -\tfrac{1}{3}\text{tr}(\boldsymbol{\sigma}) \quad \text{and} \quad \boldsymbol{s} := \boldsymbol{\sigma} + \bar{p}\boldsymbol{I} \tag{39}$$

we can define

$$\bar{p}_{\text{eff}} := -\tfrac{1}{3}\text{tr}(\boldsymbol{\sigma}_{\text{eff}}) = \bar{p} - B\,\overline{p^{w}} \tag{40}$$

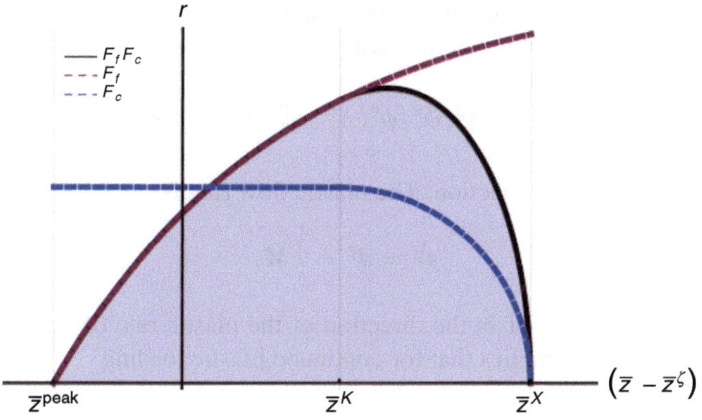

Fig. 3 A schematic of the ARENA yield surface in z-r space [54]

and

$$J_2^{\text{eff}} := \tfrac{1}{2}\boldsymbol{s}_{\text{eff}} : \boldsymbol{s}_{\text{eff}} \quad \text{where} \quad \boldsymbol{s}_{\text{eff}} := \boldsymbol{\sigma}_{\text{eff}} + \bar{p}_{\text{eff}}\boldsymbol{I} = \boldsymbol{\sigma} + \bar{p}\boldsymbol{I} = \mathbf{s}\,. \tag{41}$$

Then the ARENA yield function (sketched in Fig. 3) can be expressed as

$$f(\boldsymbol{\sigma}, B, \overline{p^w}, \overline{X}) = \beta\sqrt{J_2^{\text{eff}}} - F_f(\bar{p}_{\text{eff}})\, F_c(\bar{p}_{\text{eff}}, \overline{X}), \tag{42}$$

where

$$F_f(\bar{p}_{\text{eff}}) = a_1 - a_3 \exp[-3a_2\bar{p}_{\text{eff}}] + 3a_4\bar{p}_{\text{eff}} \tag{43}$$

and

$$F_c(\bar{p}_{\text{eff}}, \overline{X}) = \begin{cases} 1 & \text{for} \quad 3\bar{p}_{\text{eff}} \le \bar{\kappa} \\ \sqrt{1 - \left(\dfrac{3\bar{p}_{\text{eff}} - \bar{\kappa}}{\overline{X}_{\text{eff}} - \bar{\kappa}}\right)^2} & \text{for} \quad 3\bar{p}_{\text{eff}} > \bar{\kappa}\,. \end{cases} \tag{44}$$

Here a_i are material parameters, $\overline{X}_{\text{eff}}(\boldsymbol{\varepsilon}^{\text{p}}, B, \overline{p^w}) = \overline{X} - 3B\overline{p^w}$ is the shifted form of the apparent hydrostatic compressive strength ($\overline{X}/3$) of the partially saturated material, and $\bar{\kappa}$ is the branch point at which the cap function F_c starts decreasing until it reaches the hydrostatic strength point ($\overline{X}$):

$$\bar{\kappa} = 3\bar{p}_{\text{eff}}^{\text{peak}} - \left(3\bar{p}_{\text{eff}}^{\text{peak}} - \overline{X}_{\text{eff}}\right) R_c, \tag{45}$$

where $\bar{p}_{\text{eff}}^{\text{peak}}$ is the maximum hydrostatic tensile stress that the material can support and R_c is a cap ratio. Nonassociativity is modeled using the parameter β that modifies $\sqrt{J_2^{\text{eff}}}$, though we caution that nonassociativity is intrinsically unstable [70].

Yield Condition, Flow Rule, and Consistency
For continued plastic loading on the yield surface, the yield condition is

$$f\left(\boldsymbol{\sigma}, \overline{p^w_{\rm e}}, \overline{p^w_{\rm p}}, B^{\rm e}, B^{\rm p}, \phi^{\rm e}, \phi^{\rm p}, S^{\rm e}_w, S^{\rm p}_w, \overline{X}^{\rm e}, \overline{X}^{\rm p}; \boldsymbol{\varepsilon}^{\rm p}\right) = 0, \tag{46}$$

where $f(\ldots)$ is the yield function. The plastic flow rule is

$$\dot{\boldsymbol{\varepsilon}}^{\rm p} = \boldsymbol{d}^{\rm p} = \dot{\lambda}\boldsymbol{M}, \tag{47}$$

where $\boldsymbol{M}$ is the unit tensor in the direction of the plastic rate of deformation. The consistency condition implies that for continued plastic loading

$$\dot{f}(\ldots) = 0. \tag{48}$$

As the yield function depends on far more internal variables than encountered in typical engineering plasticity models, an atypically large number of terms are required to compute the yield function gradient, associated yield normal, as well as the ensemble hardening modulus governing hardening. The specific details (available in [62]) are straightforward and therefore not included in this short overview chapter.[10] The overall structure and numerical solution of ARENA's plasticity component falls in the general category of conventional numerical plasticity (cf. [60]) using recent advances in return algorithms for some of the distinctive challenges encountered in this class of model [54].

Evolution of Internal Variables
ARENA uses evolution equations for the yield surface similar to [55] or its effective-stress extension [54] with the notable exception that the porosity and saturation are tracked explicitly. For the simplified version of the ARENA *partially saturated* model discussed in this chapter, we require the internal variables to be of the form

$$\overline{p^w_{\rm p}} \equiv \overline{p^w_{\rm p}}(\varepsilon^{\rm p}_v),\quad B^{\rm p} \equiv B^{\rm p}(\varepsilon^{\rm p}_v),\quad \phi^{\rm p} \equiv \phi^{\rm p}(\varepsilon^{\rm p}_v),\quad S^{\rm p}_w \equiv S^{\rm p}_w(\varepsilon^{\rm p}_v),\quad \overline{X^{\rm p}} \equiv \overline{X}^{\rm p}(\varepsilon^{\rm p}_v)$$

so that

$$\begin{aligned} \dot{\overline{p^w_{\rm p}}} &= \dot{\lambda}\frac{d\overline{p^w_{\rm p}}}{d\varepsilon^{\rm p}_v}\,\mathrm{tr}(\boldsymbol{M}),\quad \dot{B}^{\rm p} = \dot{\lambda}\frac{dB^{\rm p}}{d\varepsilon^{\rm p}_v}\,\mathrm{tr}(\boldsymbol{M}),\quad \dot{\phi}^{\rm p} = \dot{\lambda}\frac{d\phi^{\rm p}}{d\varepsilon^{\rm p}_v}\,\mathrm{tr}(\boldsymbol{M}) \\ \dot{S}^{\rm p}_w &= \dot{\lambda}\frac{dS^{\rm p}_w}{d\varepsilon^{\rm p}_v}\,\mathrm{tr}(\boldsymbol{M}),\quad \dot{\overline{X^{\rm p}}} = \dot{\lambda}\frac{d\overline{X}^{\rm p}}{d\varepsilon^{\rm p}_v}\,\mathrm{tr}(\boldsymbol{M}). \end{aligned} \tag{49}$$

[10]To obtain a numerically tractable set of equations, several simplifying assumptions (such as neglecting elastic contributions to pore pressure) are adopted in [62], suggesting avenues appropriate for future research to justify or replace these choices.

In particular, we use the assumptions

$$\overline{p^w} = \overline{p^w_{\text{p}}}, \quad B = B^{\text{p}} \approx 1, \quad \phi = \phi^{\text{p}}, \quad S_w = S^{\text{p}}_w, \quad \overline{X} = \overline{X}^{\text{p}}, \tag{50}$$

and find that only the pore pressure evolution equation cannot be expressed in closed form. The rate equation for the pore pressure can be simplified to[11]

$$\frac{d\overline{p^w_{\text{p}}}}{d\varepsilon^{\text{p}}_v} = \frac{1}{\mathcal{B}^{\text{p}}}, \tag{51}$$

where

$$\begin{aligned} \mathcal{B}^{\text{p}} := & \frac{1}{(1-S_0)\exp\left[\overline{\varepsilon^{\text{p}}_v} - \overline{\varepsilon^a_v}\left(\overline{p^w_{\text{p}}}\right)\right] + S_0\exp\left[\overline{\varepsilon^{\text{p}}_v} - \overline{\varepsilon^w_v}\left(\overline{p^w_{\text{p}}}\right)\right]} \\ & \times \left[-\frac{(1-\phi^{\text{p}})\phi^{\text{p}}}{\phi_0}\left(\frac{S^{\text{p}}_w}{K_w\left(\overline{p^w_{\text{p}}}\right)} + \frac{1-S^{\text{p}}_w}{K_a\left(\overline{p^w_{\text{p}}}\right)}\right)\right. \\ & \left. + \frac{1-S_0}{K_a\left(\overline{p^w_{\text{p}}}\right)}\exp\left[\overline{\varepsilon^{\text{p}}_v} - \overline{\varepsilon^a_v}\left(\overline{p^w_{\text{p}}}\right)\right] + \frac{S_0}{K_w\left(\overline{p^w_{\text{p}}}\right)}\exp\left[\overline{\varepsilon^{\text{p}}_v} - \overline{\varepsilon^w_v}\left(\overline{p^w_{\text{p}}}\right)\right]\right]. \end{aligned} \tag{52}$$

In the above, S_0 is the initial saturation, ϕ_0 is the initial porosity, $\overline{\varepsilon^{\text{p}}_v}$ is the volumetric plastic strain, $\overline{\varepsilon^a_v}$ and $\overline{\varepsilon^w_v}$ are the volumetric strains in air and water, and K_a and K_w are the bulk moduli of air and water.

We only need to integrate the rate equation (51) for the pore pressure. For the evolution of porosity and the saturation, the closed form expressions are

$$\begin{aligned} \phi^{\text{p}}\left(\overline{p^w_{\text{p}}}\right) &= (1-S_0)\phi_0\exp\left[\overline{\varepsilon^{\text{p}}_v} - \overline{\varepsilon^a_v}\left(\overline{p^w_{\text{p}}}\right)\right] + S_0\phi_0\exp\left[\overline{\varepsilon^{\text{p}}_v} - \overline{\varepsilon^w_v}\left(\overline{p^w_{\text{p}}}\right)\right] \\ S^{\text{p}}_w\left(\overline{p^w_{\text{p}}}\right) &= \frac{\mathcal{C}^{\text{p}}}{1-S_0+\mathcal{C}^{\text{p}}}, \quad \mathcal{C}^{\text{p}} := S_0\exp\left[\overline{\varepsilon^a_v}\left(\overline{p^w_{\text{p}}}\right) - \overline{\varepsilon^w_v}\left(\overline{p^w_{\text{p}}}\right)\right]. \end{aligned} \tag{53}$$

The closed form expression for the drained hydrostatic compressive strength ($\overline{X}_d/3$) is

$$\overline{X}_d\left(\overline{\varepsilon^p_v}, \phi_0\right) - p_0 = p_1\left[\frac{1-\exp(-p_3)}{1-\exp\left(-p_3+\overline{\varepsilon^p_v}\right)} - 1\right]^{1/p_2}, \quad p_3 := -\ln(1-\phi_0), \tag{54}$$

[11] Details of the derivation of internal variable evolution equations can be found in [62].

where $\overline{X}_d/3$ is the drained hydrostatic compressive strength, and p_0, p_1, p_2, p_3 are model parameters.

The effective hydrostatic compressive strength for a partially saturated material is expected to be different from the drained value (we do not have any direct experimental data which supports that conjecture). We follow the approach used by Grujicic [8] and use a model of the form

$$\overline{X}_{\text{eff}}\left(\overline{\varepsilon_v^p}, \phi_0, S_w\right) - p_0 = \left[(1 - S_w) + p_1^{\text{sat}} S_w\right]\left(\overline{X}_d - p_0\right), \tag{55}$$

where $\overline{X} = \overline{X}_{\text{eff}} + 3B\overline{p^w}$ and $p_1 \times p_1^{\text{sat}}$ is the value of p_1 in a fully saturated material.

2.1.5 Density-Dependence Model

Unlike its predecessor models, ARENA accounts for the effect of sample preparation. Mechanical properties of sands and soils depend strongly on the initial density. Two otherwise identical sands of the same composition and grain morphology will, for example, exhibit significantly different behavior if one of them is shaken to compact the grains prior to subjecting it to a load.

Suppose that a sample's initial porosity ϕ_0 differs from the initial porosity ϕ_{ref} of the reference material that was used to calibrate the bulk modulus parameters and the crush curve. Following Pabst and Gregorova [71], the need to re-calibrate elastic properties for each different initial porosity is avoided by using a modulus scaling factor K_{fac}

$$K_{\text{fac}} = \exp\left[-\frac{\phi_0}{1-\phi_0} + \frac{\phi_{\text{ref}}}{1-\phi_{\text{ref}}}\right] \tag{56}$$

such that the bulk and shear moduli of the test material is

$$K_d \leftarrow K_{\text{fac}} K_d, \quad \text{and} \quad G \leftarrow K_{\text{fac}} G. \tag{57}$$

These scaled moduli are used in the partially saturated model.

The effect of sample compaction on hydrostatic strength is modeled phenomenologically and calibrated using high density dry Mason sand SHPB tests. Specifically, examination of data for high-density dry Mason sand used in SHPB tests shows that the crush curve for the high-density sand is of the same shape as the low density sand except scaled in the pressure axis. For the crush formula adopted from [55], this scaling can be achieved by simply replacing the reference-density fitting parameter p_1 as follows:

$$p_1 \leftarrow p_1 \exp\left[\rho_{\text{fac}} K_{\text{fac}}(K_{\text{fac}} - 1)\right] \tag{58}$$

where ρ_{fac} is calibrated using high density uniaxial-strain compression data and has a value between 1 and 10.

3 Parameter Fitting

The ARENA model of the previous section was developed iteratively in cycles of development based on parameter fitting from training data, to ultimately apply the fitted (and hence no longer tunable) model to validation simulations. This section describes on the fitting step, showing how the parameters of a moderately complex soil plasticity model for high-rate applications can be determined using limited—often quasistatic—data.[12]

Model parameters for ARENA were determined in eight steps:

1. Stage 1: Fit the bulk modulus model for the dry soil using low strain-rate hydrostatic compression data. Fit a shear modulus model assuming a constant Poisson's ratio (experimental data from [72–75]).
2. Stage 2: Fit the hydrostatic compressive strength model (crush curve) for the dry soil using low strain-rate hydrostatic compression data (experimental data from [72–75]).
3. Stage 3: Fit the limit surface model for the dry soil using low strain-rate triaxial compression data (experimental data from [72–75]).
4. Stage 4: Fit the rate-dependence parameters for the dry soil using high rate split Hopkinson pressure bar (SHPB) data (experimental data from [76–78]).
5. Stage 5: Fit the damage model for the dry soil using high rate SHPB data (experimental data from [76–78]).
6. Stage 6: Fit the density-dependence model for the dry soil using high rate SHPB data (experimental data from [76–78]).
7. Stage 7: Fit the fully saturated hydrostatic strength parameter for the saturated soil using SPHB data for partially saturated soil (experimental data from [76–78]).
8. Stage 8: Fit the parameter variability model using low strain-rate triaxial compression data (experimental data from [72–75]).

These steps presume the idealized case that all required data are available, spanning ranges expected in applications. As is common for complex constitutive models, some hand fitting of parameters might be needed along with engineering judgment for data extrapolation. Detailed examples of each step are provided in [62]. Only some highlights are provided here in this overview chapter.

For the bulk modulus and crush curve models, loading and unloading segments must be extracted from hydrostatic stress–strain curves. These curves for dry

---

[12]Such efforts should be seen as no more objectionable than long-accepted similar procedures in gas dynamics, where the low-rate (isothermal) pressure–volume curve along with the constant-pressure specific heat and thermal expansion properties is sufficient to infer the isentropic properties required for a purely mechanical model to accurately model acoustic waves. The key in such simplifications is to recognize that the resulting mechanical model is, by definition, limited in its application scope.

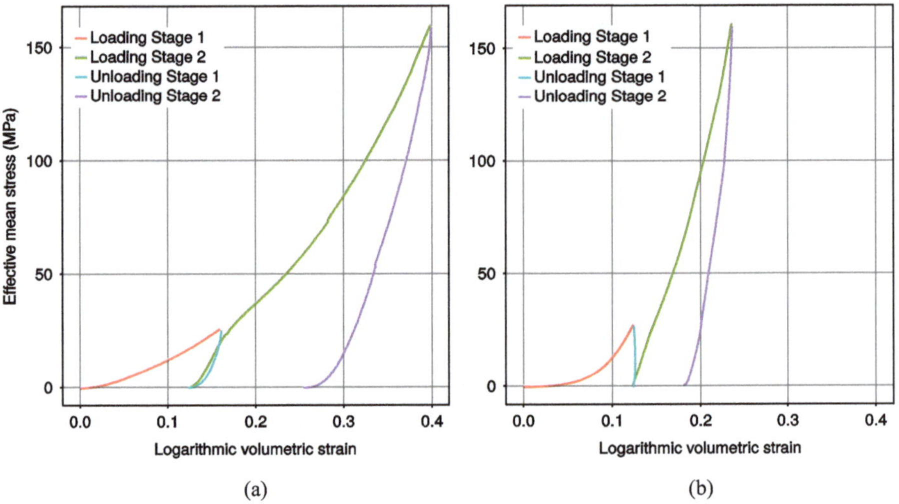

Fig. 4 Load–unload curves for sand and clay under hydrostatic compression (new data provided by collaborator John McCartney, U. Colorado, Boulder). (**a**) Dry Colorado Mason sand. (**b**) Dry Boulder clay

Colorado Mason sand and Boulder clay are shown in Fig. 4. As mentioned in the introduction, creep and/or fluid seepage (especially pronounced in the clay data) must be removed (e.g., by using a viscoelastic model) to obtain crush-curve data appropriate for high-rate loading in which these effects are negligible. Ideally, quasistatic crush data should be taken at a variety of temperatures in order to convert these effectively isothermal crush curves to the adiabatic curves needed for high-rate loading—bringing rigor to this process is an excellent avenue for future research.

Bulk modulus parameters must be fitted to the unload curves. Plastic crush (void collapse) parameters must be fitted to the loading portion of the data. Constrained optimization is typically needed to avoid inadmissible results (such as a negative bulk modulus). The elastic part of the unloading curves must be removed to obtain only the inelastic part, which is fitted to the model for porosity's dependence on pressure. Details of these fitting procedures, as well as fitted moduli/crush parameters for Colorado Mason sand and Boulder clay are provided in [62].

Following fitting procedures in [55], peak stress difference values must be extracted from a suite of triaxial compression tests conducted at a variety of lateral confining pressures. The values of the first and second stress invariants, I_1 and $\sqrt{J_2}$, at these peak states must be recorded for each experiment. The collection of these data pairs is fitted to the model's formula for pressure-dependent ultimate strength (which plays a role similar to the limit line in critical-state/cam-clay models). Details of the process and final fitted parameter values are provided in [62], resulting in the fitted strength curves in Fig. 5. In particular, for the Boulder clay data in Fig. 5b, the pressures attained in the triaxial experiments were relatively low and a manual fit to the data was needed to match SHPB test data.

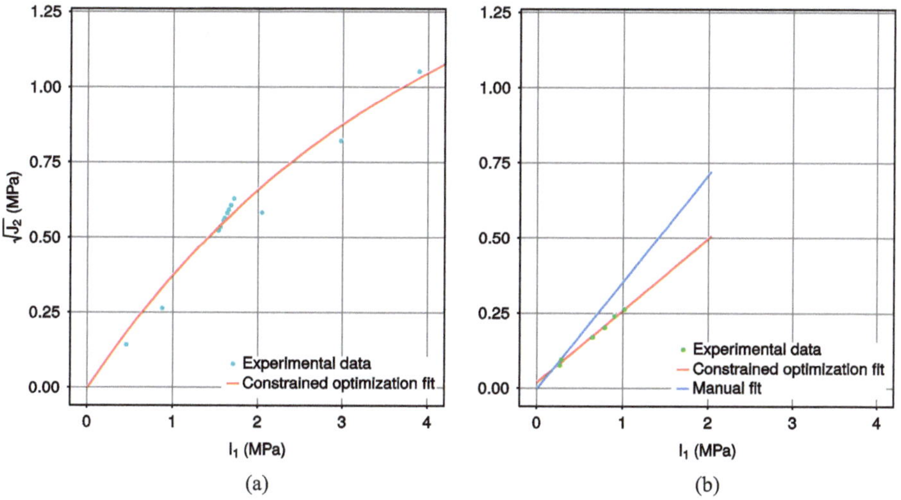

Fig. 5 Yield functions fitted to the combined triaxial loading data for dry sand and clay. Experimental data provided by John McCartney, U. Colorado, Boulder. (**a**) Dry Colorado Mason sand. (**b**) Dry Boulder clay

If the fitted rate-independent limit surface is used to simulate the response during a high-rate process such as a split-Hopkinson bar test, the predicted stresses are much lower than their observed values. This common observation is the primary evidence motivating a high-rate component in the model. The general procedure for fitting ARENA's Duvaut-Lions rate-dependence parameters is described in [60] and [55]. Details of this fitting task (and final parameters for Colorado Mason sand and Boulder clay) are provided in [62]. Resulting comparisons of model and data are shown in Fig. 6.[13] The initial rate of increase of both the radial and the axial stress is modeled well by ARENA. However, the axial stress in the sand in Fig. 6a exhibits a change in slope at approximately 10 MPa. This behavior can be replicated by increasing the initial hydrostatic strength, but at the expense of failing to match the quasistatic hydrostatic loading data.

The ARENA model is essentially a specialized cap-plasticity model (with enhancements for fluid effects and soil preparation), so its meridional yield profile is in the shape of an evolving teardrop. As depicted in Fig. 7, the movement of the stress state (blue dashed line) for experimental SHPB data is well captured with the ARENA model (red line); the sequence of evolving teardrop yield surfaces is controlled primarily by the cap hardening evolution equations. Observe that the clay fails at a much smaller stress than the sand. The colorbar in Fig. 7a shows the color of the evolving yield surface as a function of time. Notice that the initial yield

[13]Note that the ARENA model does not include damage at this stage and therefore we are only concerned with the rise part of the stress–time curves.

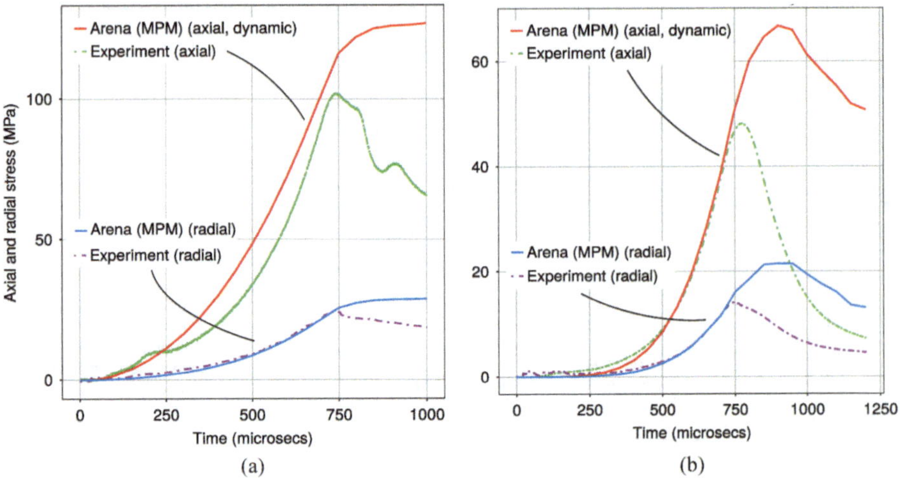

Fig. 6 Predicted and experimental stress–time curves for split-Hopkinson pressure bar uniaxial compression tests on dry sand and clay. Experimental data provided by Hongbing Lu from U. Texas at Dallas. (**a**) Dry Colorado Mason sand. (**b**) Dry Boulder clay

surface teardrop is too small to be visible in the plot and becomes large enough only after around 0.2 ms. Also note that for the Boulder clay sample, the experimental and simulated curve overlap for almost 0.5 ms (until r is around 25 MPa) during the loading phase.

Damage model parameters can also be fit using the uniaxial-strain compression data from SHPB tests. However, there are two issues that have to be considered in this case. The first is that even though the region of interest is only the unloading part of the axial and radial stress–time curves, the damage model starts to affect the stress before the peak stress is attained. The second issue is that the ARENA model does not consider the possibility of a decrease in hydrostatic compressive strength after the peak stress has been reached. These factors preclude rigorously fitting of a curve by minimizing a convex objective function. In particular, ARENA will not predict the observed decrease in mean stress with increasing volumetric strain. Therefore, it is more convenient to use a design-of-experiments approach to find an estimate of the damage model parameters. Details are provided in [62], which gives the corresponding parameter values for dry Colorado Mason sand and similar data and plots for dry Boulder clay. As seen in Fig. 8, damage provides a much better fit to the data than the non-damaging predictions in Fig. 6. The corresponding Fig. 9 shows damage collapse of the yield surface that was missing in the non-damaging simulation in Fig. 7.

It is worth pointing out a feature of the stress–time plot for the radial direction (see Fig. 8a), i.e., the damage model appears to have little effect on the predicted radial stress. Similarly, if we examine the yield surface in Fig. 9a, we observe that the damage model does not decrease the hydrostatic strength. The predicted radial

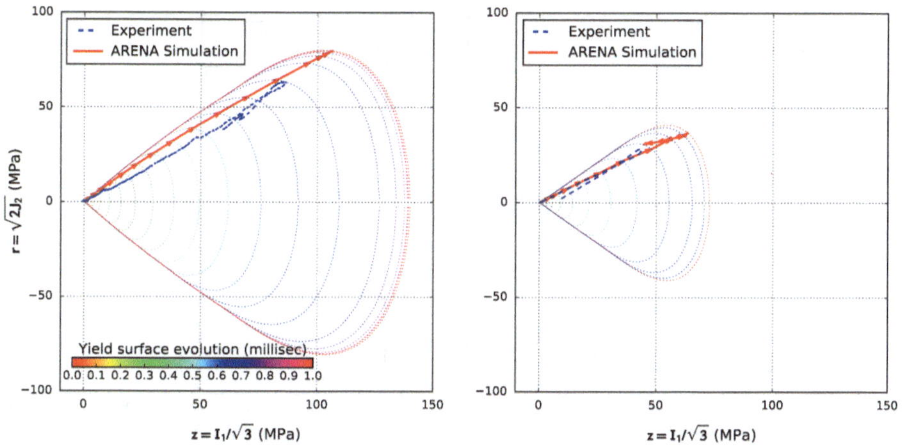

(a) (b)

Fig. 7 Dynamic stress in z-r-space—comparisons between predictions (solid) and experimental data (dashed) from split-Hopkinson pressure bar uniaxial compression tests. The teardrop shapes depict snapshots of the yield surface at a few discrete times. The end points of the arrows on the solid line depict the computed stress states at each of the discrete times for which the yield surface has been plotted. Experimental data provided by Hongbing Lu from U. Texas at Dallas. (**a**) Dry Colorado Mason sand. (**b**) Dry Boulder clay

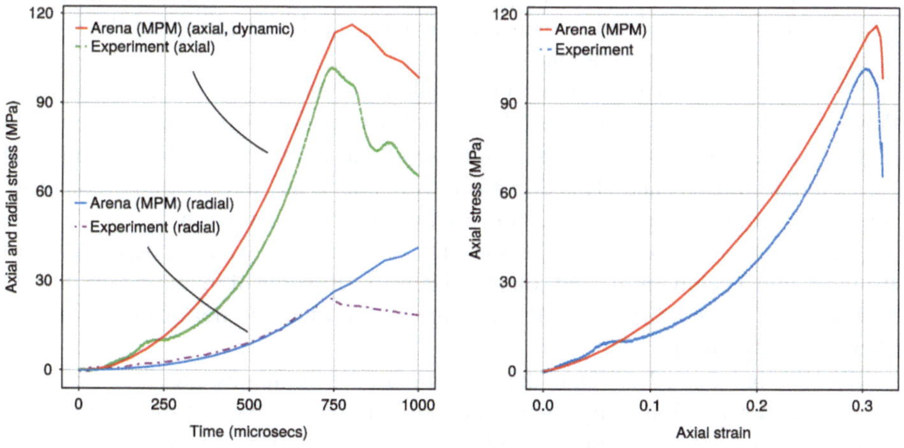

(a) Axial and radial stress vs. time (b) Axial stress vs. axial strain

Fig. 8 Dry Colorado Mason sand (**a**) stress–time curves predicted by the ARENA model (with damage) and the experimental split-Hopkinson pressure bar data, and (**b**) axial stress vs. axial strain. Experimental data provided by Hongbing Lu from U. Texas at Dallas

stress continues to rise after the sample has failed because of the path of the stress in z-r space. However, the experimental data appears to show that after failure the stress unloads almost exactly along the loading path. This effect can be predicted by

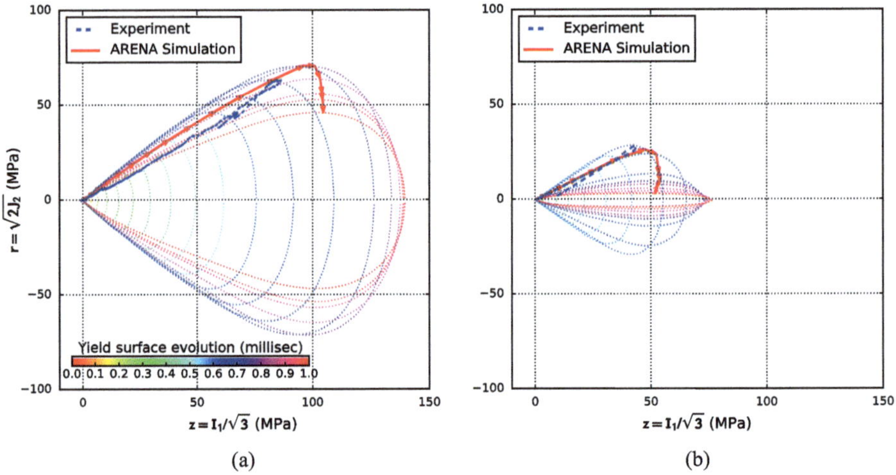

Fig. 9 Comparison of ARENA model (with damage) and the experimental split-Hopkinson pressure bar data showing yield surface evolution and stress paths in z-r-space. Experimental data provided by Hongbing Lu from U. Texas at Dallas. (**a**) Dry Colorado Mason sand. (**b**) Dry Boulder clay

ARENA only if the hydrostatic strength is allowed to drop back it a small value after failure. We are unsure whether the experimental data show a physical effect or just an artifact of the instrumentation.

Figure 9b shows a plot of the predicted stress path in z-r space (red) when damage is included in the Boulder clay model. We observe that the size of the yield surface in the r-direction decreases more rapidly as damage accumulates than that for Colorado Mason sand. This effect is mostly due to the deformation gradient vs. time curve from the experiment that is used to drive the simulation.

SHPB tests on dry Colorado Mason sand revealed a very strong dependence of the mechanical behavior of sands on initial density. Benefits of including the density scaling (Eqs. 57 and 58) are evident in Fig. 10, for which model fitting details are also provided in [62].

Recall from Sect. 2.1.4 that the ARENA model computes the pore pressure, porosity, and saturation using a simplified mixture theory. As a result, the response of a partially or fully saturated soil can be inferred from the response of the dry soil without fitting additional saturation-related parameters. However, in some situations, an extra saturation parameter (p_1^{sat}) (see Eq. 55) is needed to match the experimental data.

This extra parameter was not used for our simulations of Colorado Mason sand. The implication is that the ARENA model is able to reproduce the response of partially saturated soils even though the model parameters are derived purely from dry sand experiments. But experimental data for Boulder clay indicated that the processes involved were more complex than for sand [78] and required the additional parameter (p_1^{sat}). Resulting comparisons of model with data are shown

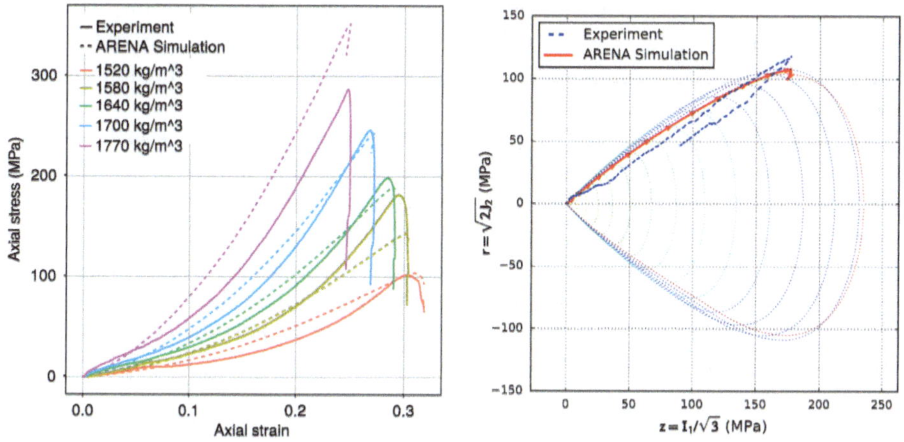

(a) Axial stress vs. axial strain (b) Dry Colorado Mason sand density = 1640 kg/m3

Fig. 10 (**a**) Axial stress vs. axial strain for sands of various initial densities: model vs. split-Hopkinson pressure bar experimental data on dry Colorado Mason sand. (**b**) Plot of stress path and yield surface in *z*-*r* space for a validation sample. Experimental data provided by Hongbing Lu from U. Texas at Dallas.

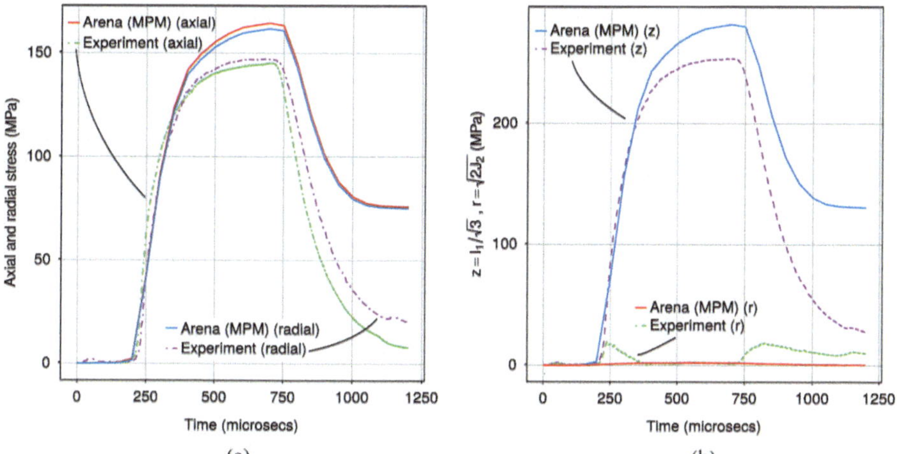

(a) (b)

Fig. 11 Comparisons of predicted axial, radial, mean, and deviatoric stress with experimental data from split-Hopkinson pressure bar uniaxial compression tests on 90% saturated Boulder clay. Experimental data provided by Hongbing Lu from U. Texas at Dallas. (**a**) Axial/radial stress vs. time. (**b**) z, r vs. time

in Fig. 11. Better fits to a particular sample can be found by tuning the crush curve parameter, but at the expense of worse fits to out-of-sample data. Note that the small difference between the radial and axial stresses are indicative of small hoop strains in the SHPB experiment. The variability of the stress difference observed in the experiments can be attributed to the lack of sensitivity of the instrumentation to small hoop strains. However, since the r-component of the stress is critical in most soil plasticity models, experimental errors in determining r should be minimized if model validation tests are to be accurate.

Statistical variation (as well as specimen scale effects) in yield parameter values from point-to-point in a soil should ideally be extracted from the parameter fitting process. This process has been applied successfully to modeling ceramics [51], but experimental data is typically incomplete, implying that unique values of the fit parameters cannot be found. Lack of data, however, does not imply no *need* for data. Accordingly, any truly systematic experimental program should report not only median observations, but also repeatability of the data as quantified by statistically significant measurement of variance and scale effects.

4 Model Behavior

This section illustrates how the internal variables in the model vary with changes in porosity and saturation. The volumetric plastic strain (which is a proxy for the void ratio) is treated as an independent variable in this section.

Figure 12 shows the computed pore pressure as a function of the volumetric plastic strain. Notice that the pressure in the air causes an increase in pore pressure even when the saturation is zero.

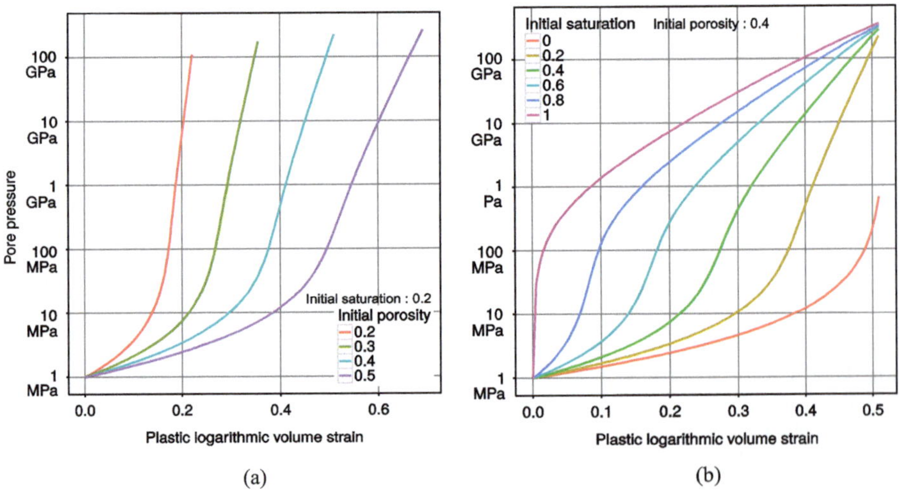

Fig. 12 Evolution of pore pressure as a function of plastic volumetric strain predicted by the ARENA model for various values of initial porosity and saturation. The pore pressure, porosity, saturation, and hydrostatic strength are evolved as internal variables in the model. The plastic volumetric strain is an input in these calculations. (**a**) Varying initial porosity. (**b**) Varying initial saturation

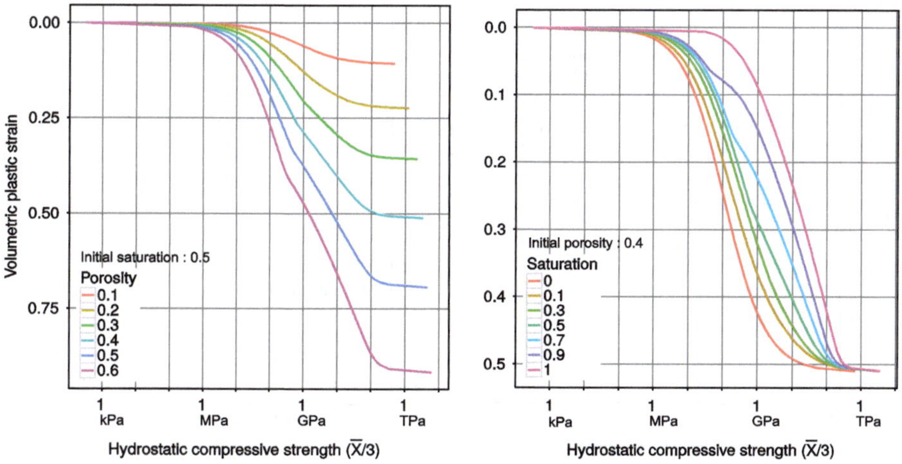

(a) Varying initial porosity (b) Varying initial saturation

Fig. 13 Predicted evolution of hydrostatic compressive strength as a function of plastic volumetric strain for several values of initial porosity and saturation. The state at which the plastic strain becomes constant corresponds to the crushing out of all pore spaces. In (**b**), an increase in saturation leads to an appropriate increase in hydrostatic strength at constant volumetric plastic strain

The effect of varying porosity and saturation on the hydrostatic compressive strength is shown in Fig. 13. The data are presented in a form similar to that used in quasistatic consolidation curves and should be compared to such curves for a better appreciation of the effects predicted by ARENA. Note that the hydrostatic compressive strength continues to increase after all voids have been crushed out and that there is a cusp in the curves due to the presence of air in the voids. In actual experiments, we would expect the pore air to go into solution in the pore water beyond some threshold compression value. ARENA does not model that process.

Figure 14 shows how the porosity computed by ARENA changes as a function of the compressive volumetric plastic strain. If the soil is dry, the porosity decreases almost linearly with increasing volumetric strain. However, as the water content is increased, ARENA predicts a drop in porosity followed by an increase as compression progresses. This is because the total plastic volumetric strain becomes less than the volumetric strain in the pore water. Whether such an effect is physically realistic can only be determined through micromechanical studies.

Plots of saturation as a function of volumetric plastic strain are shown in Fig. 15. These results correspond to intuition in that the saturation increases rapidly and then asymptotes to the fully saturated condition. But the correctness of the shapes of the saturation evolution curves needs to be confirmed with experimental or micro-simulation data. We are not aware of any micromechanical studies that have

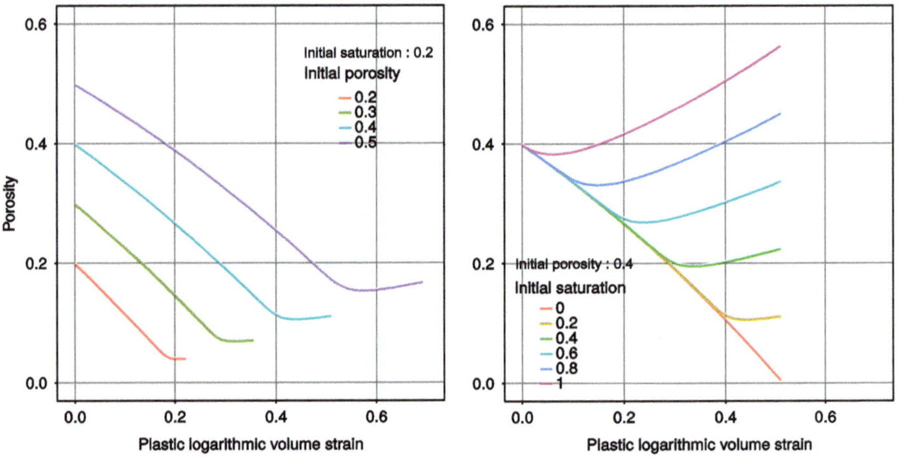

(a) (b)

Fig. 14 The evolution of porosity as a function of plastic volumetric strain (a proxy for the void ratio) for several initial values of porosity and saturation. (**a**) Varying initial porosity. (**b**) Varying initial saturation

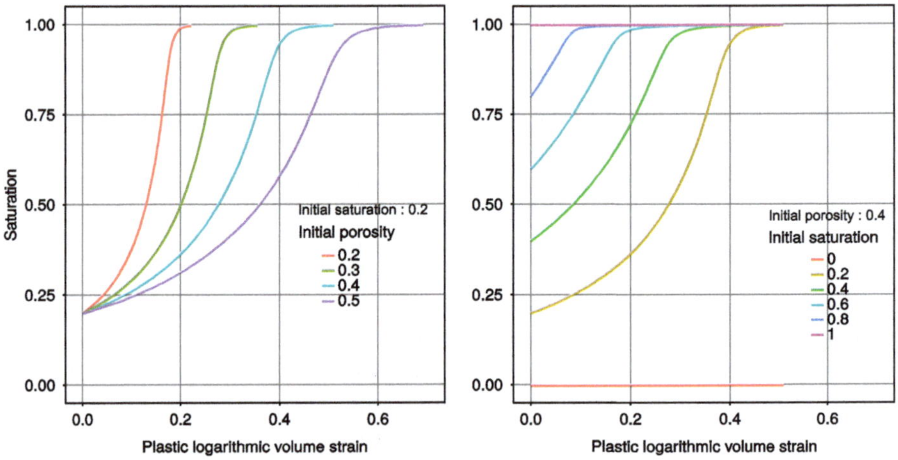

(a) (b)

Fig. 15 Evolution of saturation as a function of the plastic volumetric strain for a range of initial porosities and saturation levels. (**a**) Varying initial porosity. (**b**) Varying initial saturation

explored this issue. A finite element/discrete element coupled approach has been developed to explore some of the micromechanics of these materials but, so far, application has been limited to dry materials [79].

5 Model Verification and Validation

Typical verification tests for pressure-dependent nonlinear plasticity models for granular material include hydrostatic compression, uniaxial compression, uniaxial tension, and multiaxial strain loading paths, all of which are affine (single element) deformations. A momentum balance solver (in this case the material point method—MPM) and a complete constitutive model implementation for the momentum solver are used to compute predicted values of stresses, plastic and elastic strains, and internal variables in non-affine deformations. As summarized in [55], predicted values may then either be compared with model predictions from, for example, the bulk modulus/crush curve model, trend tests, or manufactured solutions (cf. [53]).

The full ARENA documentation [62] provides evidence of passing the following verification tests:

1. hydrostatic compressive loading–unloading of a dry sand,
2. hydrostatic loading of a fully saturated sand,
3. uniaxial compressive loading of a dry sand,
4. uniaxial compressive loading of a fully saturated sand,
5. uniaxial tensile loading of a dry sand, and
6. multiaxial loading–unloading of a dry sand.

We summarize results from two validation cases here:

1. uniaxial strain SHPB on a Mason sand (dry density 1700 kg/m^3) containing 18.4% water by weight, and
2. uniaxial strain SHPB on a Boulder clay (dry density 1300 kg/m^3) containing 12.8% water by weight.

The ARENA model requires the initial density, porosity, and saturation as inputs. The procedures for setting these and other parameters are provided in [62] along with simulation details and more detailed commentary for the following figures. Overall, the conclusion is that the ARENA model is able to quantitatively emulate the main features in the data.

Figure 16a shows axial and radial stresses extracted from SPHB tests on a Colorado Mason sand sample containing 18.4% water by weight. These stresses are compared with ARENA calculations with a MPM host code. The predicted and experimental values are remarkably similar, particularly when we take into consideration the fact that the ARENA model was calibrated with dry sand data. As has been observed for other classes of geomaterials, rate dependence in high-pressure conditions could be furthermore sensitive to the degree of confinement, providing a boost to initial strength in comparison to hydrostatic data; while such a feature was omitted from the minimal ARENA model, theory an implementation for it is available in the more general-purpose model on which ARENA is based [55].

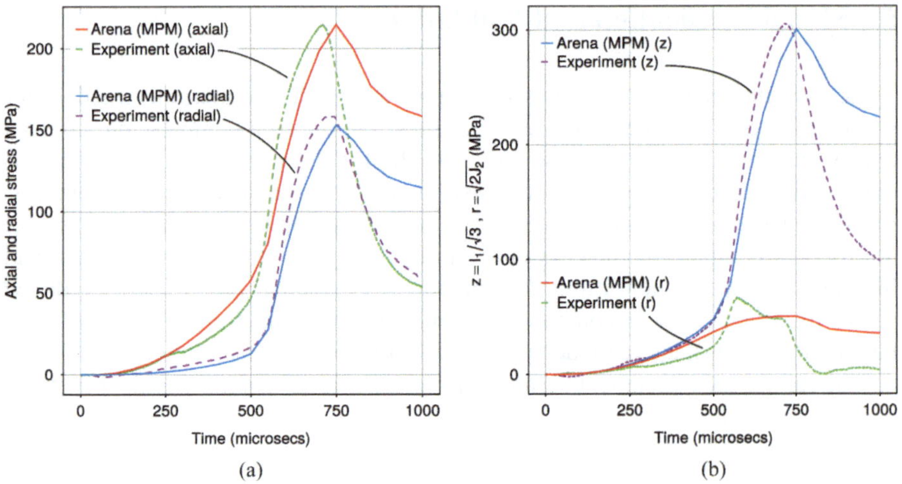

Fig. 16 Validation of ARENA using uniaxial strain split-Hopkinson pressure bar experimental data for wet Colorado Mason sand containing 18.4% water by weight. (Experimental data provided by Hongbing Lu from the University of Texas at Dallas.) (**a**) Axial/radial stress vs. time. (**b**) z and r vs. time

Consider the same data shown as isomorphic stress invariants vs. time.[14] Figures 16b and 17b reveal that that the experimental data from the r stress invariant (proportional to equivalent shear) fluctuates more than the z stress invariant (proportional to pressure), and this is believed to be attributable to noise in SHPB lateral stress measurements. *A useful visual assessment of noise in experimental data is a plot of the stress in isomorphic r vs. z or more conventional (but non-isomorphic) $\sqrt{3J_2}$ vs. $I_1/3$ (commonly called q-p) coordinates [62].*

The Colorado Mason sand plots in Fig. 16 show better correspondence between ARENA predictions and experiments than the Boulder clay plots in Fig. 17. This is partly because there is more variability in the Boulder clay data [78] and also because Boulder clay has a more complex microstructure and chemical composition

[14]These invariants, denoted r and z, do not refer to radial and axial directions of an SHPB rod. Instead, they are radial and axial directions in six-dimensional stress space, defined such that z is the projection of stress tensor onto the hydrostat, while r is the hyper-distance of the stress from the hydrostat, making these, respectively, the magnitudes of the isotropic and deviatoric parts of stress. Unlike more traditional invariants (such as equivalent stress $q = \sqrt{3J_2}$ and pressure $p = I_1/3$), the (r, z) stress invariants are isomorphic to stress space, making lengths and angles in an (r, z) plot identical to those in stress space. For axisymmetric problems, r is a multiple of the difference between axial and lateral SHPB stresses, and therefore discrepancies in r will be magnified whenever the stress difference is small in comparison to the mean stress. Other applications involving larger shearing stresses, on the other hand, would likely exhibit errors that would highlight a long-standing need for induced anisotropy (requiring greater resources to develop and calibrate than what is available in a typical engineering budget).

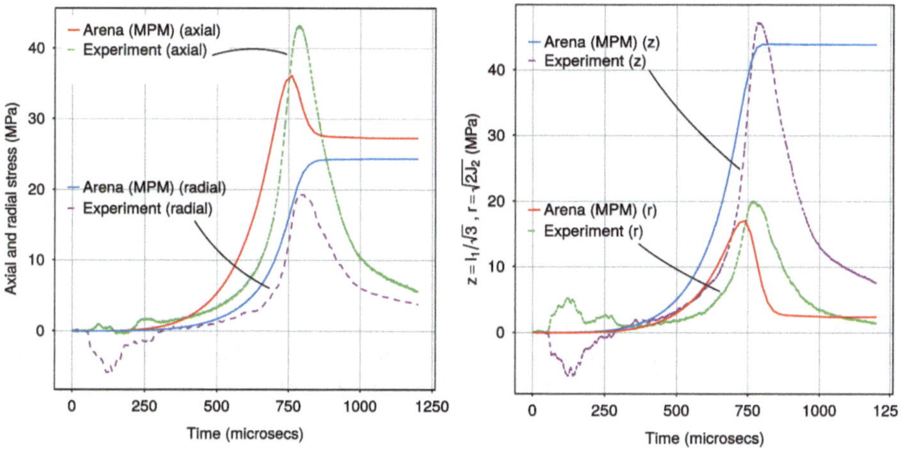

(a) (b)

Fig. 17 Validation of ARENA with wet Boulder clay split-Hopkinson pressure bar data. The clay contains 12.8% water by weight. Note: r and z stress invariants, not axial and lateral stresses. (Experimental data provided by Hongbing Lu from the University of Texas at Dallas.) (**a**) Axial/Radial stress vs. time. (**b**) z and r vs. time

than sand. We also observe that there is more noise in the Boulder clay SPHB data, making it more difficult to compare simulations with experiment.

5.1 Explosion Simulations

Of particular interest in engineering design is the situation where an explosion occurs under a vehicle. Experimental studies have shown that a V-shaped hull in the undercarriage of a vehicle is effective at reducing the effect of the blast. The following simulations show that ARENA also predicts similar results.

Consider a Boulder clay layer containing 13% water by weight inside a centrifuge rotated at a speed that leads to an acceleration of 20 g in the soil. The centrifuge has internal dimensions of $1.2 \times 1 \times 0.19$ m. A metal hull is placed 5.1 cm above the soil and an explosion is initiated inside the soil at a depth of 5.1 cm. Three hull shapes are tested: a flat hull, a rounded hull, and a V-shaped hull (the volumes of the three hulls are identical). Figure 18 shows the initial configurations of the V-hull simulation and the deformed state a short time (1.2 ms) after the explosion has been initiated.

However, if we examine the total momentum imparted to each hull, we notice that the V-shaped hull attains a smaller velocity than the flat or the round hull (see Fig. 19). In fact, the flat and the round hull have essentially the same momentum transferred to them by the explosion. This observation suggests that the ARENA

Fig. 18 Initial and deformed configurations after an explosion at 20 g in Boulder clay containing 13% water by weight. Zones of movement are visible in coloring by velocity magnitudes ranging from 0 to 150 m/s. (**a**) V-shaped hull: initial. (**b**) V-shaped hull after 1.2 ms

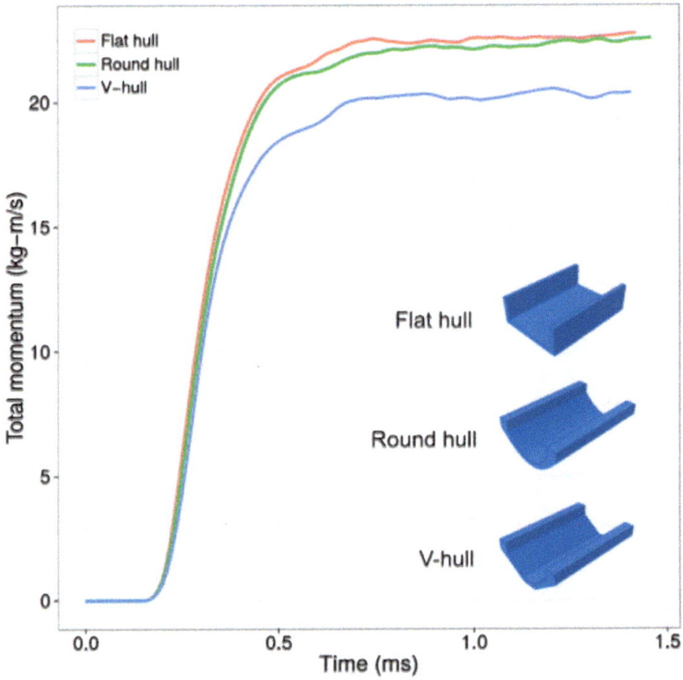

Fig. 19 The momentum transferred to the three hull shapes from an explosion in Boulder clay containing 13% water by weight

model may be used to predict the response of soils even under conditions that involve a significant amount of disaggregation.

An open-source implementation of the ARENA model can be downloaded from https://bbanerjee.github.io/ParSim/.

6 Discussion

Complex constitutive models such as ARENA are relevant primarily for computational simulations. Though progress has been made in the realm of multiscale computational models, the plateauing of clock-speeds in CPU hardware and the complexity of parallel implementations of multiscale models has meant that macroscale constitutive models are still needed for predictive simulations. The ARENA model illustrates the benefits and deficiencies of such models in simulating the high-rate loading of soils, particularity if they contain liquids.

The main benefit that a model such as ARENA provides is that it removes the need to solve mass and momentum balance equations for pore fluids while utilizing some of the underlying micromechanics. Because of the considerably different time scales of fluid motion relative to the solid matrix, and complex fluid solubility and cavitation processes during explosions, a continuum simulation of the triphasic material is extremely expensive without simplification. However, the simplification process introduces assumptions that may hold only for specific regimes of a simulation (for instance, during the compression phase after an explosion but not in the disaggregation phase).

One of the main deficiencies of the ARENA model is the assumption that the fluid phases do not displace relative to the solid phase. While this assumption may be valid for highly confined soils at large strain rates (such as in split-Hopkinson pressure bar tests), it is generally not true near the surface of the soil and particularly not valid when the soil has disaggregated due to an explosion or is displaced as ejecta after an impact event. Another major assumption in ARENA is the intrinsic pressures in the three phases are equal. This assumption simplifies the momentum equation considerably but introduces errors that can have adverse effects on the predictive capability of the model. Note that such assumptions are implicit in many models in the literature and need further research to validate.

Even if we solve the complete set of mass and momentum balance equations without assumptions such as those in ARENA, we are left with the need for a consistent constitutive model that depends on the pore pressure, porosity, and saturation. In the literature, dependence on pore pressure, porosity, and saturation is typically postulated and modeled empirically without much experimental evidence. Continuum plasticity models also typically ignore the effects of these quantities on the elastic behavior and consequently on the elastic–plastic coupling that is implied. On the other hand, if we include these quantities in the continuum model, they have to be treated as internal variables with potentially elastic and inelastic components.

The dissipation inequality of the triphasic mixture can, in principle, be used to determine constitutive equations using the Coleman–Noll approach. But we have found that these equations contain gradients of the phase pressures and porosity that make it difficult to design tractable constitutive models from thermodynamics considerations. If we include temperature (entropy) dependence and phase changes in the model, the situation becomes even more dire. With these considerations in mind, the ARENA model makes the strong assumption that the internal variables do not vary with the elastic strain. This assumption makes the stress-update algorithm much simpler than it would otherwise be even though its validity is questionable.

Even with the gross simplifications we have made in ARENA, the calibration of the model is nontrivial. Firstly, most of the experimental data that are needed for fitting model parameters are from quasistatic tests but need to be applied to high rate conditions. The Duvaut-Lions models in ARENA allow us some leeway in that process but more research is needed to determine the form of the rate-dependent models that is most appropriate for soils.

Keeping in mind the constraints and simplifying assumptions discussed above, it is remarkable that the ARENA model is able to reproduce experimental split-Hopkinson bar data as well as it does. As we saw in Sect. 5, we used parameters fitted to a dry Colorado Mason sand sample and the full machinery of the partially saturated soil model to predict the behavior of Mason sand containing 18.4% water by weight. The correspondence between our predictions and experimental data is remarkable given that we did not perform any calibration for saturation-dependence.

Some further observations about the performance of the ARENA model are listed below.

- When the model is driven using pressures or plastic strains, without solving the momentum equation, the results are reasonable for the most part. However, the ARENA model predicts an increase in porosity with increasing compressive strength (after decreasing initially) for high values of saturation. This behavior is counterintuitive and merits further exploration.
- Single particle hydrostatic, compressive, tensile verification tests show that the ARENA models have been implemented correctly in our code. Multiple strain path tests provided reasonable results but would not be verified rigorously because of the lack of an exact manufactured solution. We suggest that creating such manufactured solutions for partially saturated soil models will provide much value to the geomechanics community.
- The ARENA model as implemented in our code uses a closet point projection to the yield surface in transformed stress space. We have discovered that in some uniaxial extension tests, this projection can produce a compressive mean stress. This can possibly be mitigated by using nonassociativity.
- The damage model in ARENA was fitted using a dry Colorado Mason sand SHPB test, and matched experiments reasonably well during the loading phase. However, the simulations did not capture the observed decrease in the radial stress when the deformation gradient became constant after failure. Such a decrease would have been observed in the simulations if the hydrostatic strength in the ARENA model had been allowed to decrease after failure. We do not have any direct experimental evidence that the hydrostatic strength decreases with increasing confinement, so further study is warranted.
- Simulations of Boulder clay mixed with 40.8% water by weight suggested that the clay swells significantly (the porosity remains nearly constant, thereby reducing the saturation). Addressing this problem requires using parameter values inconsistent with calibration, suggesting a need for further research.

- In contrast to our success with Colorado Mason sand, when we used ARENA to model SHPB experiments on Boulder clay with a saturation of 30% (12.8% water by weight), we were able to match experiments only after a recalibration that suggested the need for a model enhancement that can predict hydrostatic strength decreases relative to dry clay. We recommend further research to investigate our hypothesis that compressive strength decreases with saturation because of a grain lubrication effect.

References

1. Omidvar M, Iskander M, Bless S (2014) Response of granular media to rapid penetration. Int J Impact Eng 66:60–82
2. Iskander M, Omidvar M, Bless S (2015) Behavior of granular media under high strain-rate loading. In: Rapid Penetration into Granular Media: Visualizing the Fundamental Physics of Rapid Earth Penetration. Elsevier, Amsterdam
3. Tong X, Tuan CY (2007) Viscoplastic cap model for soils under high strain rate loading. J Geotech Geoenviron Eng 133(2):206–214
4. An J, Tuan CY, Cheeseman BA, Gazonas GA (2011) Simulation of soil behavior under blast loading. Int J Geomech 11(4):323–334
5. Lu G, Fall M (2017) Modelling blast wave propagation in a subsurface geotechnical structure made of an evolutive porous material. Mech Mater 108:21–39
6. Reguerio R, Pak R, McCartney J, Sture S, Yan B, Duan Z, Svoboda J, Mun W, Vasilyev O, Kasimov N, Brown-Dymkoski E, Hansen C, Li S, Ren B, Alshibli K, Druckrey A, Lu H, Luo H, Brannon R, Bonifasi-Lista C, Yarahmadi A, Ghodrati E, Colovos J (2013) ONR MURI project on soil blast modeling and simulation. In: Song B (ed) Dynamic Behavior of Materials, vol 1, Chapter 42. Springer, Vienna, pp 341–353
7. Fan H, Li S (2017) A peridynamics-SPH modeling and simulation of blast fragmentation of soil under buried explosive loads. Comput Methods Appl Mech Eng 318:349–381
8. Grujicic M, He T, Pandurangan B, Bell WC, Cheeseman BA, Roy WN, Skaggs RR (2009) Development, parameterization, and validation of a visco-plastic material model for sand with different levels of water saturation. Proc Inst Mech Eng Pt L J Mater Des Appl 223(2):63–81
9. Suescun-Florez E, Kashuk S, Iskander M, Bless S (2015) Predicting the uniaxial compressive response of granular media over a wide range of strain rates using the strain energy density concept. J Dynam Behav Mat 1(3):330–346
10. Coussy O (2007) Revisiting the constitutive equations of unsaturated porous solids using a Lagrangian saturation concept. Int J Numer Anal Methods Geomech 31(15):1675–1694
11. Pesavento F, Schrefler BA, Sciumè G (2017) Multiphase flow in deforming porous media: a review. Arch Comput Meth Eng 24(2):423–448
12. Bishop AW, Skinner A (1977) The influence of high pore-water pressure on the strength of cohesionless soils. Philos Trans Royal Soc Lond A Math Phys Eng Sci 284(1318):91–130
13. Hassanizadeh SM, Gray WG (1979) General conservation equations for multi-phase systems: 1. Averaging procedure. Adv Water Resour 2:131–144
14. Gray WG, Hassanizadeh SM (1989) Averaging theorems and averaged equations for transport of interface properties in multiphase systems. Int J Multiphase Flow 15(1):81–95
15. Alonso EE, Gens A, Josa A, et al (1990) Constitutive model for partially saturated soils. Géotechnique 40(3):405–430
16. Hassanizadeh SM, Gray WG (1990) Mechanics and thermodynamics of multiphase flow in porous media including interphase boundaries. Adv Water Resour 13(4):169–186

17. Gray WG, Hassanizadeh SM (1991) Unsaturated flow theory including interfacial phenomena. Water Resour Res 27(8):1855–1863
18. Svendsen B, Hutter K (1995) On the thermodynamics of a mixture of isotropic materials with constraints. Int J Eng Sci 33(14):2021–2054
19. Berg CR (1995) A simple, effective-medium model for water saturation in porous rocks. Geophysics 60(4):1070–1080
20. Pietruszczak S, Pande G (1996) Constitutive relations for partially saturated soils containing gas inclusions. J Geotech Eng 122(1):50–59
21. Vlahinić I, Jennings HM, Andrade JE, Thomas JJ (2011) A novel and general form of effective stress in a partially saturated porous material: the influence of microstructure. Mech Mater 43(1):25–35
22. Meroi EA, Schrefler BA, Zienkiewicz OC (1995) Large strain static and dynamic semisaturated soil behaviour. Int J Numer Anal Methods Geomech 19(2):81–106
23. Gawin D, Schrefler BA, Galindo M (1996) Thermo-hydro-mechanical analysis of partially saturated porous materials. Eng Comput 13(7):113–143
24. Houlsby G (1997) The work input to an unsaturated granular material. Géotechnique 47(1):193–196
25. Gray WG, Schrefler BA (2001) Thermodynamic approach to effective stress in partially saturated porous media. Eur J Mech A Solids 20(4):521–538
26. Grasley ZC, Rajagopal KR (2012) Revisiting total, matric, and osmotic suction in partially saturated geomaterials. Z Angew Math Phys 63(2):373–394
27. Borja RI (2004) Cam-Clay plasticity. Part V: A mathematical framework for three-phase deformation and strain localization analyses of partially saturated porous media. Comput Methods Appl Mech Eng 193(48):5301–5338
28. Li C, Borja RI, Regueiro RA (2004) Dynamics of porous media at finite strain. Comput Methods Appl Mech Eng 193(36):3837–3870
29. Borja RI (2006) On the mechanical energy and effective stress in saturated and unsaturated porous continua. Int J Solids Struct 43(6):1764–1786
30. Song X, Borja RI (2014) Mathematical framework for unsaturated flow in the finite deformation range. Int J Numer Methods Eng 97(9):658–682
31. Fuentes W, Triantafyllidis T (2013) Hydro-mechanical hypoplastic models for unsaturated soils under isotropic stress conditions. Comput Geotech 51:72–82
32. Nedjar B (2013) Formulation of a nonlinear porosity law for fully saturated porous media at finite strains. J Mech Phys Solids 61(2):537–556
33. Lakeland DL, Rechenmacher A, Ghanem R (2014) Towards a complete model of soil liquefaction: the importance of fluid flow and grain motion. Proc Royal Soc A Math Phys Eng Sci 470(2165):20130453
34. Hamiel Y, Lyakhovsky V, Agnon A (2004) Coupled evolution of damage and porosity in poroelastic media: theory and applications to deformation of porous rocks. Geophys J Int 156(3):701–713
35. Arson C, Gatmiri B (2009) A mixed damage model for unsaturated porous media. C R Mec 337(2):68–74
36. Le Pense S, Arson C, Pouya A (2016) A fully coupled damage-plasticity model for unsaturated geomaterials accounting for the ductile–brittle transition in drying clayey soils. Int J Solids Struct 91:102–114
37. Kohler R, Hofstetter G (2008) A cap model for partially saturated soils. Int J Numer Anal Methods Geomech 32(8):981–1004
38. Papastavrou A, Steinmann P (2010) On deformational and configurational poro-mechanics: dissipative versus non-dissipative modelling of two-phase solid/fluid mixtures. Arch Appl Mech 80(9):969–984
39. Gajo A (2011) Finite strain hyperelastoplastic modelling of saturated porous media with compressible constituents. Int J Solids Struct 48(11):1738–1753

40. Madeo A, Dell'Isola F, Darve F (2013) A continuum model for deformable, second gradient porous media partially saturated with compressible fluids. J Mech Phys Solids 61(11):2196–2211
41. Gray WG, Schrefler BA, Pesavento F (2009) The solid phase stress tensor in porous media mechanics and the Hill–Mandel condition. J Mech Phys Solids 57(3):539–554
42. Gray WG, Schrefler BA (2007) Analysis of the solid phase stress tensor in multiphase porous media. Int J Numer Anal Methods Geomech 31(4):541–581
43. Nikooee E, Habibagahi G, Hassanizadeh SM, Ghahramani A (2012) The effective stress in unsaturated soils: insights from thermodynamics. In: Unsaturated Soils: Research and Applications. Springer, Berlin, pp 5–11
44. Homel MA, Guilkey JE, Brannon RM (2017) Mesoscale validation of the effective stress approach for modeling the plastic deformation of fluid-saturated porous materials. J Dyn Behav Mat 3(1):23–44
45. Itskov M (2004) On the applicability of generalized strain measures in large strain plasticity. Mech Res Commun 31(5):507–517
46. Rubin MB, Ichihara M (2010) Rheological models for large deformations of elastic-viscoplastic materials. Int J Eng Sci 48:1534–1543
47. Dienes J (1979) On the analysis of rotation and stress rate in deforming bodies. Acta Mechanica 32:217
48. Davison de St Germain J, McCorquodale J, Parker SG, Johnson CR (2000) Uintah: a massively parallel problem solving environment. In: Proceedings the Ninth International Symposium on High-Performance Distributed Computing, 2000. IEEE, Piscataway, pp 33–41. This has a generic bibtex key (just Uintah) so that it can be edited to be whatever is currently considered the main reference for Uintah, not any particular publication.
49. Sulsky D, Chen A, Schreyer H (1994) A particle method for history-dependent materials. Comput Methods Appl Mech Eng 118:179–196
50. Kamrin K, Bazant MZ (2007) Stochastic flow rule for granular materials. Phys Rev E 75:1–28 (preprint), file is Kamrin07.pdf
51. Strack O, Leavy R, Brannon R (2014) Aleatory uncertainty and scale effects in computational damage models for failure and fragmentation. Int J Numer Methods Eng 102(3–4):468–495
52. Huq F, Brannon R, Brady L (2015) An efficient binning scheme with application to statistical crack mechanics. Int J Numer Meth Eng 105:33–62
53. Kamojjala K, Brannon R, Sadeghirad A, Guilkey J (2015) Verification tests in solid mechanics. Eng Comput 2:193–213
54. Homel MA, Guilkey JE, Brannon RM (2015) Continuum effective-stress approach for high-rate plastic deformation of fluid-saturated geomaterials with application to shaped-charge jet penetration. Acta Mech 227(2):279–310
55. Brannon RM, Fuller T, Strack O, Fossum A, Sanchez J (2015) Kayenta: Theory and user's guide. Sandia National Laboratories report SAND2015-0803
56. Uzuoka R, Borja RI (2012) Dynamics of unsaturated poroelastic solids at finite strain. Int J Numer Anal Methods Geomech 36(13):1535–1573
57. Borja RI, White JA (2010) Conservation laws for coupled hydromechanical processes in unsaturated porous media: theory and implementation. In: Mechanics of unsaturated geomaterials, pp 185–208
58. Song X, Borja RI (2014) Finite deformation and fluid flow in unsaturated soils with random heterogeneity. Vadose Zone J 13(5)
59. Gray WG, Hassanizadeh SM (1991) Paradoxes and realities in unsaturated flow theory. Water Resour Res 27(8):1847–1854
60. Brannon RM (2007) Elements of phenomenological plasticity: geometrical insight, computational algorithms, and topics in shock physics. In: Shock Wave Science and Technology Reference Library. Springer, Berlin, pp 225–274
61. Mo es N, Dolbow J, Belytschko T (1999) A finite element method for crack growth without remeshing. Int J Numer Methods Eng 46(1):131–150

62. Banerjee B, Brannon R (2017) Theory, verification, and validation of the ARENA constitutive model for applications to high-rate loading of fully or partially saturated granular media. Tech. Rep. PAR-10021867-1516.v1, Parresia Research Limited and University of Utah, https://doi.org/10.13140/RG.2.2.10671.53922
63. Norris AN (2008) Eulerian conjugate stress and strain. J Mech Mater Struct 3(2):243–260
64. Brannon R, Leelavanichkul S (2010) A multi-stage return algorithm for solving the classical damage component of constitutive models for rock, ceramics, and other rock-like media. Int J Fracture 163(1–2):133–149
65. Fuller T, Brannon R (2012) On the effects of deformation-induced anisotropy on classical elastic-plastic materials. Int J Numer Anal Methods Geomech 37(9):1079–1094
66. Qian X, Gray DH, Woods RD (1993) Voids and granulometry: effects on shear modulus of unsaturated sands. J Geotech Eng 119(2):295–314
67. Berryman JG, Milton GW (1991) Exact results for generalized Gassmann's equations in composite porous media with two constituents. Geophysics 56(12):1950–1960
68. Berryman JG (2006) Effective medium theories for multicomponent poroelastic composites. J Eng Mech 32(5):519–531
69. Dvorkin J, Moos D, Packwood JL, Nur AM (1999) Identifying patchy saturation from well logs. Geophysics 64(6):1756–1759
70. Pučik T, Brannon R, Burghardt J (2015) Nonuniqueness and instability of classical formulations of nonassociated plasticity, I: case study. J Mech Mater Struct 10(2):123–148
71. Pabst W, Gregorová E (2015) Critical assessment 18: elastic and thermal properties of porous materials–rigorous bounds and cross-property relations. Mater Sci Technol 31(15):1801–1808
72. Mun W, Teixeira T, Balci M, Svoboda J, McCartney J (2016) Rate effects on the undrained shear strength of compacted clay. Soils Found 56(4):719–731
73. Mun W, McCartney JS (2017) Roles of particle breakage and drainage in the isotropic compression of sand to high pressures. J Geotech Geoenviron Eng 143(10):04017071
74. Mun W, McCartney JS (2017) Constitutive model for drained compression of unsaturated clay to high stresses. J Geotech Geoenviron Eng 143(6):04017014
75. Mun W, McCartney JS (2017) Compression of unsaturated clay under high stresses. J Geotech Geoenviron Eng 143(7):02817003
76. Lu H, Luo H, Cooper WL, Komanduri R (2013) Effect of particle size on the compressive behavior of dry sand under confinement at high strain rates. In: Dynamic behavior of materials, vol 1, Springer, Berlin, pp 523–530
77. Luo H, Du Y, Hu Z, Lu H (2015) High-strain rate compressive behavior of dry mason sand under confinement. In: Dynamic behavior of materials, vol 1, Springer, Berlin, pp 325–333
78. Luo H, Hu Z, Xu T, Lu H (2017) High-strain rate compressive behavior of a clay under uniaxial strain state. In: Dynamic behavior of materials, vol 1, Springer, Berlin, pp 117–122
79. Jensen E (2017) Hierarchical multiscale modeling to inform continuum constitutive models of soils. PhD thesis, University of Colorado at Boulder

Part II
Applications

Planetary Impact Processes in Porous Materials

Gareth S. Collins, Kevin R. Housen, Martin Jutzi, and Akiko M. Nakamura

1 Introduction

Granular and porous materials are ubiquitous in the solar system in many contexts and at a range of length scales. The growth of submicron dust (and ice) particles in a protoplanetary disk into km-scale planetesimals—the precursors to larger planetary bodies—is a complex process but, while many aspects of this growth remain enigmatic, it appears certain that planetesimals formed with very high initial porosity [1], up to 70–90% void fraction (10–30% solid fraction). High initial porosities of primordial solids are also supported by the very low inferred density of comets, believed to be among the most primitive objects in the solar system, and the highest porosities observed within the fine-grained matrix of primitive meteorites [2].

While impacts and self-gravitation—particularly when facilitated by internal heating fueled by the decay of short-lived radionuclides (e.g., ^{26}Al)—will tend to compact planetesimals as they grow into planets, it is clear that even after 4.5 billion

G. S. Collins (✉)
Imperial College London, London, UK
e-mail: g.collins@imperial.ac.uk

K. R. Housen
University of Washington, Seattle, WA, USA
e-mail: housen@aa.washington.edu

M. Jutzi
Center for Space and Habitability, Physics Institute, University of Bern, Bern, Switzerland
e-mail: martin.jutzi@space.unibe.ch

A. M. Nakamura
Graduate School of Science, Kobe University, Kobe, Japan
e-mail: amnakamu@kobe-u.ac.jp

T. J. Vogler, D. A. Fredenburg (eds.), *Shock Phenomena in Granular and Porous Materials*, Shock Wave and High Pressure Phenomena,
https://doi.org/10.1007/978-3-030-23002-9_4

years of solar system history, planetary objects (asteroids and satellites) smaller than about 500-km diameter exhibit substantial internal porosity, although the nature and scale of this porosity is unknown. The masses and volumes of several asteroids ranging in size from 1 to 500 km have been determined to sufficient accuracy to infer bulk densities as low as 1200 kg m^{-3}, implying bulk porosities as high as 70%, with a mean around 30% [3, 4]. Some of this residual porosity may be primordial, but given the violent collisional evolution of asteroids it seems likely that in many cases the present porosity is the result of either fracturing or one or more episodes of catastrophic disruption and reaggregation to produce granular "rubble-pile" asteroids [5]. The response of porous asteroids to shock is crucial, therefore, for understanding the internal structure of asteroids.

Meteorites—fragments of asteroids and other planetary surfaces that have survived impact with Earth—provide additional evidence for the presence and evolution of porosity in solar system objects [4]. Stony chondritic meteorites—those with a composition most representative of the bulk solar system, often regarded as remnant planetary building blocks—have typical bulk porosities of 10–20%: considerable, but substantially lower than inferred for asteroids that might represent their parent bodies. The nature of this porosity is still being elucidated: in some cases it appears to be primordial [6], but in others a fracture origin seems likely, presumably caused by impacts on the parent asteroid [4]. In either case, as chondritic meteorites are composed of a large volume fraction of mm-scale chondrules—nonporous, solidified melt droplets—the porosity of the fine-grained matrix between the chondrules is substantially larger than the bulk porosity, and in some samples is as high as 70% [2].

Porosity can also represent a volumetrically minor, but important component of large planetary bodies. Although porosity here is confined to shallow depths, below which high internal pressures and temperatures from self-gravity have removed internal porosity, its location at the surface implies it has an important role in regulating surface processes and fluxes of fluids and energy. Again, the nature and length scale of porosity can vary. On the Moon, the uppermost layer is a thin veneer of fine-grained soil, called regolith. This has a typical porosity of 35% and thickness of 5–15 m [7], produced by continual impact gardening by micrometeorites. Below this lies a thicker layer of fractured rock, often referred to as the "megaregolith". Recent gravity measurements by NASA's GRAIL mission indicate a bulk crustal density of 10% lower than the expected grain density implying significant porosity in the megaregolith crust [8], which may extend as deep as 30 km [9]. The correlation of spatial variations in crustal porosity with impact craters suggests that shock processes play a pivotal role in the formation and evolution of crustal porosity [10, 11].

The abundance and ubiquity of porous and granular materials in the solar system, and in particular their position at the interface between space and planet, makes knowledge of their influence on shock and impact processes imperative for understanding and interpreting the myriad of solar system environments. The fact that porous and granular materials respond very differently to shock than nonporous materials is well known to shock physicists, but planetary scientists are only beginning to appreciate its ramifications in many contexts [12]. The enhanced shock attenuation and localization of shock heating exhibited by porous materials has several important implications for the evolution of planetary materials and surfaces. Here we summarize some of those implications to highlight how advances in understanding shock and impact processes in porous granular materials have direct application in planetary science.

2 The Role of Porosity in Planetary Impact Processes

Planetary impacts involve the collision of two bodies—an impactor and a target—at high speed. When the target is much larger than the impactor an impact crater is formed and the mass that escapes the system is negligible. As the impactor size approaches that of the target body, cratering gives way to distortion or disruption of the entire target body, and a significant portion of the impactor and target mass can escape the system. From the perspective of energy conservation, the kinetic energy of the impactor (assuming the target is at rest) is partitioned into kinetic energy of ejected target and impactor material, thermal energy of the heated target and impactor and a change in the gravitational potential energy of the target. If either the target or impactor is porous, this can dramatically affect the partitioning of energy in either cratering or disruption scenarios (Fig. 1).

Compaction of pore space absorbs energy from the shock wave, partitioning a greater fraction of the impact energy into thermal energy of the compacted impactor or target than would occur if the material was nonporous (Fig. 1). If the target material is porous, the expanding shock wave also attenuates more rapidly (Fig. 2), implying not only that more heating occurs, but also that this heating is more localized: thermal energy is deposited closer to the impact site. At the same time, the more rapid attenuation of shock pressure and particle velocity in porous materials serves to reduce the momentum of the cratering flow (Fig. 2) and with it ejection speeds, ejected mass, momentum and energy, crater dimensions and the related changes in gravitational potential energy. In the following sections, we discuss each of these mechanisms in detail.

2.1 *Crater Formation in Porous Materials*

The formation of impact craters has been studied since the early 1960s, sparked mainly by the birth of the space program. Explosion craters have been modeled and studied even longer, primarily for military purposes. Most of these studies, whether experimental or numerical, have focused on targets composed of common geologic materials, such as soils and rock, thought to be reasonable proxies for the surfaces of the terrestrial planets. However, as noted above, it has become increasingly clear that those materials are not good analogues for the minor bodies of the solar system.

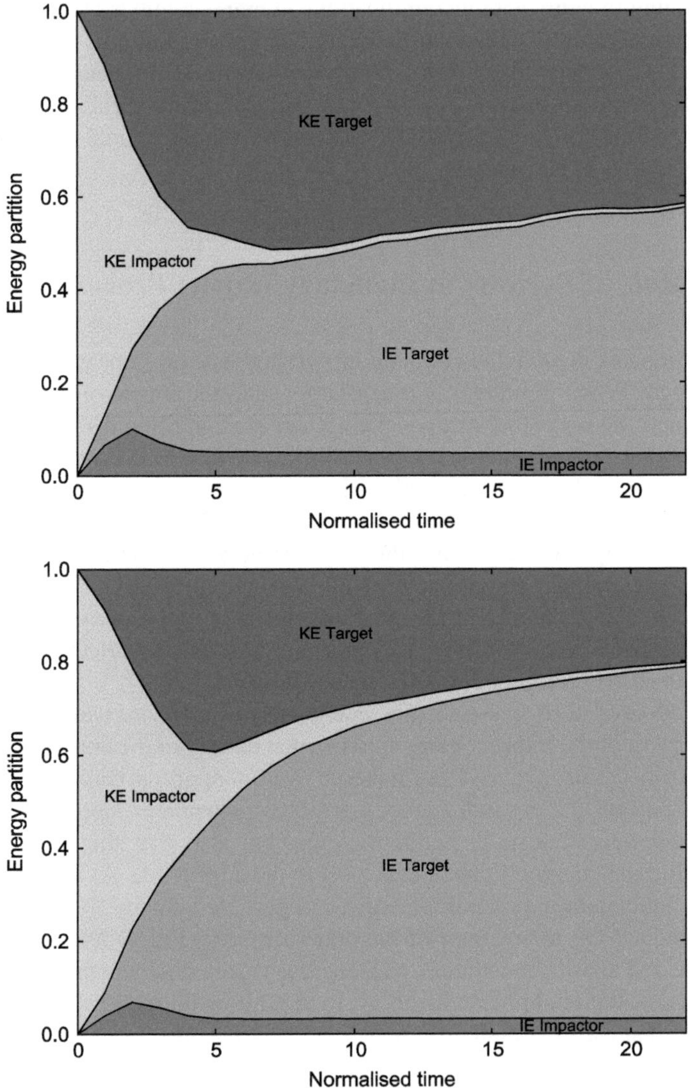

Fig. 1 Comparison of energy partitioning in the early stages of impact cratering in a nonporous (top) and porous (bottom) material. Shown are the proportions of the kinetic energy (KE) of the impactor that remain in this reservoir or are partitioned into kinetic energy of the target or internal energy (IE) of the impactor and target. Time is normalized by a/U (where a is the impactor radius and U is the impactor speed), which is the time taken for the impactor to travel its own radius. The results were derived from iSALE shock physics calculations of a 100-m diameter nonporous impactor colliding vertically at 10 km/s with a flat nonporous (top) and 33% porous (bottom) target. Strength effects are neglected

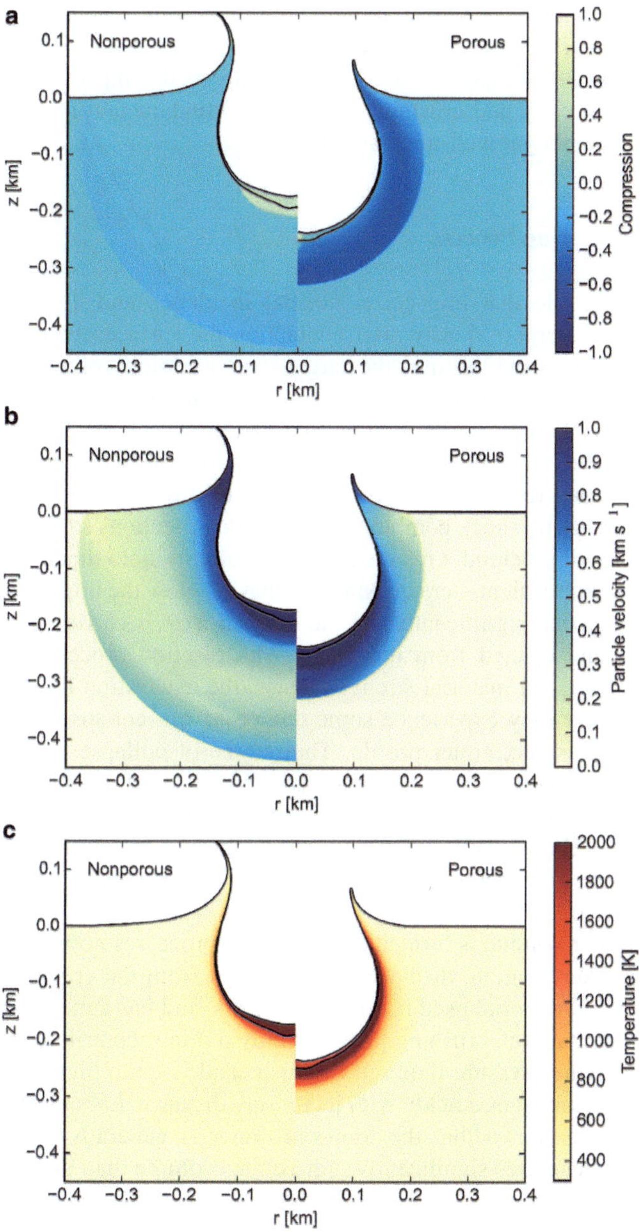

Fig. 2 Comparison of shock effects and attenuation in the early stages of impact cratering in a nonporous and porous material. The results derived from shock physics calculations of a 100-m diameter nonporous impactor colliding vertically at 10 km/s with a flat nonporous (left) and 33% porous (right) target. Shown are the state 0.08 s after impact ($Ut/a = 8$; see Fig. 1 for definition of terms) in terms of (**a**) material compression ($\rho_0/\rho - 1.$; where ρ and ρ_0 are the density and initial density, respectively); (**b**) particle velocity; and (**c**) temperature. Strength effects are neglected

Terrestrial rocks and soils do not exhibit the highly porous structures that have been inferred from observations of asteroids and meteorites [3, 13].

In this section we discuss how target porosity affects the impact cratering process. We show that porosity and crushing strength are fundamental material properties whose effects are not yet well-understood.

2.1.1 The Cratering Process

Consider a projectile that impacts at normal incidence and high speed into an infinite half-space target. Shortly after contacting the surface, the impactor energy and momentum are coupled into the target over a region comparable in size to the impactor. This initial compression creates a region of very high pressure and a roughly hemispherical shock that propagates from the impact point (Fig. 2). Depending on the initial porosity of the target and the conditions of the impact, the pressure in this region can be high enough to crush the target material. The outwardly propagating shock combines with tensile reflections from the free surface of the target, leaving behind a residual particle velocity field that eventually forms an evolving—i.e., transient—crater. Material that is below the impact point is driven downward while that significantly off the vertical axis is carried along upwardly curving paths and ejected from the crater. The ejection process continues until either gravity forces or material strength arrests the excavation flow. At this point, the transient crater may experience some degree of rim collapse or floor rebound to form the final, static, crater profile. The degree of collapse and floor rebound depends on the properties of the target material, gravity and the size of the transient crater. Floor rebound can produce complex crater shapes with central peaks and ring structures. Rebound is not considered here because it generally only occurs in very large craters, where high lithostatic pressures squeeze out any significant target porosity.

The final crater volume is formed from the four processes noted above: crushing in the high-pressure region, displacement, ejection from the crater and collapse.[1] Sand is a common material used in lab experiments, and has a moderate porosity of ~35%. For that material, crushing provides only a minor contribution to the crater volume because the pressure required to crush sand is quite high and because the shock pressure diminishes quickly with increasing distance. Laboratory experiments in sand [17] have shown that the transient crater is generally bowl-shaped and that excavation generates significantly more crater volume than does displacement. Collapse of the transient crater generally reduces the final crater volume because material from the uplifted crater rim slides back into the crater and the material bulks as it slumps.

[1] A fifth process, spallation, can be important, but is not relevant to the present discussion. Spallation occurs when the compressive shock reflects from the target surface in tension. If the tensile stress exceeds the strength of the target material, thin plate-shaped fragments are ejected from the periphery of the crater [14–16].

The situation is quite different for highly porous targets. Geological materials with high porosity tend to crush easily [18]. The reduced strength, as well as the low density of porous targets, reduces the decelerating force on the impactor, allowing it to penetrate deep into the target [19, 20]. The outwardly propagating impact shock crushes a much larger volume of material compared to an impact in sand. The voids in the target are permanently compressed, creating a deep transient crater by compaction. If the cohesion of the initial target material is large compared to lithostatic forces, the transient crater does not collapse, leaving a narrow deep final crater. "Such carrot-shaped" craters are observed in porous targets made from sintered glass beads [19, 20], in cohesive pumice [21, 22], as well as extremely porous targets such as aerogels [23, 24]. Such craters were also formed in snow by surviving fragments of the Chelyabinsk meteorite [25].

If the target strength is small compared to lithostatic forces, e.g., in cohesionless materials, the late-time flow field causes the transient crater to expand beyond the initial cavity created by shock compaction. Gravity eventually arrests the expansion and the transient crater collapses. In contrast to the case for sand, the deep penetration of the impactor causes much of the flow field to be directed either radially outward or down [26]. As a result, much less material is ejected from the crater compared to sand targets.

Figure 3 illustrates how crater formation depends on the target porosity. The images in the upper part of the figure are from laboratory quarter-space impact experiments, in which the crater formation is viewed in cross section [18]. In each case, the impactor was a polyethylene cylinder impacting vertically at $\sim$1800 m/s into a target with very little cohesion. The results for four target porosities are shown. Each image in the upper row shows the profile of the final crater. The black curve is the profile of the transient crater just prior to collapse, determined from high-speed videos of the events. The left image is for pure sand whose porosity is 43% and crush strength is 41 MPa.[2] The transient crater did not collapse, so it coincides with the final crater profile. The next three images to the right show the effect of increasing target porosity, and decreasing crush strength. As the crush strength goes down, the projectile penetrates deeper into the target, reaching a depth about equal to the maximum depth of the transient crater.

The illustration at the bottom of the figure summarizes how these craters form. As the impactor penetrates the target, the shock compression permanently compacts material into a bulb-shaped cavity that is lined with the newly crushed material. The residual particle velocity (i.e., the flow field) left behind after the shock passes causes the cavity to continue to expand until it is arrested by gravity. In a lab

[2]The crush strength is not as well defined as, say, the tensile strength of a brittle material. In actuality porous geological materials crush over a wide range of pressure, with more void space being eliminated as pressure increases. This has been described using the so-called P-α model [27] or a modified version referred to as the ϵ-α model [28]. We refer to a single crush strength measure here as a way to simply characterize the material's resistance to compaction. The specific values of crush strength cited refer to the stress that compresses the material 50% of the way to zero porosity [18].

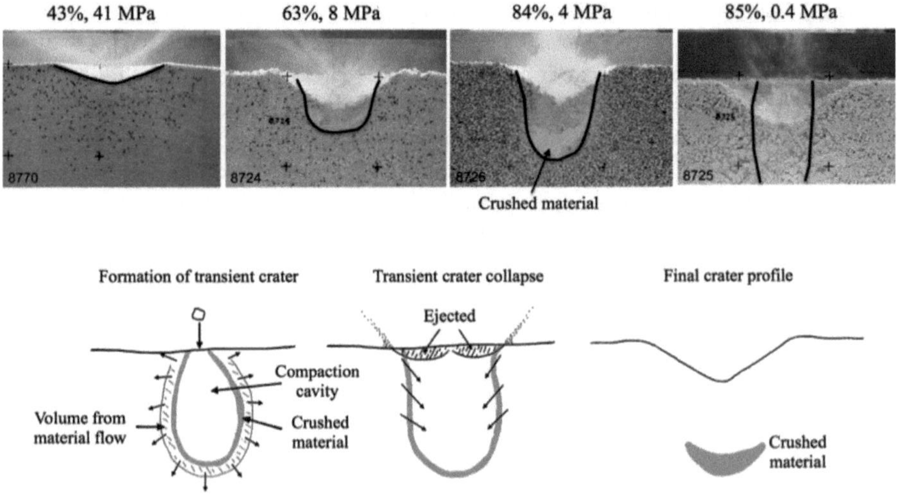

Fig. 3 The upper part shows cross sectional views of the final craters produced in laboratory impacts into targets of various porosities and crushing strengths, whose values are shown immediately above the images. The dark line shows the observed profile of the transient crater. The impactor penetrates deepest in the most porous targets. The sketch in the lower part of the figure illustrates the formation and subsequent collapse of the transient crater. Based on results from [18]

experiment at a gravity of 1G, the growth due to the flow is considerable and makes up a significant part of the transient crater volume. At larger scales, the greater lithostatic forces limit the post-compaction growth. This effect is simulated in the lab by performing the impact experiments on a centrifuge. In fact, the experiments shown in Fig. 3 were conducted at a centrifuge acceleration of 150G, where the lithostatic forces are about the same as for a large crater on a 100 km asteroid. For this condition, the expansion of the cavity due to material flow is limited. The middle sketch at the bottom of the figure shows the transient crater at the onset of collapse, which occurs in the experiments because the lithostatic stress beneath the crater exceeds the low cohesion of the target material. As collapse occurs, the crushed material lining the transient crater collects at the bottom and is buried beneath the crater by surrounding uncrushed material. This highly crushed material is seen in the middle two images, just above the bottom of the transient crater profile. It is below the field of view in the rightmost image.

As noted, impacts into highly porous targets produce flow fields that are generally directed down into the target, rather than curving upward as for an impact in sand. As a result, only a small mass of material is ejected from the crater, as illustrated in the middle sketch in Fig. 3. This effect is more significant for targets with lower crush strengths. The result of this is that less material is ejected from targets with lower crush strengths, as illustrated by the lack of ejecta at the rims of the two craters on the right side of Fig. 3. This mechanism has been used to explain the apparent lack of ejecta around large craters on asteroid Mathilde [29] and Hyperion [30].

2.1.2 Crater Scaling

The purpose of crater scaling relations, or "laws", is to relate the outcome of an impact event, e.g., the crater size, ejecta speeds, formation time, to the initial conditions of the impact, including the impactor size, speed, material properties, the target material properties, gravity, etc. Crater scaling laws usually make use of the simplifications afforded by dimensional analysis and so are expressed in terms of dimensionless variables. This method has been used since the 1950s to study the formation of explosion and impact craters in the lab, in the field, or numerically. An important development occurred during the 1980s when it was recognized that, for many aspects of cratering, the explosive charge or the impactor is approximated well as a point source [31]. This explained many of the power-law relations observed in cratering experiments.

The validity of point-source scaling has been demonstrated through many experimental and numerical studies, though principally for materials with either low porosity (water, rock, metals) or moderate porosity (sand, gravel, glass beads). On the other hand, the fact that an impactor can penetrate deeply into a porous target raises the question of whether the energy and momentum are deposited as a point source, or over a more extended region [20, 22]. We consider point-source scaling here and present some evidence for its applicability to highly porous materials. However, given the current lack of relevant data we also note that additional studies are needed to fully understand the scaling of craters formed in porous targets. For brevity, the scaling relationships are simply stated and we do not consider large craters for which the final and transient crater sizes can differ significantly. Additional details and derivations are available in [18, 31, 32].

The diameter, D, of the final crater produced by an impact depends upon the gravitational acceleration, g, and the various properties of the impactor and the target listed in Table 1. In the point-source theory, D is not a function of the separate impactor properties, but instead depends only on the single measure $aU^{\mu}\delta^{\nu}$ (known as the "coupling parameter") where the material dependent exponent μ is constrained to the range of 1/3 to 2/3, referred to as energy and momentum scaling, respectively, because the point source measure is proportional to the impactor momentum or energy in those two limiting cases. Experimentally, μ has been found to be about 0.4 for dry sand and 0.55–0.6 for nonporous materials such as water or metals. From the theory, μ is expected to be at the low end of the range for highly porous materials. The density exponent ν in the coupling parameter has been measured to be about 0.4.

Point-source scaling relations take the form of power-laws in two limiting cases. In the first, the final crater size is determined by a target strength measure (e.g., cohesion or crush strength) and is independent of gravity. A lab crater produced in cohesive pumice is an example. In the second case, target strength is small compared to the shear strength induced by gravity: $\rho g h \tan(\phi)$, where ρ is the target density, g is gravity, h is a characteristic length scale such as the crater depth and ϕ is the friction angle of the target material. These two limiting cases produce three regimes

Table 1 Scaling relations for crater diameter and volume

Regime	Crater diameter	Crater volume
Cohesion	$D\left(\frac{\rho}{m}\right)^{1/3}=\left(\frac{c}{\rho U^2}\right)^{-\frac{\mu}{2}}\left(\frac{\delta}{\rho}\right)^{\frac{3\nu-1}{3}}f(n)$	$\frac{\rho V}{m}=\left(\frac{c}{\rho U^2}\right)^{-\frac{3\mu}{2}}\left(\frac{\delta}{\rho}\right)^{3\nu-1}f(n)$
Gravity	$D\left(\frac{\rho}{m}\right)^{1/3}=\left(\frac{ga}{U^2}\right)^{-\frac{\mu}{2+\mu}}\left(\frac{\delta}{\rho}\right)^{\frac{6\nu-2-\mu}{3(2+\mu)}}f(n,\phi)$	$\frac{\rho V}{m}=\left(\frac{ga}{U^2}\right)^{-\frac{3\mu}{2+\mu}}\left(\frac{\delta}{\rho}\right)^{\frac{6\nu-2-\mu}{(2+\mu)}}f(n,\phi)$
Compaction	$D\left(\frac{\rho}{m}\right)^{1/3}=\left(\frac{Y_c}{\rho U^2}\right)^{-\frac{\mu}{2}}\left(\frac{\delta}{\rho}\right)^{\frac{3\nu-1}{3}}f(n)$	$\frac{\rho V}{m}=\left(\frac{Y_c}{\rho U^2}\right)^{-\frac{3\mu}{2}}\left(\frac{\delta}{\rho}\right)^{3\nu-1}f(n)$

Variable definitions

Crater parameters	Impactor parameters	Target parameters
V, volume	a, radius	c, cohesion
D, diameter	U, speed	Y_c, crush strength
$\pi_v=\frac{\rho V}{m}$	δ, density	ρ, density
$\pi_D=D\left(\frac{\rho}{m}\right)^{1/3}$	m, mass	ϕ, friction angle
		n, porosity

of cratering in which the crater size is determined by either the cohesive strength of the target c, by gravity or by the crush (compaction) strength Y_c.

Table 1 shows the scaling laws for the three regimes. The relationships are expressed in terms of dimensionless ratios of the variables shown in the table. The scaled crater diameter, $\pi_D = D(\rho/m)^{1/3}$ is, to within a factor, the ratio of crater size to impactor size. The scaled crater volume, $\pi_V = \rho V/m$, is the "crater mass" normalized by the impactor mass. In the cohesion-dominated regime, these scaled variables depend on the ratio of the cohesive strength to the dynamic pressure of the impact, ρU^2, the ratio of impactor and target densities, and porosity. In the gravity regime they depend on the ratio $\pi_2 = ga/U^2$, essentially an inverse Froude number, often referred to as the gravity-scaled size of the impact.[3] Additional dependencies on the density ratio, the target porosity and friction angle are also shown. The third (compaction) regime has the same form as the cohesion regime, with the cohesive strength c replaced by the crush strength of the target Y_c. We note that the specific measure of cohesive or crush strength that c and Y_c terms represent is not yet known [33].

Crater size measurements are usually plotted in terms of the scaled crater volume or size vs π_2 because a common question is how do those quantities depend on the size scale of the impact? Some important trends are illustrated by the data in Fig. 4, which shows π_V and π_D as a function of π_2. Materials with zero cohesion and large crush strength (or low porosity) exhibit the monotonic downward sloping power-law of pure gravity scaling. Dry sand exhibits this behavior, as do other cohesionless materials, e.g., water [34], glass beads [35], and numerical simulations of cohesionless soil [36].

[3]The original definition of π_2 included a multiplicative factor of 3.22, which is not used here.

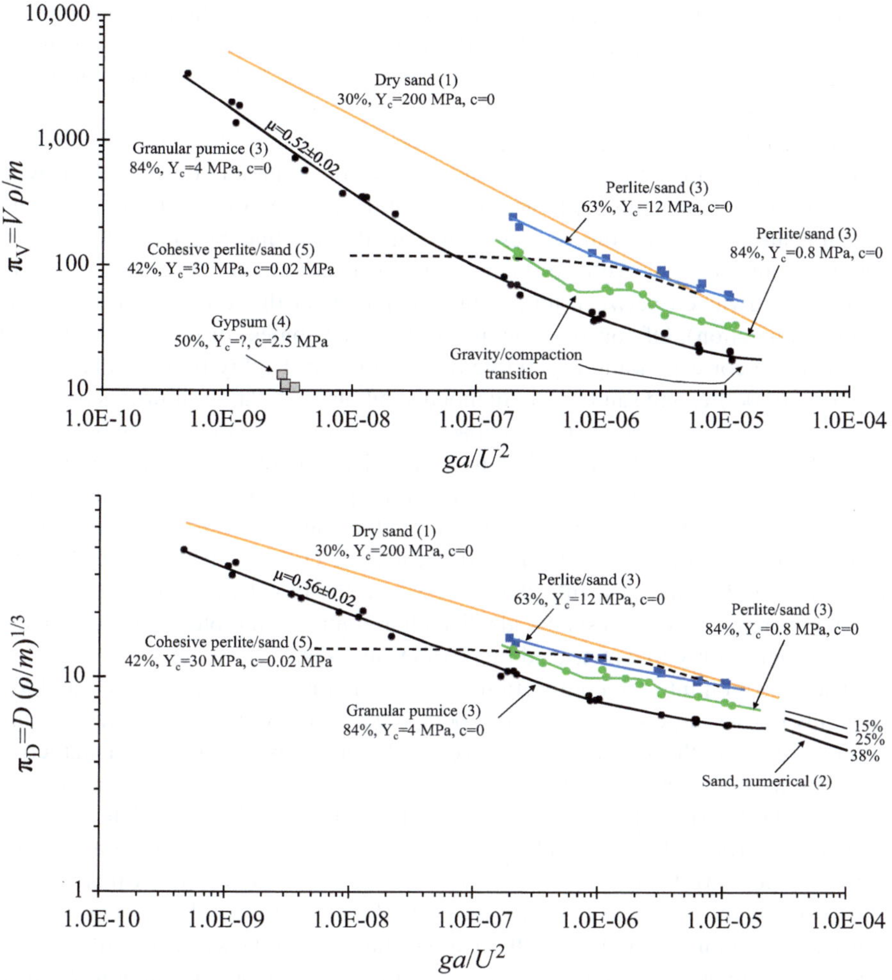

Fig. 4 Experimental and numerical results for the scaled crater volume and crater diameter for impacts into porous materials. Each set of data are labelled by the target porosity, crush strength and cohesion. Data sources: (1): [34]; (2): [36]; (3): [18]; (4): [37]; (5): [38]

Impacts into materials with high cohesion show a different, horizontal trend because π_V and π_D are independent of the impactor size.[4] At sufficiently large size scale, i.e., when π_2 is large, the shear strength induced by the lithostatic overburden is large compared to the target cohesion. The flat line then transitions into the downward-sloping line (on a log–log plot) of the gravity regime. Impacts into moderately porous ($n = 42\%$) cohesive mixtures of perlite and sand (dashed

[4]This is strictly true only when the target strength does not itself depend on size or timescales.

line in Fig. 4) exhibit this behaviour. In the cohesion regime, an increase in cohesion reduces cratering efficiency, shifting the trend line down, as exemplified by the results of [37] for high cohesion gypsum targets ($n = 50\%$) that lie well below the dashed line.

At even larger size scales in porous targets, most of the crater volume is determined by compaction because gravity arrests the material flow quickly, leaving a transient crater only slightly larger than the initial compaction cavity. The cratering then transitions into the compaction regime, another flat line. So, the trend expected for a porous material with cohesion is a downward-sloping line in the gravity regime, sandwiched between horizontal asymptotes at the left (cohesion) and the right (compaction). Of course if the material is cohesionless or has very high crush strength, the corresponding asymptote does not occur. Highly porous cohesionless mixtures of perlite and sand (open diamond symbols, $n = 84\%$) show this transition from the gravity regime to the compaction regime. π_V initially slopes down to the right, as expected, but then flattens to a horizontal compaction asymptote near $\pi_2 = 6 \times 10^{-7}$. This transition also marks the point at which ejecta blankets no longer appear, another signal of a transition to the compaction regime [18]. Further to the right π_V again turns downward because of changes in the target material properties as the increased lithostatic stresses in the target cause it to crush under its own weight. This suggests the idealized horizontal asymptote in the compaction regime applies only to a restricted range of event sizes.

The transition into the compaction regime and cratering efficiency (π_V and π_D) in highly porous targets are sensitive to both target porosity and the crushing strength (more precisely, the entire crush curve). Laboratory experiments and numerical simulations both show that higher target porosity causes a decrease in π_V and π_D (Fig. 4). In particular, numerical simulations that are able to isolate the effect of porosity from other material parameters show a moderate reduction in crater size with porosity from zero to $n = 67\%$ (Fig. 5). The effect of crush strength is illustrated by the results in Fig. 4 for the perlite/sand mixture with 84% porosity and those for the granular pumice. Both materials have about the same porosity, density and friction angle. Yet, the granular pumice transitions to the compaction regime at larger π_2 and has a lower compaction asymptote than does the perlite/sand. This is a result of the five times larger crush strength of the granular pumice compared to the perlite/sand, which dramatically reduces the size of the compaction cavity.

Despite the emergence of clear trends, at present there is no single set of data, either from lab or numerical experiments that spans all three cratering regimes. Our understanding of crater scaling in porous materials therefore remains incomplete. For example, while highly porous materials show a power-law relation between π_V (also π_D) and π_2, not all such materials exhibit the expected power-law exponent consistent with a μ-value close to the momentum-scaling limit. Whether this relates to differences in friction coefficient or some other material property is not yet known. What is clear is that additional work should be done to explore the entire range of scaling regimes for porous target materials.

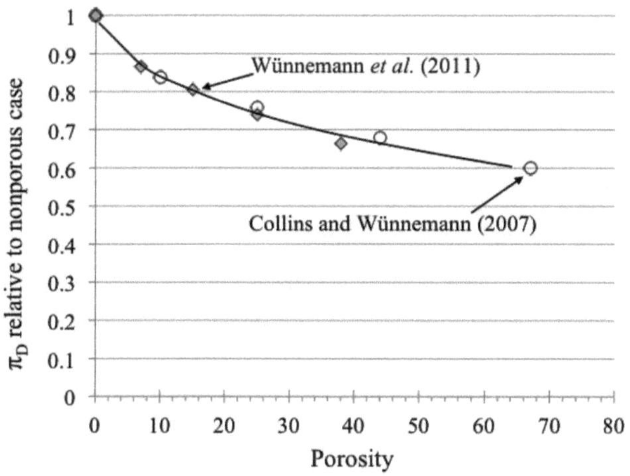

Fig. 5 Results from numerical simulations of impacts show how the scaled crater size depends on the target porosity

2.2 *Collisional Disruption of Porous Bodies*

Not all impacts result in crater formation. When the impactor mass is comparable to the mass of the target body the collision affects the entire object and can result in global-scale distortion or potentially disruption, followed by reaccretion or dispersal of the fragments. Internal porosity in the target body increases its resistance to collisional disruption and reduces momentum transfer efficiency by virtue of enhanced shock attenuation and reduced particle velocity. In this section we describe current understanding of these effects based on laboratory experiments and numerical modelling.

2.2.1 Collision Regimes

To characterize the outcome of a collision, it is helpful to distinguish between erosive and accretionary collisions [39]. The first type is defined as events that lead to a net reduction of the mass of the target body. This includes cratering impacts as well as disruptive collisions. Disruptive collisions lead to the destruction of the target and to a significant loss of mass. At large scales, they are possibly followed by the gravitational reaccumulation of a part of the ejected material on the main remaining body as well as among the smaller fragments [40, 41]. The second type are accretionary collisions that lead to a net increase of the target's mass.

The ratio v_{imp}/v_{esc} is a crucial parameter that determines in which regime a collision takes place [39]. Here v_{imp} is the impact velocity and v_{esc} is the mutual escape velocity of the bodies involved. An overview of various collision regimes is presented in Fig. 6. According to this simplified picture, which ignores the effects of

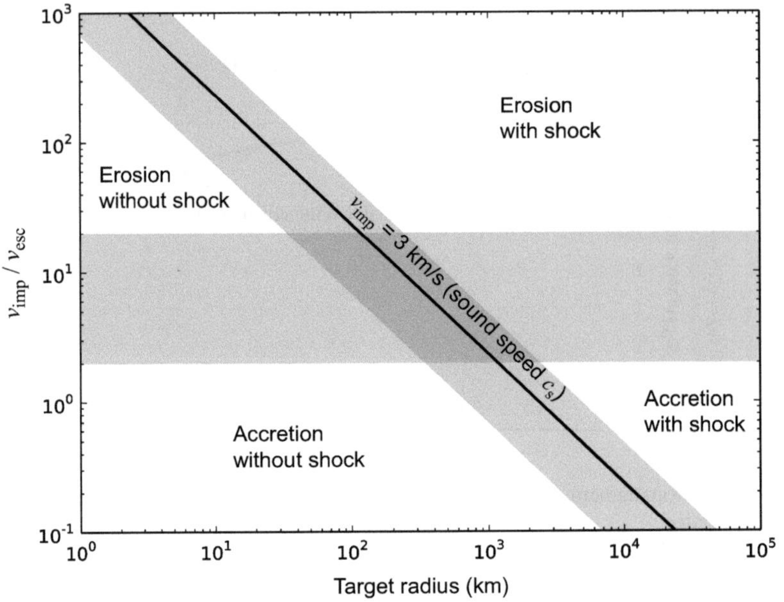

Fig. 6 Collision regimes, distinguished by the ratio v_{imp}/v_{esc} as function of target size. In this simplified picture, where only "gravitational sticking" is considered, collisions transition from accretionary to erosive when the impact velocities become higher than a few times v_{esc} (horizontal grey band). The solid line of constant velocity corresponds to the typical sound speed of rocks, marking the transition between sub-sonic and super-sonic impacts. Adapted from [39]

"cohesional sticking" at very small scales as well as the influence of impact angle, collisions transition from accretionary to erosive when the impact velocities become higher than a few times v_{esc}.

At small scales, there is a large range of v_{imp}/v_{esc} with impact velocities below the sound speed, which means that accretionary collisions do not involve shocks (at very small scales, this is even true for erosive events). At large scales, on the other hand, accretionary collisions take place with velocities above the sound speed, involving strong shocks.

Current impact velocities in the main belt are typically $v_{imp} \sim 2$–8 km/s, leading to $v_{imp}/v_{esc} \gg 1$ which means that asteroid collisions today take place in an erosive regime.

2.2.2 Catastrophic Disruption Threshold

Disruptive collisions are characterized by their specific impact energy, traditionally defined as $Q = 0.5 m_p v_{imp}^2 / M_T$, where m_p, v_{imp} and M_T are the mass and speed of the projectile and the mass of the target, respectively (more recent definitions also take into account the mass of the projectile; e.g., [42]). For small target bodies, the catastrophic disruption threshold Q_D^* is the same as the "shattering threshold" Q_S^*, defined as the specific impact energy that results in a largest intact fragment M_{lr} containing 50% of the original target's mass. In this so-called strength regime it is the target body's tensile strength—its resistance to shattering—that modulates disruption. For large target bodies, on the other hand, the gravitational binding energy of the body, rather than the tensile strength controls the mass of the body post collision. In this "gravity regime", the catastrophic disruption threshold Q_D^* is defined as the specific impact energy that results in a modified (potentially reaccumulated) target body with a M_{lr} equivalent to 50% of the original target's mass. In this case, Q_D^* can be much larger than Q_S^* and the largest intact fragment may be much smaller than the largest reaccumulated fragment.

Values of Q_D^* and Q_S^* have been estimated using both laboratory and numerical hydrocode experiments.

2.2.3 Experiments and Numerical Modelling of Disruptive Collisions

Laboratory impact disruption experiments of porous targets have been performed for a variety of target materials [43, 44]. Disruption by a projectile with a velocity higher than 1 km/s by two-stage light-gas gun has been conducted for targets of weakly glued pebble aggregates and pre-shattered fragments of cement mortar targets [45], sintered glass beads [20, 46, 47], gypsum [48–50], pumice [21, 44] and chondrites [51, 52]. Numerical impact disruption experiments have also been performed for both porous and nonporous targets [53]. By considering impacts on target bodies of a wide range of sizes these experiments spanned both the strength regime where $Q_D^* = Q_S^*$ and the gravity regime where $Q_D^* \geq Q_S^*$.

Figure 7a shows the experimentally and numerically determined disruption threshold (Q_D^*) as a function of target size. To the left of the diagram, in the strength regime where $Q_D^* = Q_S^*$, the disruption threshold decreases with increasing size scale. This is a consequence of both a reduction in tensile strength and a decrease in strain rate with increasing target size. Above a critical target body size (~200 m) and to the right of the diagram, in the gravity regime, the disruption threshold shows the reverse trend, increasing with body size. In this case, large impacts where $Q_D^* \geq Q \geq Q_S^*$ completely shatter the object, but are unable to separate the fragments sufficiently for disruption.

The experimental data which lie exclusively in the strength regime to the left side of the diagram are for head-on collisions, while curves of numerical simulation are for 45° oblique impacts onto basalt-like (nonporous) and pumice-like (porous) bodies. The higher disruption thresholds in the strength regime determined by the numerical simulation than those of experiments are probably because of the oblique incidence. In this regime, experiments and numerical simulations are in agreement in showing that more porous targets tend to have higher disruption threshold, because of their more efficient shock absorption.

The slopes of the relationships between disruption threshold and target diameter in the strength regime are different for different materials and different methods,

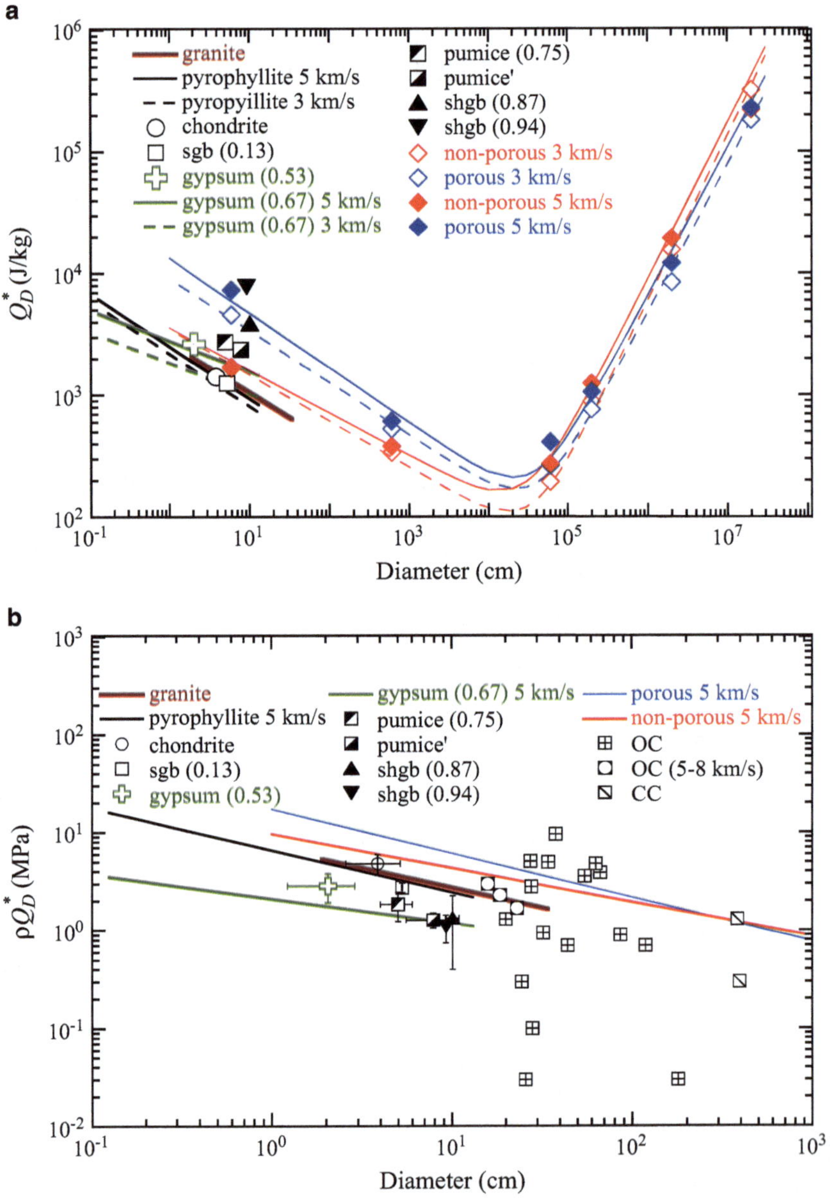

Fig. 7 Experimentally and numerically determined disruption threshold Q^*_D. Disruption thresholds (**a**) in unit of energy per unit mass, or energy density, and (**b**) in unit of energy per unit volume, or pressure, for granite [54], pyrophyllite [50], chondrites (ordinary chondrite and Allende carbonaceous chondrite, [51]), sintered glass bead (sgb) [47], normal gypsum [49], high-porosity gypsum [50], pumice of 0.75 porosity [44], another pumice (pumice') [21], sintered hollow glass bead (shgb) [20]. The curves of numerical simulation are for nonporous (basalt-like material) and porous (pumice-like material) bodies at 45° oblique impacts [53]. Also shown are fireball data [55]. Numbers in parentheses show porosity

i.e., experimental and numerical. The slope is given as $\frac{-9\mu}{(2m-3)}$ in the scaling theory of [54], where μ is the velocity exponent from point-source theory, introduced previously. Point-source theory is valid if the impactor is small relative to the target body and the impact velocity is high compared to material sound speed. As described previously, the value of μ for nonporous materials is in the range of 0.55–0.60, whereas for moderately porous materials, such as dry sand, μ is known to have lower value $\mu \sim 0.40$ [56]. The parameter m describes the flaw-size distribution in the target [54, 57]. It is a material parameter and dependent on the homogeneity of the target [58]; therefore, the different slopes shown in Fig. 7a may be consequence of different degree of homogeneity of the target materials.

Although Fig. 7a shows that more porous bodies are harder to disrupt in the context of the disruption threshold, Q^*_D, with the unit of energy per mass, Fig. 7b shows that disruption threshold in the unit of energy per volume (or, equivalently, stress) ρQ^*_D is similar for nonporous and porous materials. The threshold in this unit is called impact strength [59]. As highly porous materials are generally weaker than less porous materials, the effect of porosity and strength tends to offset one another when the disruption threshold is strength-dominated. Hence, the kinetic energies required to disrupt nonporous and porous bodies of the same size are similar in strength-dominated collisions [21]. Also shown in Fig. 7b are the dynamic strengths of ordinary chondrite (OC) and carbonaceous chondrite (CC) meteorites as inferred from fireball lightcurves [55]. These dynamics strengths can be orders-of-magnitude lower than the measured tensile strength of intact meteorite fragments, suggesting that some objects enter Earth's atmosphere already pervasively fractured.

The outcome of a disruptive collision is a complex function of material properties. In addition to porosity, other material properties such as cohesive and tensile strength as well as friction angle have an important effect on the overall impact strength of a body (i.e., its disruption threshold). The modeling results shown in Fig. 7a were obtained using a very simple von Mises shear strength model combined with a tensile failure model and only the effect of pore crushing was investigated in this study. More sophisticated strength models reproduce the behavior of rocky materials more accurately by accounting for the pressure-dependent shear strength of fractured material. Using such models, the threshold for catastrophic disruption Q^*_D was recently explored numerically at large scales for objects with various material properties [60], to identify the effects of porosity (crushing) as well as material strength (cohesion and friction). For this, the following material rheologies were investigated:

1. no friction, no crushing, no cohesion
2. friction, no crushing, no cohesion
3. friction and crushing, no cohesion
4. friction, crushing and cohesion

The first case is a fluid, with no porosity or resistance to deformation; the second represents a cohesionless material with internal friction but no porosity; the third is a cohesionless material with both internal friction and crushable porosity, like sand;

and the last is a cohesive material with internal friction and crushable porosity, like pumice. For the two cases with crushing, a crush curve with parameters for pumice was applied [61].

As illustrated in Fig. 8, the effects of pore crushing and friction are quite dramatic, even at the 100 km scales considered in this study. It was found that the catastrophic disruption energy Q^*_D is ~5–10 times higher when friction is included. The disruption threshold further increases by a factor ~2–3 when the energy dissipation by compaction (pore crushing) is taken into account. The results imply that although the surfaces of many asteroids and comets have a strength comparable to snow and orders-of-magnitude weaker than hard rock, their shock-absorbing ability implies they can withstand more violent collisions than dense monolithic rocky bodies.

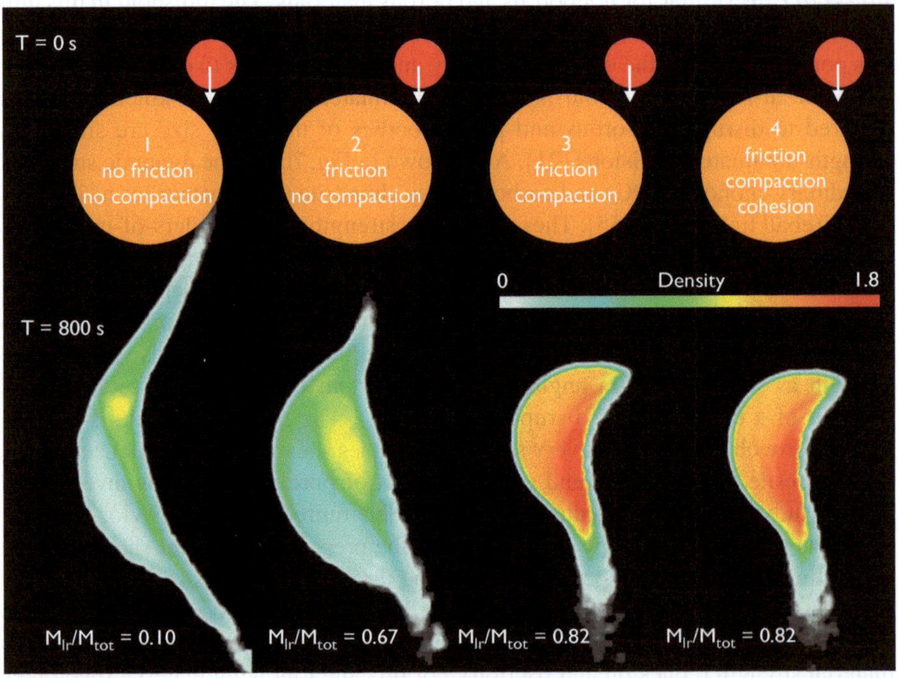

Fig. 8 3D SPH simulations of collisions between targets with a radius R_t = 100 km and projectiles with R_p = 27 km using a relative velocity of 3 km/s and a 45° impact angle. Shown are cross-sections through the mid-plane. Different rheologies are investigated, as described in the text. The degree of disruption, the size of the largest remnant (M_{lr}/M_{tot}) and the increase in density strongly depend on the target properties. From [60]

2.3 *Ejection Speeds and Momentum Transfer*

The ejection of material by impact events is an important mechanism in the collisional evolution of asteroids and the modification of planetary surfaces. The deposition of excavated material is a primary driver of the geological evolution of the surfaces of all airless, low-to-moderate porosity bodies. Owing to efficient shock wave absorption, craters in highly porous materials can form mostly by compaction with a negligible amount of ejected material, as indicated by experiments and observations (e.g., [29, 62, 63]). In this regard, porosity plays a key role in determining how asteroids evolve due to collisions.

Asteroids not only collide with each other but also, occasionally, with Earth. One of the more technologically feasible ways to deflect a hazardous asteroid is direct impact by a spacecraft. The spacecraft imparts its momentum to the asteroid, but also generates a crater and ejects surface material. Some of that material escapes the asteroid, providing an additional contribution to the momentum transfer. Therefore, the effectiveness of the impact deflection method depends on the mass–velocity distribution of the ejecta, an example of which is shown in Fig. 9. The figure shows the total mass of ejecta, normalized by the impactor mass, with speed greater than some value v in the form of point-source scaling. The main point of the figure is that ejecta mass decreases as the target porosity increases, which means porous asteroids could be harder to deflect than nonporous ones.

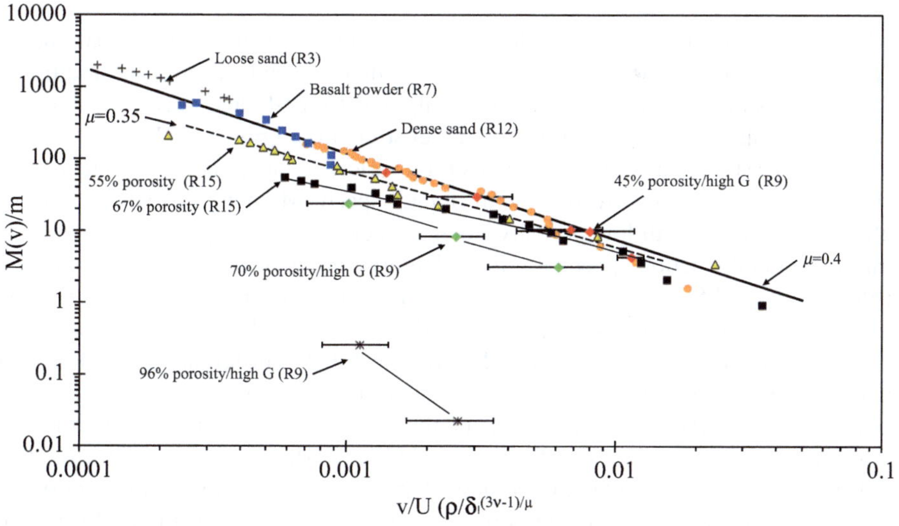

Fig. 9 Mass of material ejected faster than a given velocity. Increasing target porosity leads to an decrease of ejection speeds. From [63]. Data sources: (R3): [17]; (R7): [64]; (R9): [38]; (R12): [63]; (R15): [65]

The effectiveness of the impact deflection method is characterized by a parameter, β, defined as the change in the asteroid momentum divided by the spacecraft momentum. If no ejecta escape, $\beta = 1$. However, in general, the mass of crater ejecta can be many times that of the impactor (Fig. 9). If much of that escapes, β can be significantly larger than 1.

How large does β need to be to deflect an asteroid? The answer depends on the momentum change that must be imparted to the asteroid, and on the momentum that can be delivered by the impact. For illustration, suppose a 200 m asteroid is discovered roughly a decade before it would strike Earth. Assuming a spherical shape and a density of 2000 kg/m^3, the asteroid mass is 8.4×10^9 kg. For the given lead time, a velocity change of about 0.01 m/s is needed [66], which translates to a momentum change of 8.4×10^7 N-s. For a deflection mission, a typical impactor has a mass of several tons ($\sim 5 \times 10^3$ kg) and a speed of several to 10 km/s (say 7000 m/s), giving a spacecraft momentum of 3.5×10^7 N-s. This means $\beta = 8.4 \times 10^7/3.5 \times 10^7 = 2.4$. That is, the spacecraft would have to impart 2.4 times its momentum in order to deflect the asteroid. Could a spacecraft impact be reasonably expected to achieve this?

The answer to this question comes from experiments, both in the lab and on the computer, and on scaling theory. Point-source crater scaling theory [67] suggests that $\beta - 1$, i.e., the momentum transfer due to the ejecta, varies as a power of the impact speed U

$$\beta - 1 \sim U^{3\mu - 1}, \tag{1}$$

where the exponent μ is, again, the velocity exponent of the coupling parameter. As such, β should increase directly with impact speed. There are additional dependences on impactor size, target density and strength, and gravity, depending on whether the impact is controlled by target strength or gravity [67].

Figure 10 shows some measurements for a variety of target materials. The results show the expected increase of β with impact speed. Momentum transfer is most efficient in targets with low porosity, such as rock, where damping of the shock is minimized and ejection speeds are high. Highly porous materials, such as pumice (porosity of 80–85%), tend to have a lower value of β. One notable exception is the impact into cohesive pumice reported by Flynn et al. [21]. Their measured value of β is nearly as large as for rock. The reason for this difference is unclear, but underscores the present lack of understanding of how porosity—and different types of porosity—affects impact processes and the need for further studies. Nevertheless, returning to our deflection example, the data in Fig. 10 show that $\beta = 2.4$ (or $\beta - 1 = 1.4$) would be achieved by the hypothetical spacecraft if the threatening asteroid is rocky, but not if it has significant porosity. We note that this is a simple illustrative example. The result of a deflection mission depends on the specific details of the target asteroid, the impactor and the warning time before impact. However other studies of deflection have similarly shown that high target porosity makes deflection harder. For impact speeds of several km/s ejecta velocity distributions have been used to show that β is close to 1 for highly porous asteroids and as high as 3–13

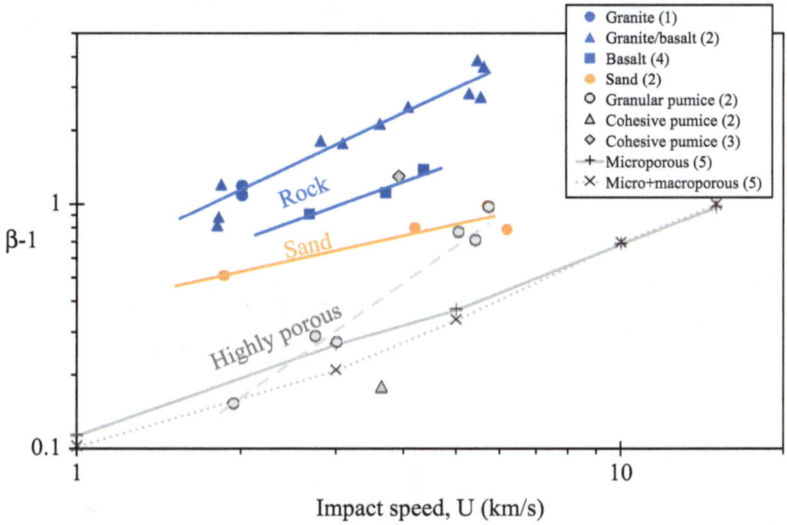

Fig. 10 Laboratory (closed symbols) and numerical results (open symbols) for the momentum multiplication factor, β. Data sources: (1): [70]; (2): [71]; (3): [21]; (4): [72]; (5): [69]

for rocky objects [67, 68]. Numerical simulations [69] yield β in the range of 1.5–1.7 for target porosity in the range of 50–70%, and 2.3 for 15% porosity. Therefore target porosity is a key factor in the design of asteroid deflection missions.

The effect of different degrees and scales of target porosity on β was investigated numerically for a range of impact conditions [69]. Two kinds of porous target structures were considered: (a) homogeneous: microporous only, and (b) heterogeneous: both micro- and macro-porous. The microporosity of the material had a porosity of 50% for both structures and was accounted for using a continuum p-α porosity model. In the case of structure (b), macroscopic cracks with a size scale of ~0.3 m are randomly distributed in the target (in addition to the microporosity), with a resulting total macroscopic void fraction of 10%. Figure 11 shows the outcome of a 10 km/s impact on such targets. The results of the calculations of β for different impact velocities are shown in Fig. 10.

Overall, results of this study verify that the momentum multiplication factor β is small for impacts on porous targets (with ~50% porosity), even for very high impact velocities ($\beta < 2$ for $v_{imp} \leq 15$ km/s). This is consistent with experimental results, as discussed above. The effect of the target inhomogeneities (macroporosity) on the momentum multiplication factor β was found to be quite small. At low impact velocities, the amount of momentum transferred is slightly smaller using structure (b) than structure (a). However, these differences disappear at high velocities. This suggests that the exact form of the modeling of porosity (explicit or implicit) does not matter for any phenomena on a length scale large compared to the size scale of that porosity.

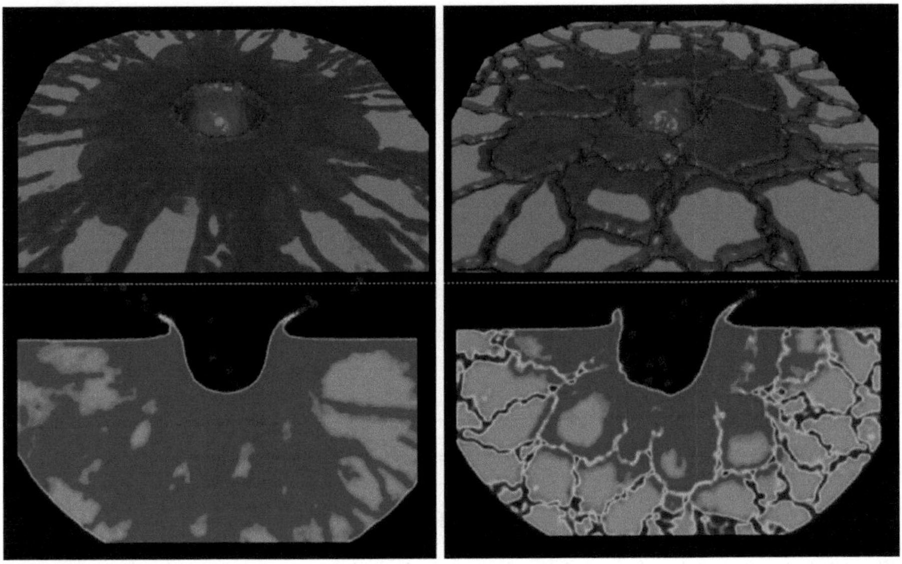

Fig. 11 Left: Impact at 10 km/s on a microporous target; the top figure shows the target from above, while the bottom figure shows a vertical slice. Right: same for a target containing both microporosity and macroporosity. Dark grey zone show the damage resulting from the impact. From [69]

2.4 Collateral Impact Effects: Compaction and Heating

As well as producing a crater and ejecta, or deforming an entire body, impacts can produce a range of collateral effects in the material processed by the shock wave. These include: compaction and lithification, shock metamorphism and phase changes, including melting and vaporization. In this section, we briefly summarize some of these important effects and how they are influenced by porosity.

2.4.1 Compaction and Shock Effects at the Meso- and Macro-Scale

In porous materials, an important impact effect is compaction: the permanent crushing of pore space and densification of the compacted material. This may result in a pulverized, fine-grained granular material or if temperatures in the compacted material are high enough, sintering and lithification or even melting [17, 73]. Because porosity has a length scale, compaction has localized effects at the length scale of the porosity as well as bulk effects at larger scales.

Among planetary materials, the shock response of quartz and sandstone is the most studied [74–78]. For nonporous quartzo-feldspathic rocks, seven shock stages, 0, 1a, 1b, 2, 3, 4 and 5, are distinguished based on a progressive sequence of shock effects—produced experimentally and observed in nature—from fracturing, planar fractures and planar deformation features (PDFs) at low pressures (0, 1a, 1b); through to crystallographic phase transformations (stishovite/coesite) and diaplectic glass at intermediate pressures (2, 3); and then whole rock melting (4, 5). A similar classification scheme has been developed for meteorites [79].

As compaction of porous rocks produces additional waste heat and localized pressure amplifications around closing pores, higher initial porosity tends to shift shock-stage transitions to lower pressures. However, a quantitative calibration of the effect of initial porosity on shock stage is lacking for many common geological materials. Quartz is a notable exception. A crude calibration was derived for quartz sandstone based on observations of shock deformation in sandstones at Meteor Crater, Arizona [74, 75]. This was recently revised, based on a series of shock recovery experiments on Seeberger sandstone [78, 80]. The boundary between shock stage 1a and 1b is defined by the complete closure of pore space; higher pressure boundaries are defined by a progressive increase in the volume fraction of diaplectic glass and SiO_2 melt that replaces intense fracturing.

Porosity also introduces additional heterogeneity into a material, which results in a heterogeneous shock response. Rapid closure of pore space during shock compaction leads to localized pressure amplification (of approximately four times ambient shock pressure) and hot spots that have been observed in natural samples [75], shock recovery experiments [80], and numerical modelling [81]. This "mesoscale" response is particularly significant for meteoritic material, which is often comprised of a bimodal mixture of large (mm-scale) nonporous igneous inclusions—chondrules—surrounded by a porous matrix of much finer grains.[5] Recent numerical simulations [2, 82] have shown that compaction of such material, which is accommodated primarily by crushing of the matrix, results in a bimodal temperature and strain distribution (Fig. 12). The compacted matrix is strongly deformed and heated, while the strong, nonporous chondrules remain largely unscathed. As analysis of meteorites has tended to focus on the chondrules for establishing shock effects, it is possible that important evidence of impact processing in the early solar system has been overlooked. Indeed, recent analytical interrogation of inter-chondrule matrix of the primitive Allende meteorite has provided evidence for impact compaction in a meteorite classified as shock stage 1 (low shock) [83].

2.4.2 Impact Heating, Melting and Vaporization

An important collateral effect of impact cratering at the meso- and macro-scale is the deposition of thermal energy, which raises the temperature of rocks beneath the crater and can result in phase changes, including melting and vaporization. Long cooling timescales associated with large impacts imply the thermal anomaly can be an important heat source and a driver for subsurface fluid flow for long periods after impact. Molten ejecta and vapour can be transported great distances from the impact

[5]See discussion of mesoscale modeling in chapter by Vogler and Fredenburg.

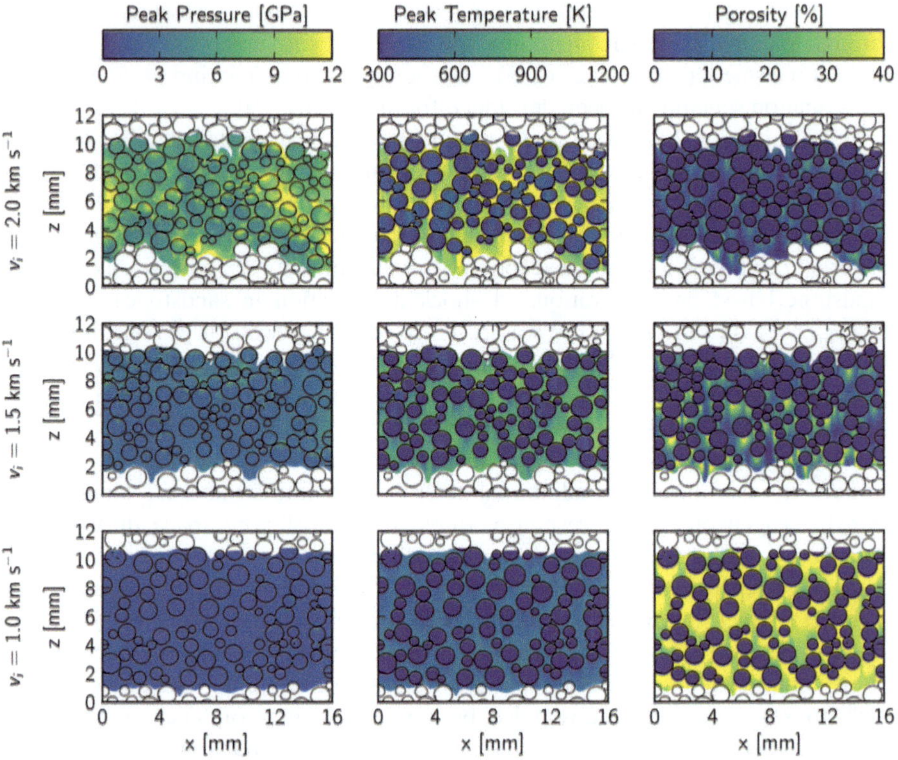

Fig. 12 The spatial variation of peak pressure, peak temperature and porosity within a bimodal mixture of nonporous chondrules and 70% porous matrix (volume ratio 3:7) subjected to planar shocks of varying intensity [82]. Results for three different impact speeds (v_i) are shown; see [82] for details. In these 2D mesoscale numerical simulations, the chondrules are modelled explicitly as nonporous dunite discs, whereas the matrix grains are modelled as a continuum using the ϵ-α porosity compaction model [28, 84]

site and pose an environmental hazard if produced in sufficiently large quantities. For example, the 66 Ma Chicxulub impact in the Gulf of Mexico released ~5000 gigatons of vaporized impactor and target material into the atmosphere, including carbon dioxide and sulphate aerosols. The Cretaceous–Palaeogene (K-Pg) mass extinction event immediately following the impact has been attributed to global cooling caused by dust, soot and, in particular, aerosols injected into the atmosphere by the impact [85].

Porosity in the target material affects impact heating—and consequently melt and vapour production—in two ways. First, as permanent crushing of pore space generates additional waste heat compared to compression of nonporous materials, post-shock temperatures in porous materials are higher for the same shock pressure [12]. Or, to say this another way, porous materials require a much smaller degree of shock compression to experience a given temperature increase or to undergo a thermodynamic phase transformation, such as melting or vaporization. On the other hand, the more rapid attenuation of expanding shock waves in porous materials [31, 86] implies that at a given distance from the impact site, the shock pressure will be lower in a porous material compared to a nonporous material. The volume of material shocked to a given shock pressure will also be less in porous targets than in nonporous targets. Thus, whether more or less material is melted, vaporized or heated to a given post-shock temperature, as a consequence of porosity depends on the competition between these two effects. High porosity results in more extreme, but also more localized shock heating; low porosity results in more modest shock heating, but distributed over a larger volume.

The effect of porosity on shock melting has been investigated experimentally [80, 87] and using numerical modelling [88, 89]. Numerical modelling estimates of the critical shock pressure required for melting as a function of porosity have been derived for quartz (sandstone) [88] and forsterite/dunite [89] (a common proxy for chondritic material). These suggest that 20% porosity is sufficient to reduce the shock pressure for melting by one half and 50% porosity reduces the shock pressure for melting to between one-eighth and one-tenth. Natural observations and laboratory experiments provide support for these estimates. Based on studies of shocked Coconino sandstone at Meteor Crater, Arizona, with a preimpact porosity of 20%, and comparisons to the Hugoniot for quartz, Kieffer [74] inferred a lowering of the shock pressure for melting of Coconino sandstone to ~30 GPa from a nominal value of 60 GPa required to melt solid quartz. More recently, shock recovery experiments of 25–30% porous sandstone samples suggested 80% melting by volume at a shock pressure of only 17.5 GPa [80].

As most planetary bodies smaller than 500-km diameter have substantial internal porosity [3, 13], the lower threshold for melting in porous rocks has important implications for small-body collisions during planet formation and in the current asteroid belt. As the maximum shock pressure achieved in such impacts is directly related to the collision speed, significant melting may occur in slower collisions if the initial porosity is higher. Numerical simulations [89] suggest that while melting is minor in collisions between nonporous objects at speeds below 10 km/s, significant melting is expected at speeds greater than 7 km/s for an initial porosity of 20% and at speeds greater than 5 km/s for an initial porosity of 50% (Fig. 13). For collisions between bodies of different sizes, such as those with an impactor-to-target mass ratio below 0.1, significant localized heating occurs in the target body. At impact velocities as low as 5 km/s, the estimated mass of melt is nearly double the mass of the impactor, and the mass heated by 100 K is nearly five times greater [89].

Melt (and vapour) production in porous rocks has been investigated using observations and numerical modelling [88]. The total volume of impact melt can sometimes be estimated at terrestrial impact structures (Fig. 14) allowing a comparison of melt production in targets of different composition. However, the challenge of recognizing impact-generated melt rocks in sedimentary rock targets, as well as determining initial porosity and water content, has hampered efforts to determine whether craters in porous rocks show a melt deficiency or excess

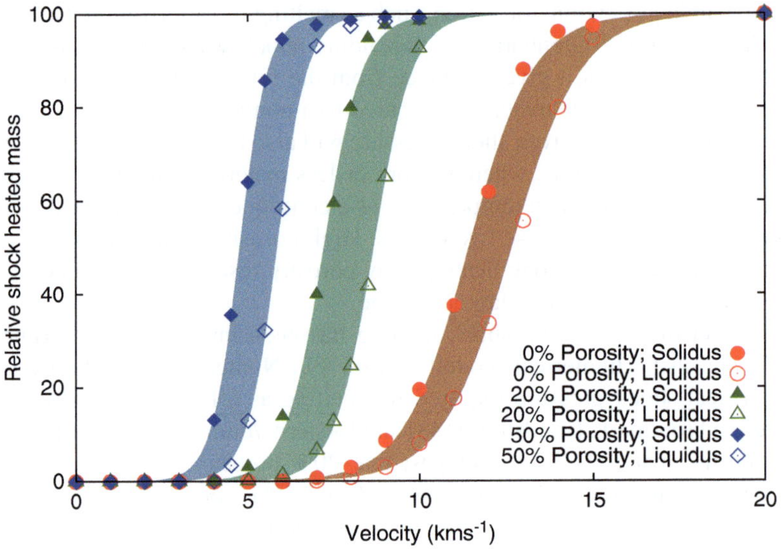

Fig. 13 The effect of velocity on the fraction of material melted in head-on collisions between two asteroids of equal size. The average collision velocity in the asteroid belt today is 5 km/s. At this speed, negligible melt production is expected if the asteroids are nonporous, but melting could be significant for asteroids with typical porosities of 20–50% [89]

compared to craters in nonporous crystalline rocks [90]. For volatile-rich porous targets the story is complex and the degree of saturation likely plays an important role in determining whether more or less melt is produced in a porous target as well as the nature of the end product. For dry porous and nonporous silicate rocks numerical simulations suggest that porosity enhances impact melt production [88]. In these simulations, the effect of dry porosity on lowering of the critical pressure for melting is reduced but not totally diminished by faster shock wave decay in porous materials. The results for nonporous rocks agree well with estimated melt volumes at terrestrial impact craters in volatile-poor target rocks. For dry porous targets, the results suggest that a moderate initial porosity of 25% increases melt volume by approximately a factor of two (Fig. 14). Similar numerical simulations of impacts in water-ice targets, however, show the opposite trend [91]. In this case, melt production appears to be reduced by initial porosity, suggesting that, for ice, enhanced shock wave attenuation dominates over the reduction in shock pressure required for melting. Further work is required to establish the reason for this difference and constrain the effect of porosity on melt production in water–rock mixtures. The sensitivity of impact vapor production to target and impactor porosity, and implications for large terrestrial impacts such as the one that formed the Chicxulub crater, is another area that requires further work.

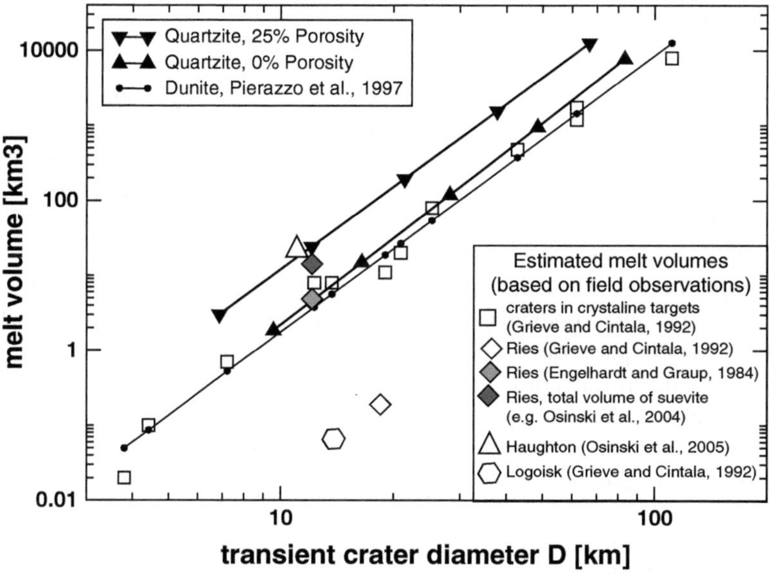

Fig. 14 Comparison between observed melt volumes at various terrestrial craters of different sizes and estimates derived from shock physics calculations for nonporous dunite and 0, 25% porous quartzite. See [88] for details. Crater sizes are estimates of the transient crater diameter at the preimpact surface. Larger melt volumes are expected in craters formed in dry porous rocks. After [88]

2.4.3 Fracturing and Dilatancy

Impacts can generate porosity as well as destroy it. The principal gravity anomaly of most terrestrial impact craters is negative as a result of fracturing by the shock wave and shear bulking during crater formation [92, 93]. Gravity measurements by NASA's GRAIL mission indicate the presence of substantial porosity in the upper kilometers of the lunar crust [8, 9] and the correlation of spatial variations in crustal density with impact craters suggests that the porosity is impact-generated [10, 94]. Numerical simulations of impacts that account for porosity generation and destruction show that the balance between these effects depends on both impact size and preimpact porosity, with compaction dominating in large events in porous targets and vice versa [11]. Confining pressure also appears to have an important control on porosity generation [93], which explains why the process appears to be more effective on the Moon than Earth and suggests high crustal porosity is likely to be common among minor planets. As porosity has a fundamental control on fluid and heat transfer, impact processes must play a pivotal role in the formation and evolution of planetary crusts.

3 Conclusions

The outcome of a planetary collision depends to a large degree on the porosity of the target material. Porosity affects the size of the resulting crater, the presence and morphology of ejecta blankets, the speed of the ejected material, the sensitivity of the target object to deformation and disruption and a range of collateral effects from shock metamorphism, melting and vaporization to the competition between compaction and distension.

Through these mechanisms, target porosity also has a significant effect on the evolution of planetary bodies, their internal structures, and their surfaces. The size distributions of impact craters and the surface ages deduced from those distributions are sensitive to crustal porosity, as are impact melt and vapour production and fracturing-related density and permeability changes. At global scales, internal porosity can increase an object's resilience to disruption and collisional lifespan, reduce the mass and speed of escaping ejecta, and make it harder to deflect off course.

Some effects of porosity are intuitive: for example, more effective shock absorption and higher shock temperatures. But other consequences are not, and have only been exposed by detailed spacecraft measurements, innovative analog experiments, and sophisticated numerical modelling. For example, the fact that light and often fragile porous materials appear to be more resilient to the cratering process than heavier, stronger nonporous rocks is counter intuitive. As is the fact that a highly porous hazardous asteroid on collision course with Earth can be much harder to divert than a stronger, denser object of the same mass.

While the principal effects of porosity on impact processes are now quite well known, many details are lacking and we are only just beginning to appreciate their many implications. For example, a universal crater scaling relation for porous materials is yet to be established and accurate shock calibration only exists for quartz-dominated rocks. Further impact experiments, numerical models and observations of natural craters are all required to address these gaps. Moreover, few studies have explored the role of the type and size scale of porosity on impact processes. For instance, it is likely that inter-granular porosity in a cohesionless soil will influence shock propagation and cratering in different ways to vesicular porosity in a well-bonded igneous rock. In this case, mesoscale numerical simulations combined with space-and-time-resolved laboratory experiments, for example, using dynamic X-ray radiography, will be invaluable for relating particle- or pore-scale effects to the bulk response. Finally, this review has focussed on the effects of *target* porosity on impact processes; the influence of *impactor* porosity is largely unexplored and certainly warrants further work.

Since the beginning of the space program, a major driver of planetary impact research has been human and robotic spacecraft discovery. The next decades will see several pioneering space missions to asteroids, including NASA's OSIRIS-REx, Psyche and Lucy and JAXA's Hayabusa 2. As well as providing new tests of our understanding of impacts on porous bodies, these missions will also doubtless raise further exciting questions that will drive the science of planetary impacts forward.

References

1. Blum J, Wurm G (2008) The growth mechanisms of macroscopic bodies in protoplanetary disks. Ann Rev Astron Astrophys 46(1):21–56. https://doi.org/10.1146/annurev.astro.46.060407.145152
2. Bland PA, Collins GS, Davison TM, Abreu NM, Ciesla FJ, Muxworthy AR, Moore J (2014) Pressure–temperature evolution of primordial solar system solids during impact-induced compaction. Nat Commun 5. https://doi.org/10.1038/ncomms6451
3. Britt DT, Yeomans D, Housen K, Consolmagno G (2002) Asteroid density, porosity, and structure. In: Asteroids III, Bottke Jr WF, Cellino A, Paolicchi P, Binzel RP (eds) University of Arizona Press, Tucson, pp 485–500
4. Consolmagno GJ, Britt DT, Macke RJ (2008) The significance of meteorite density and porosity. Chemie der Erde - Geochem 68(1):1–29. https://doi.org/10.1016/j.chemer.2008.01.003
5. Asphaug E, Ryan EV, Zuber MT (2002) Asteroid interiors. In: Asteroids III, Bottke Jr WF, Cellino A, Paolicchi P, Binzel RP (eds) University of Arizona Press, Tucson, pp 463–484
6. Sasso MR, Macke RJ, Boesenberg JS, Britt DT, Rivers ML, Ebel DS, Friedrich JM (2009) Incompletely compacted equilibrated ordinary chondrites. Meteorit Planet Sci 44(11):1743–1753. https://doi.org/10.1111/j.1945-5100.2009.tb01204.x
7. McKay DS, Heiken G, Basu A, Blanford G, Simon S, Reedy R, French BM, Papike J (1991) The Lunar Regolith. In: Heiken GH, Vaniman DT, French BM (eds) Lunar source book. Cambridge University Press, Cambridge
8. Wieczorek MA, Neumann GA, Nimmo F, Kiefer WS, Taylor GJ, Melosh HJ, Phillips RJ, Solomon SC, Andrews-Hanna JC, Asmar SW, Konopliv AS, Lemoine FG, Smith DE, Watkins MM, Williams JG, Zuber MT (2013) The crust of the moon as seen by GRAIL. Science 339(6120):671–675. https://doi.org/10.1126/science.1231530
9. Besserer J, Nimmo F, Wieczorek MA, Weber RC, Kiefer WS, McGovern PJ, Andrews-Hanna JC, Smith DE, Zuber MT (2014) GRAIL gravity constraints on the vertical and lateral density structure of the lunar crust. Geophys Res Lett 41(16):5771–5777. https://doi.org/10.1002/2014GL060240
10. Soderblom JM, Evans AJ, Johnson BC, Melosh HJ, Miljković K, Phillips RJ, Andrews-Hanna JC, Bierson CJ, Head JW, Milbury C, Neumann GA, Nimmo F, Smith DE, Solomon SC, Sori MM, Wieczorek MA, Zuber MT (2015) The fractured Moon: production and saturation of porosity in the lunar highlands from impact cratering. Geophys Res Lett 42(17):2015GL065022. https://doi.org/10.1002/2015GL065022
11. Milbury C, Johnson BC, Melosh HJ, Collins GS, Blair DM, Soderblom JM, Nimmo F, Bierson CJ, Phillips RJ, Zuber MT (2015) Preimpact porosity controls the gravity signature of lunar craters. Geophys Res Lett 42(22):2015GL066198. https://doi.org/10.1002/2015GL066198
12. Sharp TG, de Carli PS (2006) Shock Effects in Meteorites. In: Meteorites and the Early solar system II, pp 653–677
13. Carry B (2012) Density of asteroids. Planet Space Sci 73(1):98–118. https://doi.org/10.1016/j.pss.2012.03.009
14. Melosh HJ (1984) Impact ejection, spallation, and the origin of meteorites. Icarus 59(2):234–260. https://doi.org/10.1016/0019-1035(84)90026-5
15. Polanskey CA, Ahrens TJ (1990) Impact spallation experiments: fracture patterns and spall velocities. Icarus 87(1):140–155. https://doi.org/10.1016/0019-1035(90)90025-5
16. Holsapple KA, Housen KR (2013) The third regime of cratering: spall Craters. In: Lunar and planetary science conference, vol 44, p 2733
17. Stöffler D, Gault DE, Wedekind J, Polkowski G (1975) Experimental hypervelocity impact into quartz sand: distribution and shock metamorphism of ejecta. J Geophys Res 80(29):4062–4077. https://doi.org/10.1029/JB080i029p04062
18. Housen K, Sweet W, Holsapple K (2018) Impacts into porous asteroids. Icarus 300:72–96. https://doi.org/10.1016/j.icarus.2017.08.019

19. Michikami T, Moriguchi K, Hasegawa S, Fujiwara A (2007) Ejecta velocity distribution for impact cratering experiments on porous and low strength targets. Planet Space Sci 55(1–2):70–88. https://doi.org/10.1016/j.pss.2006.05.002
20. Okamoto T, Nakamura AM, Hasegawa S (2015) Impact experiments on highly porous targets: Cavity morphology and disruption thresholds in the strength regime. Planet Space Sci 107:36–44. https://doi.org/10.1016/j.pss.2014.08.008
21. Flynn GJ, Durda DD, Patmore EB, Clayton AN, Jack SJ, Lipman MD, Strait MM (2015) Hypervelocity cratering and disruption of porous pumice targets: implications for crater production, catastrophic disruption, and momentum transfer on porous asteroids. Planet Space Sci 107:64–76. https://doi.org/10.1016/j.pss.2014.10.007
22. Okamoto T, Nakamura AM (2017) Scaling of impact-generated cavity-size for highly porous targets and its application to cometary surfaces. Icarus 292:234–244. https://doi.org/10.1016/j.icarus.2017.01.007
23. Hörz F, Bastien R, Borg J, Bradley JP, Bridges JC, Brownlee DE, Burchell MJ, Chi M, Cintala MJ, Dai ZR, Djouadi Z, Dominguez G, Economou TE, Fairey SAJ, Floss C, Franchi IA, Graham GA, Green SF, Heck P, Hoppe P, Huth J, Ishii H, Kearsley AT, Kissel J, Leitner J, Leroux H, Marhas K, Messenger K, Schwandt CS, See TH, Snead C, Stadermann FJ, Stephan T, Stroud R, Teslich N, Trigo-Rodríguez JM, Tuzzolino AJ, Troadec D, Tsou P, Warren J, Westphal A, Wozniakiewicz P, Wright I, Zinner E (2006) Impact features on stardust: implications for Comet 81p/Wild 2 Dust. Science 314(5806):1716–1719, https://doi.org/10.1126/science.1135705
24. Niimi R, Kadono T, Arakawa M, Yasui M, Dohi K, Nakamura AM, Iida Y, Tsuchiyama A (2011) In situ observation of penetration process in silica aerogel: Deceleration mechanism of hard spherical projectiles. Icarus 211(2):986–992. https://doi.org/10.1016/j.icarus.2010.11.005
25. Luther R, Artemieva N, Ivanova M, Lorenz C, Wünnemann K (2017) Snow carrots after the Chelyabinsk event and model implications for highly porous solar system objects. Meteorit Planet Sci 52(5):979–999. https://doi.org/10.1111/maps.12831
26. Housen KR, Holsapple KA (2012) Craters without ejecta. Icarus 219(1):297–306. https://doi.org/10.1016/j.icarus.2012.02.030
27. Carroll MM, Holt AC (1972) Static and dynamic pore-collapse relations for ductile porous materials. J Appl Phys 43(4):1626–1636
28. Wünnemann K, Collins GS, Melosh HJ (2006) A strain-based porosity model for use in hydrocode simulations of impacts and implications for transient crater growth in porous targets. Icarus 180(2):514–527. https://doi.org/10.1016/j.icarus.2005.10.013
29. Housen KR, Holsapple KA, Voss ME (1999) Compaction as the origin of the unusual craters on the asteroid Mathilde. Nature 402(6758):155–157. https://doi.org/10.1038/45985
30. Thomas PC, Armstrong JW, Asmar SW, Burns JA, Denk T, Giese B, Helfenstein P, Iess L, Johnson TV, McEwen A, Nicolaisen L, Porco C, Rappaport N, Richardson J, Somenzi L, Tortora P, Turtle EP, Veverka J (2007) Hyperion's sponge-like appearance. Nature 448(7149):50–56. https://doi.org/10.1038/nature05779
31. Holsapple KA, Schmidt RM (1987) Point source solutions and coupling parameters in cratering mechanics. J Geophys Res 92(B7):6350–6376. https://doi.org/10.1029/JB092iB07p06350
32. Holsapple KA (1993) The scaling of impact processes in planetary sciences. Ann Rev Earth Planet Sci 21:333–373
33. Güldemeister N, Wünnemann K, Poelchau M (2015) Scaling impact crater dimensions in cohesive rock by numerical modeling and laboratory experiments. In: Osinski GR, Kring DA (eds) Large meteorite impacts and planetary evolution V. Geological Society of America
34. Schmidt RM, Housen KR (1987) Some recent advances in the scaling of impact and explosion cratering. Int J Impact Eng 5(1–4):543–560. https://doi.org/10.1016/0734-743X(87)90069-8
35. Yamamoto S, Wada K, Okabe N, Matsui T (2006) Transient crater growth in granular targets: an experimental study of low velocity impacts into glass sphere targets. Icarus 183(1):215–224. https://doi.org/10.1016/j.icarus.2006.02.002

36. Wünnemann K, Nowka, D, Collins, GS, Elbeshausen, D, Bierhaus, M (2011) Scaling of impact crater formation on planetary surfaces – insights from numerical modeling. In: Proceedings of the 11th Hypervelocity Impact Symposium 2010. Fraunhofer Verlag, Stuttgart/Freiburg, p 828
37. Yasui M, Arakawa M, Hasegawa S, Fujita Y, Kadono T (2012) In situ flash X-ray observation of projectile penetration processes and crater cavity growth in porous gypsum target analogous to low-density asteroids. Icarus 221(2):646–657. https://doi.org/10.1016/j.icarus.2012.08.018
38. Housen KR, Holsapple KA (2003) Impact cratering on porous asteroids. Icarus 163(1):102–119. https://doi.org/10.1016/S0019-1035(03)00024-1
39. Asphaug E, Collins G, Jutzi M (2015) Global scale impacts. In: Michel P, Demeo FE, Bottke WF (eds) Asteroids IV, University of Arizona Press, Tucson, pp 661–678. arXiv: 1504.02389
40. Michel P, Benz W, Tanga P, Richardson DC (2001) Collisions and gravitational reaccumulation: forming asteroid families and satellites. Science 294:1696–1700. https://doi.org/10.1126/science.1065189
41. Michel P, Benz W, Richardson DC (2003) Disruption of fragmented parent bodies as the origin of asteroid families. Nature 421:608–611
42. Leinhardt ZM, Stewart ST (2012) Collisions between Gravity-dominated Bodies. I. Outcome Regimes and Scaling Laws. Astrophys J 745:79. https://doi.org/10.1088/0004-637X/745/1/79, 1106.6084
43. Holsapple K, Giblin I, Housen K, Nakamura A, Ryan E (2002) Asteroid impacts: laboratory experiments and scaling laws. In: Asteroids III, Bottke Jr WF, Cellino A, Paolicchi P, Binzel RP (eds) University of Arizona Press, Tucson, pp 443–462
44. Nakamura AM, Hiraoka K, Yamashita Y, Machii N (2009) Collisional disruption experiments of porous targets. Planet Space Sci 57(2):111–118. https://doi.org/10.1016/j.pss.2008.07.027
45. Ryan EV, Hartmann WK, Davis DR (1991) Impact experiments 3: Catastrophic fragmentation of aggregate targets and relation to asteroids. Icarus 94(2):283–298. https://doi.org/10.1016/0019-1035(91)90228-L
46. Love SG, Hörz F, Brownlee DE (1993) Target porosity effects in impact cratering and collisional disruption. Icarus 105(1):216–224. https://doi.org/10.1006/icar.1993.1119
47. Setoh M, Nakamura AM, Michel P, Hiraoka K, Yamashita Y, Hasegawa S, Onose N, Okudaira K (2010) High- and low-velocity impact experiments on porous sintered glass bead targets of different compressive strengths: outcome sensitivity and scaling. Icarus 205(2):702–711. https://doi.org/10.1016/j.icarus.2009.08.012
48. Nakamura A, Suguiyama K, Fujiwara A (1992) Velocity and spin of fragments from impact disruptions. Icarus 100(1):127–135. https://doi.org/10.1016/0019-1035(92)90023-Z
49. Okamoto C, Arakawa M (2009) Experimental study on the collisional disruption of porous gypsum spheres. Meteorit Planet Sci 44:1947–1954. https://doi.org/10.1111/j.1945-5100.2009.tb02004.x
50. Nakamura AM, Yamane F, Okamoto T, Takasawa S (2015) Size dependence of the disruption threshold: laboratory examination of millimeter-centimeter porous targets. Planet Space Sci 107:45–52. https://doi.org/10.1016/j.pss.2014.07.011
51. Flynn GJ, Durda DD (2004) Chemical and mineralogical size segregation in the impact disruption of inhomogeneous, anhydrous meteorites. Planet Space Sci 52(12):1129–1140. https://doi.org/10.1016/j.pss.2004.07.010
52. Cintala MJ, Hörz F (2008) Experimental impacts into chondritic targets, part i: Disruption of an l6 chondrite by multiple impacts. Meteorit Planet Sci 43(4):771–803. https://doi.org/10.1111/j.1945-5100.2008.tb00684.x
53. Jutzi M, Michel P, Benz W, Richardson DC (2010) Fragment properties at the catastrophic disruption threshold: the effect of the parent body's internal structure. Icarus 207(1):54–65. https://doi.org/10.1016/j.icarus.2009.11.016
54. Housen KR, Holsapple KA (1999) Scale effects in strength-dominated collisions of rocky asteroids. Icarus 142(1):21–33. https://doi.org/10.1006/icar.1999.6206
55. Popova O, Borovicka J, Hartmann WK, Spurný P, Gnos E, Nemtchinov I, Trigo-Rodríguez JM (2011) Very low strengths of interplanetary meteoroids and small asteroids. Meteorit Planet Sci 46(10):1525–1550. https://doi.org/10.1111/j.1945-5100.2011.01247.x

56. Housen KR, Holsapple KA (1990) On the fragmentation of asteroids and planetary satellites. Icarus 84(1):226–253. https://doi.org/10.1016/0019-1035(90)90168-9
57. Jutzi M, Benz W, Michel P (2008) Numerical simulations of impacts involving porous bodies: I. Implementing sub-resolution porosity in a 3d SPH hydrocode. Icarus 198(1):242–255. https://doi.org/10.1016/j.icarus.2008.06.013
58. Weibull W (1939) A statistical theory of the strength of materials. Ingvetensk Akad Handl 151:1–45
59. Davis DR, Ryan EV (1990) On collisional disruption - experimental results and scaling laws. Icarus 83:156–182. https://doi.org/10.1016/0019-1035(90)90012-X
60. Jutzi M (2015) SPH calculations of asteroid disruptions: the role of pressure dependent failure models. Planet Space Sci 107:3–9. https://doi.org/10.1016/j.pss.2014.09.012
61. Jutzi M, Michel P, Hiraoka K, Nakamura AM, Benz W (2009) Numerical simulations of impacts involving porous bodies: II. Comparison with laboratory experiments. Icarus 201(2):802–813. https://doi.org/10.1016/j.icarus.2009.01.018
62. Housen KR, Holsapple KA (2003) Impact cratering on porous asteroids. Icarus 163:102–119. https://doi.org/10.1016/S0019-1035(03)00024-1
63. Housen KR, Holsapple KA (2011) Ejecta from impact craters. Icarus 211(1):856–875. https://doi.org/10.1016/j.icarus.2010.09.017
64. Hartmann WK (1985) Impact experiments. Icarus 63(1):69–98. https://doi.org/10.1016/0019-1035(85)90021-1
65. Cintala MJ, Berthoud L, Hörz F (1999) Ejection-velocity distributions from impacts into coarse-grained sand. Meteorit Planet Sci 34(4):605–623. https://doi.org/10.1111/j.1945-5100.1999.tb01367.x
66. Ahrens TJ, Harris AW (1992) Deflection and fragmentation of near-Earth asteroids. Nature 360(6403):429–433. https://doi.org/10.1038/360429a0
67. Holsapple KA, Housen KR (2012) Momentum transfer in asteroid impacts. I. Theory and scaling. Icarus 221(2):875–887. https://doi.org/10.1016/j.icarus.2012.09.022
68. Cheng AF, Michel P, Jutzi M, Rivkin AS, Stickle A, Barnouin O, Ernst C, Atchison J, Pravec P, Richardson DC (2016) Asteroid impact and deflection assessment mission: kinetic impactor. Planet Space Sci 121:27–35. https://doi.org/10.1016/j.pss.2015.12.004
69. Jutzi M, Michel P (2014) Hypervelocity impacts on asteroids and momentum transfer I. Numerical simulations using porous targets. Icarus 229:247–253. https://doi.org/10.1016/j.icarus.2013.11.020
70. Walker JD, Chocron S, Durda DD, Grosch DJ, Movshovitz N, Richardson DC, Asphaug E (2012) Momentum enhancement from large impacts into granite. In: Asteroids, Comets, Meteors 2012, LPI Contributions, vol 1667, p 6086
71. Housen KR, Holsapple KA (2015) Experimental measurements of momentum transfer in hypervelocity collisions. In: Lunar and planetary science conference, lunar and planetary science conference, vol 46, p 2894
72. Yanagisawa M, Hasegawa S (2000) Momentum transfer in oblique impacts: implications for asteroid rotations. Icarus 146:270–288. https://doi.org/10.1006/icar.2000.6389
73. Wünnemann K, Zhu MH, Stöffler D (2016) Impacts into quartz sand: Crater formation, shock metamorphism, and ejecta distribution in laboratory experiments and numerical models. Meteorit Planet Sci pp 1762–1794. https://doi.org/10.1111/maps.12710
74. Kieffer SW (1971) Shock metamorphism of the Coconino Sandstone at Meteor Crater, Arizona. J Geophys Res 76(23):5449–5473. https://doi.org/10.1029/JB076i023p05449
75. Kieffer SW, Phakey PP, Christie JM (1976) Shock processes in porous quartzite: transmission electron microscope observations and theory. Contrib Mineral Petrol 59(1):41–93. https://doi.org/10.1007/BF00375110
76. Grieve RAF, Langenhorst F, Stöffler D (1996) Shock metamorphism of quartz in nature and experiment: II. Significance in geoscience*. Meteorit Planet Sci 31(1):6–35. https://doi.org/10.1111/j.1945-5100.1996.tb02049.x

77. Stöffler D, Grieve RAF (2007) Impactites, a proposal on behalf of the IUGS subcommission on the systematics of metamorphic rocks. In: Fettes D, Desmons J (eds) Metamorphic rocks–a classification and glossary of terms. Cambridge University Press, Cambridge, pp 82–92
78. Kowitz A, Güldemeister N, Schmitt RT, Reimold WU, Wünnemann K, Holzwarth A (2016) Revision and recalibration of existing shock classifications for quartzose rocks using low-shock pressure (2.5–20 GPa) recovery experiments and mesoscale numerical modeling. Meteorit Planet Sci 51(10):1741–1761. https://doi.org/10.1111/maps.12712
79. Stöffler D, Keil K, Scott ERD (1991) Shock metamorphism of ordinary chondrites. Geochim Cosmochim Acta 55(12):3845–3867. https://doi.org/10.1016/0016-7037(91)90078-J
80. Kowitz A, Güldemeister N, Reimold WU, Schmitt RT, Wünnemann K (2013) Diaplectic quartz glass and SiO2 melt experimentally generated at only 5 GPa shock pressure in porous sandstone: Laboratory observations and meso-scale numerical modeling. Earth Planet Sci Lett 384:17–26. https://doi.org/10.1016/j.epsl.2013.09.021
81. Güldemeister N, Wünnemann K, Durr N, Hiermaier S (2013) Propagation of impact-induced shock waves in porous sandstone using mesoscale modeling. Meteorit Planet Sci 48(1):115–133. https://doi.org/10.1111/j.1945-5100.2012.01430.x
82. Davison TM, Collins GS, Bland PA (2016) Mesoscale modeling of impact compaction of primitive solar system solids. ApJ 821(1):68. https://doi.org/10.3847/0004-637X/821/1/68
83. Forman LV, Bland PA, Timms NE, Collins GS, Davison TM, Ciesla FJ, Benedix GK, Daly L, Trimby PW, Yang L, Ringer SP (2016) Hidden secrets of deformation: impact-induced compaction within a CV chondrite. Earth Planet Sci Lett 452. https://doi.org/10.1016/j.epsl.2016.07.050
84. Collins G, Melosh H, Wünnemann K (2011) Improvements to the ϵ-α porous compaction model for simulating impacts into high-porosity solar system objects. Int J Impact Eng 38(6):434–439. https://doi.org/10.1016/j.ijimpeng.2010.10.013
85. Schulte P, Alegret L, Arenillas I, Arz JA, Barton PJ, Bown PR, Bralower TJ, Christeson GL, Claeys P, Cockell CS, Collins GS, Deutsch A, Goldin TJ, Goto K, Grajales-Nishimura JM, Grieve RAF, Gulick SPS, Johnson KR, Kiessling W, Koeberl C, Kring DA, MacLeod KG, Matsui T, Melosh J, Montanari A, Morgan JV, Neal CR, Nichols DJ, Norris RD, Pierazzo E, Ravizza G, Rebolledo-Vieyra M, Reimold WU, Robin E, Salge T, Speijer RP, Sweet AR, Urrutia-Fucugauchi J, Vajda V, Whalen MT, Willumsen PS (2010) The Chicxulub asteroid impact and mass extinction at the Cretaceous-Paleogene boundary. Science 327(5970):1214–1218. https://doi.org/10.1126/science.1177265
86. Stewart ST, Ahrens TJ (1999) Porosity effects on impact processes in solar system materials. In: Lunar and planetary science conference
87. Hörz F, Cintala MJ, See TH, Le L (2005) Shock melting of ordinary chondrite powders and implications for asteroidal regoliths. Meteorit Planet Sci 40(9–10):1329–1346. https://doi.org/10.1111/j.1945-5100.2005.tb00404.x
88. Wünnemann K, Collins GS, Osinski GR (2008) Numerical modelling of impact melt production in porous rocks. Earth Planet Sci Lett 269(3–4):530–539. https://doi.org/10.1016/j.epsl.2008.03.007
89. Davison TM, Collins GS, Ciesla FJ (2010) Numerical modelling of heating in porous planetesimal collisions. Icarus 208(1):468–481. https://doi.org/10.1016/j.icarus.2010.01.034
90. Osinski GR, Grieve RAF, Collins GS, Marion C, Sylvester P (2008) The effect of target lithology on the products of impact melting. Meteorit Planet Sci 43(12):1939–1954. https://doi.org/10.1111/j.1945-5100.2008.tb00654.x
91. Kraus RG, Senft LE, Stewart ST (2011) Impacts onto H2o ice: Scaling laws for melting, vaporization, excavation, and final crater size. Icarus 214(2):724–738. https://doi.org/10.1016/j.icarus.2011.05.016
92. Pilkington M, Grieve RAF (1992) The geophysical signature of terrestrial impact craters. Rev Geophys 30(2):161–181. https://doi.org/10.1029/92RG00192

93. Collins GS (2014) Numerical simulations of impact crater formation with dilatancy. J Geophys Res Planets p 2014JE004708. https://doi.org/10.1002/2014JE004708
94. Bierson CJ, Phillips RJ, Nimmo F, Besserer J, Milbury C, Keane JT, Soderblom JM, Zuber MT (2016) Interactions between complex craters and the lunar crust: analysis using GRAIL data. J Geophys Res Planets 121(8):2016JE005090. https://doi.org/10.1002/2016JE005090

Recent Insights into Penetration of Sand and Similar Granular Materials

Mehdi Omidvar, Stephan Bless, and Magued Iskander

1 Introduction

This chapter is concerned with kinetic energy driven penetration into soils and soil-like granular materials. Historically this is a topic of great interest to military and civil engineers. The response of granular materials to dynamic loading also has implications for many other branches of engineering and planetary physics. Kinetic energy driven penetration refers to movement of projectiles following high-speed impact. For the phenomena discussed in this paper, there are no forces on the penetrator except the force by which the target resists penetration.

The insights that are the subject of this chapter have mainly arisen from recent experiments involving optical diagnostics to measure penetrator deceleration and the deformation fields that arise around a penetrating projectile. The in situ diagnostics have been largely made possible by advances in two measurement methods: (1) high-speed photography techniques using refractive index-matched transparent soils [1, 2] and (2) photonic Doppler velocimetry (PDV) to measure velocity of a high-speed penetrator [3].

M. Omidvar
Department of Civil and Environmental Engineering, Manhattan College, New York, NY, USA

S. Bless (✉) · M. Iskander
Civil and Urban Engineering Department, New York University, New York, NY, USA

T. J. Vogler, D. A. Fredenburg (eds.), *Shock Phenomena in Granular and Porous Materials*, Shock Wave and High Pressure Phenomena,
https://doi.org/10.1007/978-3-030-23002-9_5

2 Introduction to Mechanical Behavior of Granular Materials

Penetration, by definition, provokes mechanical failure of the target medium. High-speed penetration necessarily also invokes high stresses and high strain rates. The behavior of target materials under extreme loading affects the penetration process. Certain features of the mechanical response of granular media to loading are strain rate dependent. Some of these features include particle crushing under high stresses, shear localization, and pore fluid pressure buildup, among others [4].

Penetration mechanics is a relatively mature field of study. However, the majority of published work in this area is concerned with engineering materials: metals, ceramics, composites, and concrete. On the other hand, granular materials exhibit modes of behavior that are not found or are relatively inconsequential in these more common materials:

- The uniaxial stress–strain behavior of sand at low strains is highly sensitive to relative density. Relative density is defined on a scale of zero to one, where zero is the lowest possible packing density, and one is the greatest. For large strains loosely packed and densely packed granular media exhibit similar behavior, whereas for small strains their behaviors can be quite different.
- Uniaxial loading at large strains results in grain fracture. The onset of grain fracture produces yield-like behavior. But the onset of yielding is usually gradual due to mesoscale heterogeneity. Yielding is also often associated with large changes in void content as some grains crush or comminute and release fragments into interstitial cavities.
- Granular materials possess no or negligible cohesive strength. For example, the concept of spall strength does not really apply to granular materials.
- Shearing of granular materials results in volumetric strains. These materials can be contractive and dilatant in shear, depending on the initial packing density. This is contrary to the assumptions of conventional plasticity. Both initially loose and dense granular media eventually reach a similar critical void fraction at which the medium is very susceptible to shear bands.
- Granular materials possess large amounts of collapsible porosity that results in relatively large load-unload hysteresis, and hence energy absorption.
- Granular materials possess relatively low acoustic velocities—generally less than a few 100 m/s. Thus, supersonic penetration is much more important in these materials than in conventional materials.
- Strength in granular materials, at least for small deformations, is due to friction. Frictional behavior is often described by a Mohr–Coulomb framework. This means that the shear strength is highly dependent on the stress normal to the shear direction. The behavior is governed by the so-called *effective stresses*, which represent the buoyant force carried by the soil skeleton, normalized by the gross cross-sectional area in question.

- Granular materials can often display a phase-change like transition from jammed to unjammed states. Relative movement of particles without damage does not occur in jammed states.
- Densely packed granular materials have a tendency for shear dilatancy. In contrast, loosely packed granular materials contract when sheared. However, under high strain rate loading, even loosely packed media can shift from a contractive response to a dilative response. In saturated granular media, i.e., when the pore space is occupied by water, the large bulk modulus of pore water resists pore compression or expansion. The result is a tendency for increase in pore water pressure with shear in loosely packed media, and a tendency for reduction of pore water pressure in densely packed media. The size of the pore space—related to the mean size and size distribution of the particles—and the rate of loading determine whether there is a buildup of pore water pressure or reduction of pore water pressure under shear. It is important to distinguish between dilation due to negative hoop strain and shear dilatancy; the former occurs during cavity expansion, while the latter occurs when granular media are sheared.

There are also mesoscale phenomena that are unique to granular materials. Force chains are one of these: Often compressive loads are only carried through a limited number of contact points, with the result that grain stresses are highly heterogeneous. Load-carrying grains are often arranged in fluctuating linear chains, although with frequent branches, termed *force chains*. These have been observed during low-speed penetration tests by Clark et al. [5] and high-speed penetration tests in sand reported by Borg [6]. Likewise, the strain-softening behavior associated with grain fracture very often leads to shear strain localization. Large shear in sands leads to localization. For a more extensive discussion of force chains, see, for example, Kondic et al. [7].

Some of the aforementioned phenomena are beyond the capabilities of conventional numerical modeling, particularly when high strain rates are encountered. Thus, understanding penetration of granular materials requires a somewhat unconventional conceptual framework, which is much less well developed than for conventional engineering materials. For this reason, the approach taken in this chapter will be mainly phenomenological, observational, and empirical.

3 Poncelet Framework for Penetration Mechanics

Impact between a target and projectile produces shock waves in both objects. Shock amplitudes can be computed by conventional impedance-matching using the Hugoniot of the target and projectile (see, for example, Myers [8]). However, for deeply penetrating impacts that are the focus of this article, impact transients have relatively little effect on penetration depth because their durations are usually short, of order projectile diameter divided by the shock velocity.

Beyond the impact phase, penetration velocity is relatively constant. For the general case, in which the penetrator length shortens during penetration, it follows from geometrical arguments that

$$^{dP}\Big/_{dL} = {}^{u}\Big/_{(v-u)} \tag{1}$$

where P is penetration, L is the length of the penetrator, v is projectile velocity, and u is penetration velocity. Newton's third law requires that the stress in the target be the same as the stress in the projectile. This requirement gives rise to the so-called Tate equation (Tate [9]) usually written as

$$\frac{1}{2}\rho_p(v-u)^2 + Y = \frac{1}{2}\rho u^2 + R \tag{2}$$

Here, R and ρ are the strength and density of the target, Y and ρ_p are the strength and density of the projectile. This equation applies to penetrators that lose mass as they penetrate, so $v > u$, and the stress that decelerates the rear of the projectile is the yield stress. Equation (2) has several solution regimes depending on the relative values of R and Y. For example, when $R >> Y$ there is no penetration; the penetrator collapses on the surface of the target. The case of when R and Y are comparable applies to many practical studies of penetration in engineering materials—the projectile experiences failure as it penetrates. When $Y >> R$ the penetrator does not deform as it penetrates, although it does decelerate. That is the most common case for engineering problems involving soil penetration.

Penetration of soils and other geomaterials is conventionally discussed in terms of the Poncelet equation. This is a version of (2) in which the projectile is rigid. As a rigid projectile penetrates a granular media, it imparts momentum on a mass of soil, dm. The momentum, P, can be written as $P = dmv = \rho dV \times v = \rho Av^2 dt$. The time rate of momentum change, i.e., dP/dt, results in the force on the projectile as it displaces the soil. Therefore, the force on the projectile is proportional to $F = dP/dt = \rho Av^2$. The effect of the actual projectile nose geometry is accounted for by a coefficient known as the drag coefficient, C. The resulting equation of resistance to penetration can be written as

$$F = -M\frac{dv}{dt} = C\rho Av^2 \tag{3}$$

where A is the projectile cross-sectional area, C, termed the Poncelet drag, is effectively a drag coefficient (or half the drag coefficient as it normally used in fluid mechanics). Eq. (3) describes the velocity-dependent resistance to penetration. In soils there is also a frictional resistance to penetration, which is a function of the mean stress at the penetration front. In order to account for this additional frictional resistance, referred to herein as the bearing resistance, R, Eq. (3) is generalized as follows

$$F = -M\frac{dv}{dt} = A\left(C\rho v^2 + R\right) \tag{4}$$

R plays similar roles in Eqs. (2) and (4), that of a shear strength with little or no dependence of strain rate. However, high strain rate triaxial compression tests have shown a rate dependence of shear strength for sand [4]. This framework is for linear penetration. Achieving stable linear penetration in sand is often difficult because of instabilities that arise since even a small angle of attack can result in strong off-axis forces on the penetrator nose. There has been some success treating trajectory stability in terms of conventional aerodynamics [10]. Rod projectiles can also be stabilized using cones and fins, which function as they do in fluid penetration by applying a restoring force [11]. Discrete element numerical models suggest that instability is promoted in targets with larger mean diameter particles (relative to the projectile diameter), more uniform size distribution, and more randomly distributed porosity. Higher porosity results in more stable penetration, as the lower penetration resistance in these targets reduces projectile mushrooming [12].

Once a high-speed high aspect ratio projectile becomes unstable, off-axis forces are so large that rapid breakup often ensues [11, 13]. Severe deformation of strong projectiles is seen to occur at speed of about 500 m/s (for example, above citations plus Schneider and Stilp [14]). Analysis of trajectories once they have become unstable is very difficult and not of great practical interest; see, for example, Satapathy [15] for a discussion on the mechanics of yawed projectiles in a ductile material. Most studies have focused on attempts to predict conditions that lead to stability, and the penetration mechanics of stable projectiles. Rigid penetrator behavior is a necessary, though not sufficient, criterion for trajectory stability. The rest of this chapter deals with rigid penetrators and applies predominantly to stable trajectories.

Equation (4) can be integrated to give the relation between penetration velocity, v, impact velocity, v_0, and the penetration depth, x

$$x = \left(\frac{M}{2AC\rho}\right)\ln\frac{v_0^2 + \frac{R}{\rho C}}{v^2 + \frac{R}{\rho C}} \tag{5}$$

$$v = \left[\left(v_0^2 + \frac{R}{\rho C}\right)\exp\left(-\frac{2\rho C A x}{M}\right) - R/\rho C\right]^{1/2} \tag{6}$$

Terminal penetration depth is obtained by setting $v = 0$ in Eq. (5),

$$P = \left(\frac{M}{2AC\rho}\right)\ln\left[\frac{\rho C v_0^2}{R} + 1\right] \tag{7}$$

The form of the velocity versus depth relationship predicted by Eq. (6) (for relatively constant values of C and R) is shown in Fig. 1. The penetration as

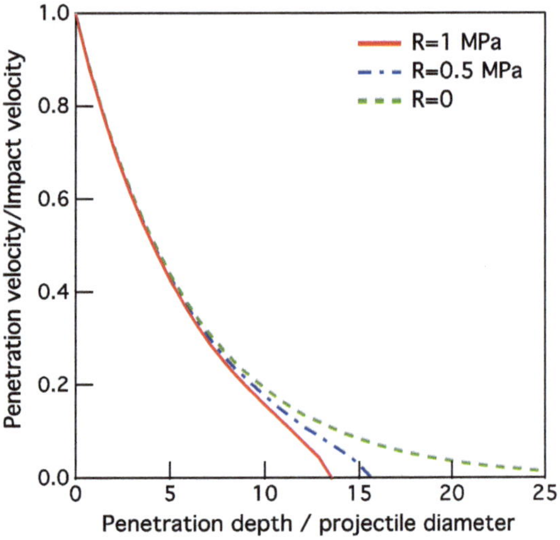

Fig. 1 General form of normalized velocity vs. normalized penetration depth predicted by the Poncelet equation (Computed for a steel sphere impacting medium dense sand target at impact velocity of 300 m/s)

a function of distance initially has positive curvature, but there is an inflection point when the second term comes to dominate the first term in Eq. (4). For R finite, when the projectile stops, $dV/dx = -\infty$. This is expected since $dv/dx = (dv/dt)(dt/dx) = (dV/dt)/v$.

In the Poncelet Eq. (3), the C term is normally associated with inertia transferred to the target medium—it represents the force required to accelerate target material laterally out of the path of the penetrator. The R term is associated with strength. This is similar to the bearing stress in geomechanics, which is discussed in Sect. 5. Penetration resistance, R in these equations, is normally taken as velocity independent. However, that may be problematic in sand because, as mentioned in the previous section, shear strength arises from friction which depends on normal stress, which in turn should be proportional to velocity squared. This aspect of velocity dependence has yet to receive an analytical treatment, although the effect of linear pressure hardening is discussed in Bless et al. [16]. The increase in stress associated with higher velocities can also lead to failure of individual sand grains which probably also effects shear resistance, and hence penetration resistance.

Penetration resistance has been computed from cavity expansion analysis by many authors for a variety of materials and geometries. For granular materials such as soils, the most often used solution is that of Forrestal et al. [17]. This solution has been applied to natural soils, but apparently not to laboratory-scale experiments in which the mechanical properties of the granular material were well characterized.

4 Early Empirical Framework for Penetration of Soils

Early analysis of penetration into sand was entirely empirical. Young's equation is perhaps the most widely used approach, which gives reasonable results for large-scale ordnance penetration. The equations were developed in the 1960s and 1970s by Sandia National Laboratories. The updated results of those studies are revised and summarized by Young [18], to include layered soils, marine sediments, ice, frozen soil, weathered rock, and modified ballistic penetrators. Young's equations are written as

$$P = 0.3\ S\ N\left(\frac{W}{A}\right)^{0.7} \ln\left(1 + 2{v_o}^2 \times 10^{-5}\right) K_s \qquad v_o < 200\ fps \tag{8}$$

$$P = 0.00178\ S\ N\left(\frac{W}{A}\right)^{0.7} (v_o - 100)\ K_s \qquad v_o \geq 200\ \text{fps} \tag{9}$$

where P is the terminal depth of penetration, W is the weight of the projectile (lbs), A is the cross-sectional area of the projectile (in^2), v_o is the impact velocity (ft/s), and S is the soil penetrability number which represents soil properties. Young [18] recommended S-numbers of 2–4; 4–6; and 6–9 for dense sand, sands without cementation, and moderately dense to loose sands, respectively. The nose performance coefficient is obtained for ogive and conical nose shapes using the following equations:

$$N = 0.18(\text{CRH} - 0.25)^{0.5} + 0.56 \qquad \text{(ogive nose shape)} \tag{10}$$

$$N = 0.25^{L_n}/_{d} + 0.56 \qquad \text{(conical nose shape)} \tag{11}$$

where the caliber radius head (CRH) is the ratio of the radius of the nose circle to the diameter of the projectile, L_n is the nose length, and d is the diameter. Equations (10) and (11) are applicable for heavy projectiles ($m \geq 27$ kg (60 lbs)). For lightweight projectiles ($2 < m < 27$ kg) Young suggests that penetration (P) be multiplied by a mass scaling term (K_s) given by the following equation.

$$\begin{aligned} K_s &= 0.2(W)^{0.4} \quad W < 60\ \text{lbs} \\ K_s &= 1 \qquad\quad\ \ W > 60\ \text{lbs} \end{aligned} \tag{12}$$

Note that these equations employ fractional exponents of the physical dimensions and density of the penetrator, so the fitting parameters contain fractional dimensions. While this is common practice in engineering empirical formulas, it is not a common feature of physics-based models. The equations therefore do not fit laboratory data with low mass projectiles well. Young's equations were developed for heavy projectiles ($m > 2$ kg), and the discrepancy is perhaps due to a dependence on aspect ratio that is missing in the Young formulation. To correct this problem Guzman et al. [19] suggested a modified mass scaling factor, K_s', for lightweight (lab scale) projectiles, where

$$K_s{}' = 4(W)^{0.4} \quad (13)$$

It should be pointed out that since Young's equation is purely empirical, the fitting parameters should not be assigned physical significance, unlike the Poncelet parameters, where some physical meaning can be attributed to C and R.

5 Determination of Resistance Force by Direct Measurement of Velocity

Several methods have been used to obtain direct measurement of resistance force during penetration. These methods can be divided into two main categories. The first method relies on measurement of displacements and double differentiation to produce deceleration. Newton's second law of motion is then invoked to derive the time history of resisting force to penetration. The displacement record used in this approach is obtained from several means, including on-board accelerometers [20, 21], flash x-ray photography [22], and high-speed photography of index-matched transparent granular media [1, 23–26].

The second method to obtain resistance to penetration is through direct measurement of projectile velocity by means of a photon Doppler velocimeter. This latter method is the most important recent advance for laboratory-scale studies and is advantageous because of the direct measurement of velocity that it produces. Details of the application of photonic Doppler velocimetry (PDV) to measurement of velocity in penetration into sand can be found in Peden et al. [3]. Results of their penetration tests into sand and calculation of resisting force to penetration using the PDV method are presented here. In these studies, 14-mm diameter spherical projectiles were impacted into sand targets. Velocity as a function of time was measured. Data were converted to $v(x)$ using integration to compute distance. Use of spheres results in rapid deceleration which is relatively easy to measure and reduces trajectory instability. On the other hand, being relatively low in mass, they do not provide data for deep penetration. Four different sand conditions were tested, covering two different relative densities of Ottawa sand, plus saturated and dry porosity. Each test condition was repeated at least twice; scatter between similar tests was negligible.

A spectrogram of the velocity-time record for impact into dense dry and dense saturated Ottawa sand, impacted at 302 and 304 m/s, respectively, are shown in Fig. 2. In the case of dry sand, the intensity maximum associated with the projectile is clearly observed almost until it stops. In saturated material, the signal was sometimes obscured at early times due to ejecta but at later times the peak associated

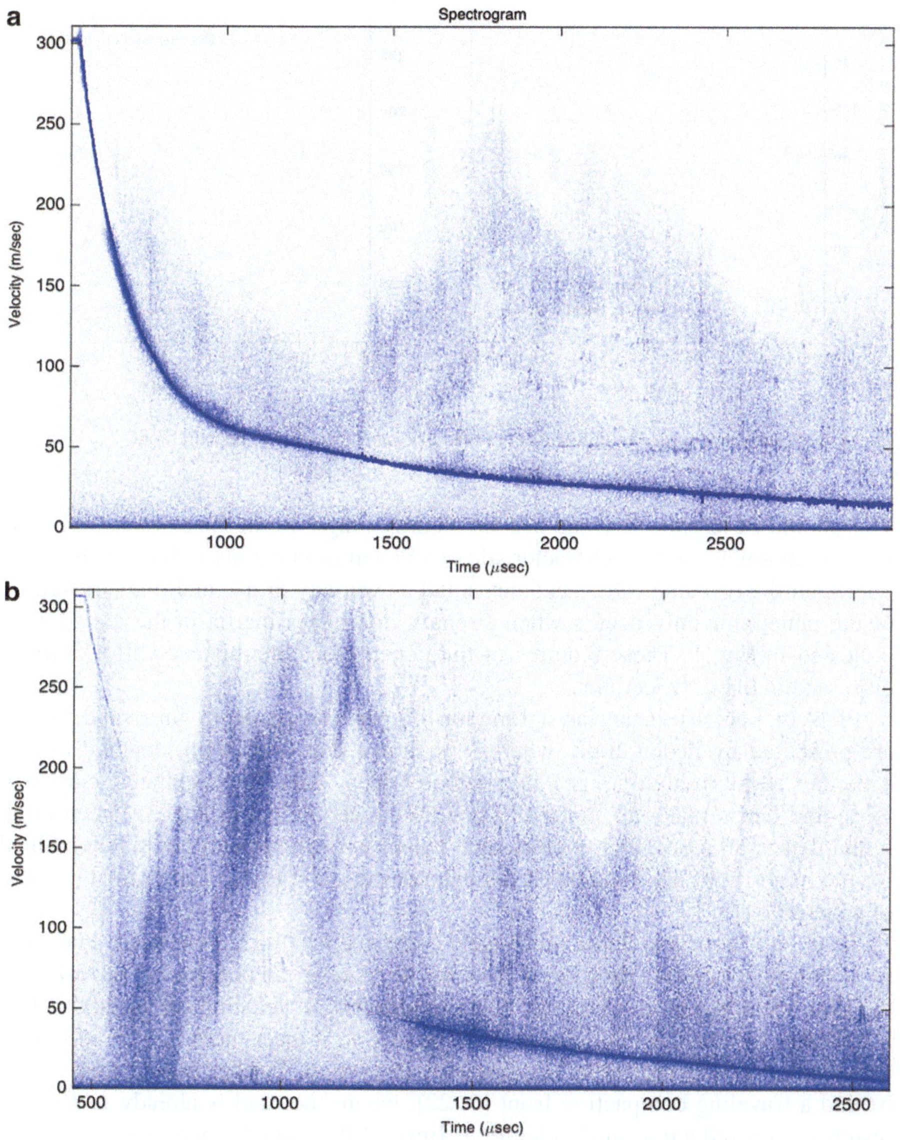

Fig. 2 Deceleration of a sphere in dense dry silica sand (**a**) and dense saturated sand (**b**) from PDV signal

with the projectile was resolvable in spite of the scatter from sand in the cavity. The record for dry material in Fig. 2 indicates a very rapid and small initial drop in velocity. This is presumably due to shock response. An impedance-match solution using the sand Hugoniot data from Chapman et al. [27], and assuming a duration of the stress of radius/sound velocity in steel, gives a velocity drop of 13 m/s in 11 μs.

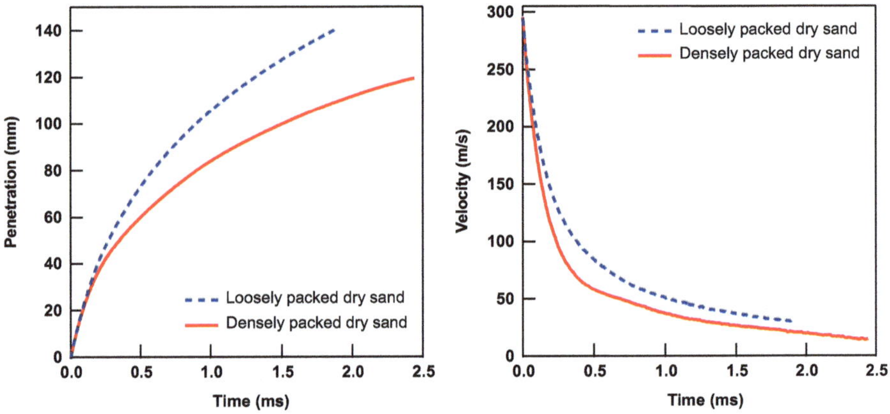

Fig. 3 Penetration test results in loose and dense dry sand from PDV measurements

This should be slightly less and last for slightly longer in the case of a sphere. The long tail at low velocity is characteristic of a penetration mainly influenced by drag, for which the resisting force becomes small as velocity drops, and final arrestation of the penetrator only occurs when strength dominates inertia in the equation as indicated in Fig. 1. These features of the penetration time history will be further discussed in the next sections.

Plots of velocity decay as a function of penetration depth into sand targets are presented by Peden et al., where it is shown that shots with identical impact velocities result in highly repeatable velocity decay data. Representative curves of loose and dense materials under dry conditions are shown in Fig. 3. Peden et al. found that in all cases, highly repeatable velocity measurements could be obtained for the majority of the duration of the penetration event (up to approximately 80% of penetration).

Tests into loose and dense silica sand were analyzed in a Poncelet framework, by assuming a constant drag coefficient for the part of the penetration recorded by the PDV method, i.e., from impact down to penetration velocities of approximately 30–50 m/s, where the PDV signal was lost. Upon impact there is a rapid rise in deceleration as the projectile sets particles in motion. Further penetration occurs behind a traveling compaction front [6, 22], where the sand is already in motion, resulting in a reduction in deceleration. Beyond this initial stage, the flow in the sand around the projectile changes only slowly (in the projectile frame of reference). A quadratic fit, as predicted by the Poncelet analysis, adequately matches the acceleration-velocity data for penetration into both loose and dense sand targets. This is shown in Fig. 4. It can be seen that a constant Poncelet drag of 0.77 and 1.05 fits the data with reasonable accuracy.

The deceleration data from PDV measurements of high-speed impact produced highly reliable deceleration data from impact down to approximately 50 m/s of penetration speed. Below this threshold the velocity record was often obscured by ejecta blocking the path of the PDV sensors. In order to better understand

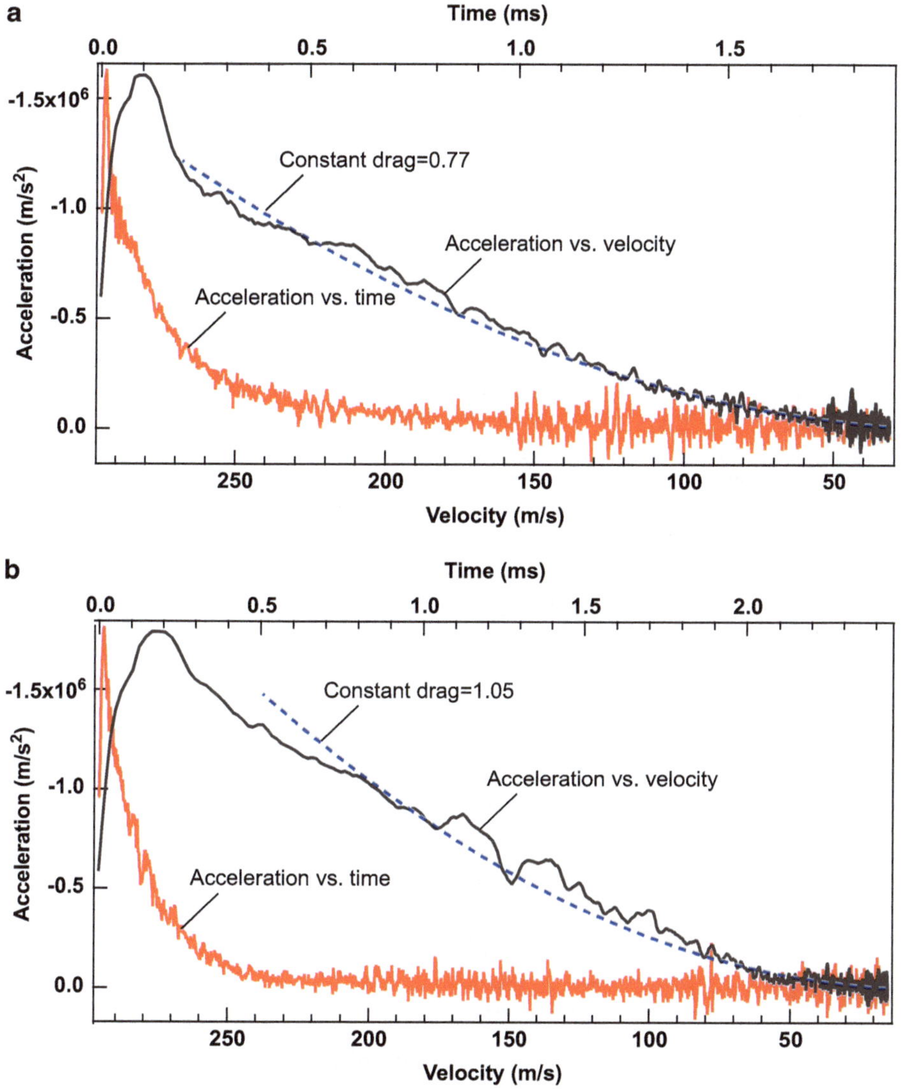

Fig. 4 Acceleration history of spheres in sand: (**a**) loose sand, (**b**) dense sand. Acceleration time histories were calculated by differentiation of PDV velocity-time histories. A quadratic fit was then used to find average drag coefficients

penetration resistance characteristics of the projectile at lower penetration velocities, Omidvar et al. [23] performed experiments using hemispherical-nose long rods at impact velocities of approximately 80 m/s into targets with relative densities similar to the high-speed impact tests. The rods had hemispherical noses so that they would have similar drag coefficients to spheres, disregarding side friction along the

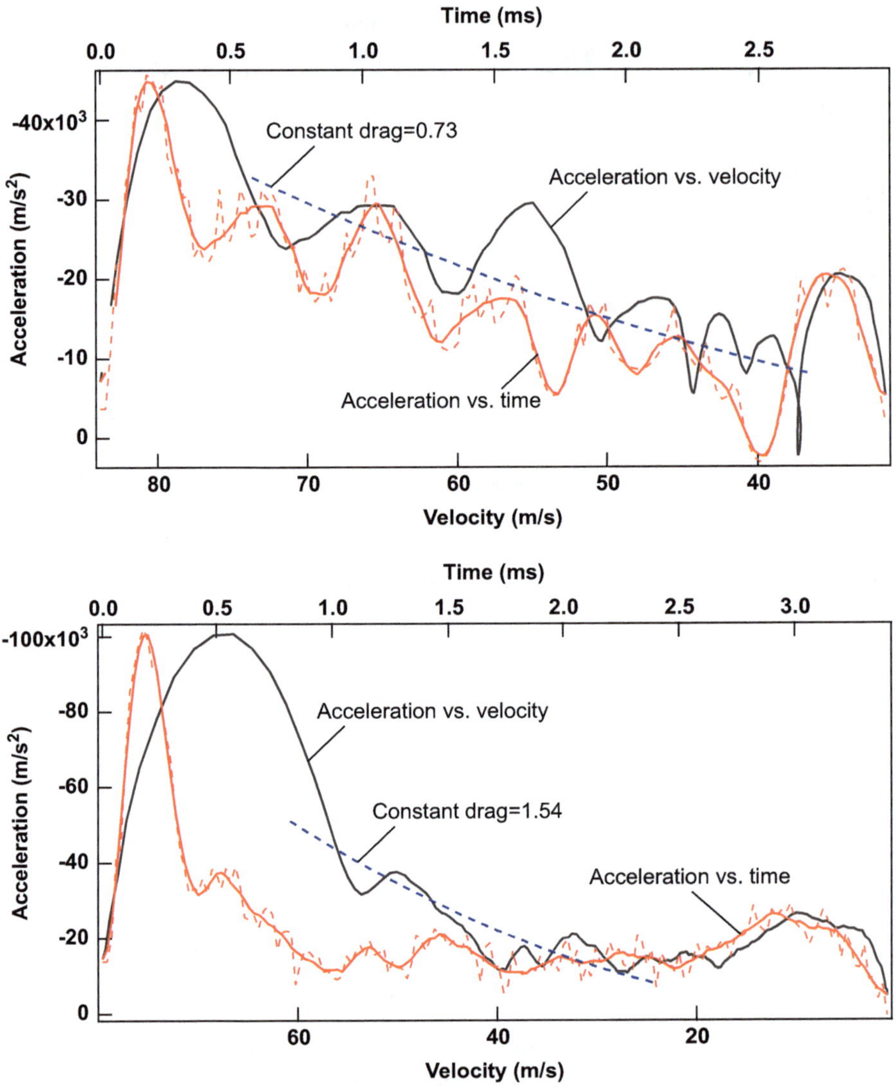

Fig. 5 Deceleration data for rods penetrating dense sand at relatively low velocity. Upper: loose sand. Lower: dense sand. Velocity- and acceleration time histories were calculated through single and double differentiation of penetration time histories, respectively. A Savitzky-Golay filter was used to remove high frequencies resulting from differentiation

afterbody of the rods. Their data for deceleration, obtained by double differentiation of distance-time measurements, are noisier than the PDV data. However, the same trends as observed at higher speeds were evident. Figure 5 shows results for deceleration. In all cases, the higher the density, the higher the Poncelet drag. A summary of the tests performed at low and high velocities along with the drag coefficients for each test is presented in Table 1.

Table 1 Summary of low velocity and high velocity tests in sand

Test ID	V_0 (m/s)	Density (kg/m^3)	Void ratio	Projectile shape	Projectile mass (g)	Projectile diameter (mm)	C
SDD1421	299.2	1817	0.46	Sphere	11.17	14	1.1
S513	79.7	1830	0.45	Rod-hemispherical nose	16.03	10	1.5
SLD1428	297.4	1587	0.67	Sphere	11.17	14	0.7
S512–2	83.9	1670	0.59	Rod-hemispherical nose	16.03	10	0.6
SWD1417	302.5	1819	0.46	Sphere	11.17	14	0.7
SWL1417	303.0	1600	0.66	Sphere	11.17	14	0.6
S521	77.3	1660	0.60	Rod-hemispherical nose	16.03	10	1.1
S522	73.9	1830	0.45	Rod-hemispherical nose	16.03	10	2.1

From an engineering point of view, in the initial stages of penetration velocity decay is perhaps the most useful way to examine the data. This is because of the concern for damage to the penetrator itself and because such records can be extrapolated to predict final penetration depth. However, it should be pointed out that the final depth of penetration is problematic for several reasons: in practice there is usually quite a lot of scatter in depth data because in the final stage of penetration there is increased sensitivity to the soil strength. Moreover, the penetrator is moving slowly at low velocities and damage to it or structures it may strike is minimal. For applications that are mainly concerned with accurate predictions of final penetration depth in full-scale field testing, elevated gravity subscale tests using a geotechnical centrifuge (which can model overburden effects) or accurate numerical models should be relied on instead of 1-g laboratory-scale models. An example of elevated stress tests in sand can be found in Taylor et al. [28], in which small-scale centrifuge penetration test results were used to calibrate and modify Young's [18] empirical correlations.

6 Effects of Pore Saturation on Penetration Resistance

In the tests reported in previous sections, the pore space consisted of air. Replacing the pore space with water affects the physics of penetration, and the resulting projectile-soil interactions. To examine this effect, penetration tests were performed in Ottawa sand under wet conditions by introducing water into the samples. Tests were performed at both high (approximately 300 m/s) and low (approximately 80 m/s) impact velocities [3]. In PDV tests on dense saturated sand, it was found that the ejecta from the impact obscured the path of the incident beams, and velocity measurements could only be obtained for approximately 50% of the penetration event. Despite the ejecta obscuring the path of the laser beams, it was found through repeating the tests into dense saturated sand four times that the velocity measurements for the initial 50% of the penetration were highly repeatable.

The effect of saturation in quartz sand depends on the instantaneous penetration velocity. Contrasting behaviors were observed in high-speed and low-speed penetration. At high penetration speed, saturation reduces penetration resistance, resulting in lower drag coefficients in the wet samples compared to dry samples. In contrast, penetration in wet sand under low impact velocities apparently increases resistance to penetration.

In order to explain the observed contrasting behavior under high and low velocity penetration, it is important to note that, depending on penetration rate, the following phenomena may occur at the micro and meso levels during penetration into the soil:

- Negative (i.e., tensile) hoop strains ahead of the projectile due to cavity expansion.
- Shear-induced dilation or contraction in the soil ahead of the projectile, depending on the initial density of the soil target.
- Rapid increase in pore water pressure in soils with shear-induced contractive response, which can lead to loss of stiffness and strength, ultimately resulting in a limit state known as liquefaction. Effective stresses, i.e., interparticle forces, are negligible in this limit state.
- Particle fracture or comminution which contributes to changes in porosity and may also affect the rate of dissipation of pore pressures due to plugging of pore space.

At high velocity, where drag originates from inertia, the drag coefficient of the fluid-saturated sand is strongly influenced by the drag coefficient of water, which is significantly lower than that of sand at the Reynolds numbers associated with penetration events (for cm-size objects Reynolds number ranges from 10^5 to 10^6). For these Reynolds numbers, the Poncelet drag on a sphere in water should be <0.2. It could be even less for a spherical-nose rod or if turbulent flow takes place behind the projectile. This explains the lower penetration resistance in wet sand at high penetration velocities. Under low velocity penetration, it can be seen from Fig. 6 that the effect of saturation is manifested at approximately $0.5–0.4V_0$, below which there is a decrease in the slope of velocity-time history in dry sand, but not in wet sand. One explanation for this observed behavior is that at low velocities, where bearing stress begins to dominate resistance to penetration, the particles have a tendency for shear-induced dilation or contraction. Presence of water in the pores suppresses relative movement of particles; pore creation is difficult because water resists dilation. The result is that cavitation can occur, which largely reduces the shear resistance of soil to further penetration in this velocity regime. This hypothesis is strengthened by the observation that saturation has less effect on low-speed penetration into loose sand, where movement of the particles is less restricted. It should be noted, however, that the above explanations are based on the limited tests reported in this study. Further examination of the effects of pore fluid on penetration dynamics is required to assess global validity of these explanations.

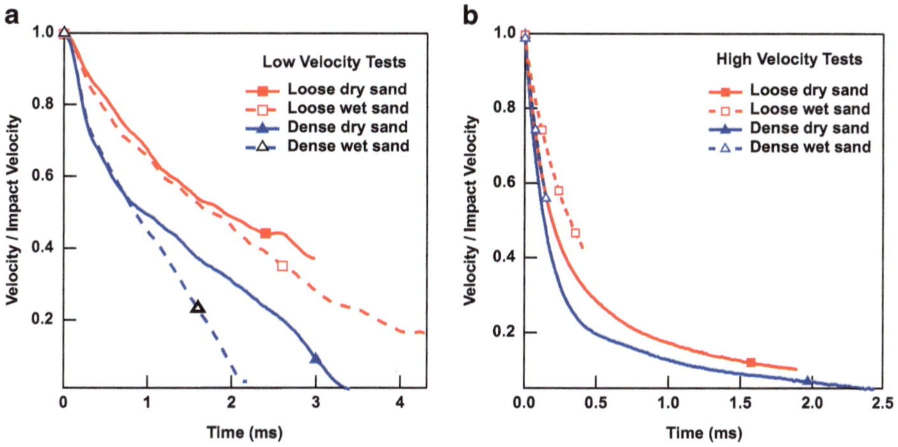

Fig. 6 Effect of pore saturation on penetration into sand under (**a**) low (~80 m/s) impact velocity and (**b**) high impact velocity (~300 m/s)

7 In Situ Observations of Soil-Projectile Interaction

The velocity and displacement measurements cited in previous sections have been complemented by studies using transparent sand surrogates. This is a relatively new technology based on rendering granular materials transparent by saturation with an index-matching pore fluid, usually a blend of mineral oils. Recent investigations of dynamic behavior using this technique can be found in [24–26, 29]. Studies of particle motion around penetrating projectiles have shown that the zone of influence of the penetrator only extends a few diameters beyond the cavity. The in situ data in the following figures are all obtained for a soil simulant made from oil-saturated fused silica [24–26]. Incremental soil displacement fields in the horizontal and vertical directions from typical tests are shown in Figs. 7 and 8, respectively. Details of the image analysis techniques used to produce these figures can be found in [26]. It can be seen that penetration of the blunt nose projectile produces greater horizontal and vertical displacements in the soil. The zone of influence of the blunt projectile is also larger than that of the conical nose projectile. Comparison of the horizontal and vertical displacement fields reveals that once the projectile passes a given depth, there is little additional horizontal displacement, while there is noticeable vertical displacement near the cavity.

Incremental displacements were integrated along a representative horizontal section at a depth of $6D$ from the target surface. The results revealed that there are considerably larger vertical displacements ahead of the projectile in the case of the blunt projectile compared to the conical nose projectile. Comparison was also made between cumulative displacements resulting from rapid penetration and quasi-static penetration of the blunt nose projectile. The results showed that while horizontal displacements are comparable in the two cases, vertical displacements ahead of the

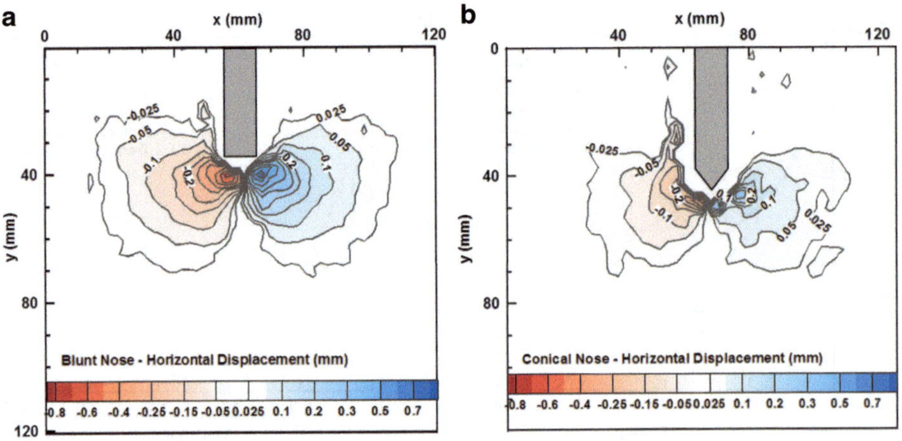

Fig. 7 Incremental horizontal displacement field at a penetration depth of approximately 4D in soil simulant: (**a**) blunt nose projectile; (**b**) conical nose projectile

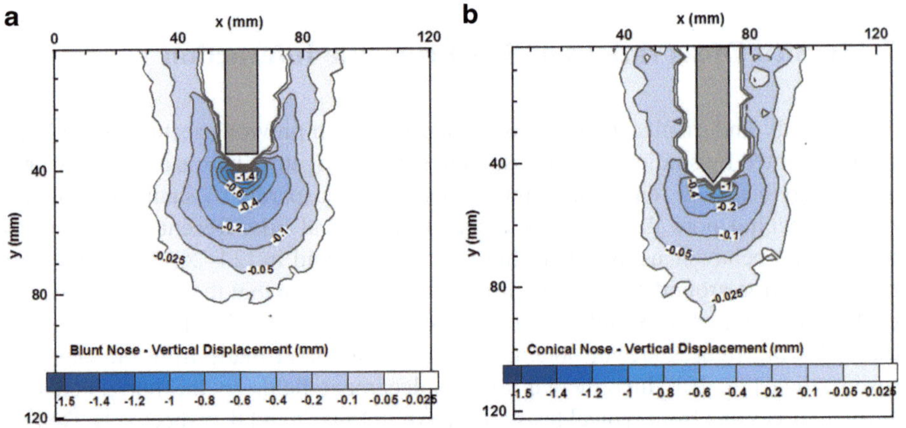

Fig. 8 Incremental vertical displacement field at a penetration depth of approximately 4D in soil simulant at a velocity of 12–14 m/s: (**a**) blunt nose projectile; (**b**) conical nose projectile

projectile are larger in the case of dynamic penetration, by approximately 60%. It can therefore be concluded that the majority of the momentum transfer to the particles during dynamic penetration occurs in the direction of the penetration axis.

Shear strains were computed from the incremental digital image correlation (DIC) analyses described in previous sections. The DIC grid was first discretized into finite elements. DIC displacements then represented nodal displacements. Shape functions were used to compute element displacement and deformation gradients. A finite strain formulation was then employed to compute the right Cauchy–Green strain, from which logarithmic strains were calculated. Next, the Biot strain matrix was used to find the principal engineering strains. Finally, shear

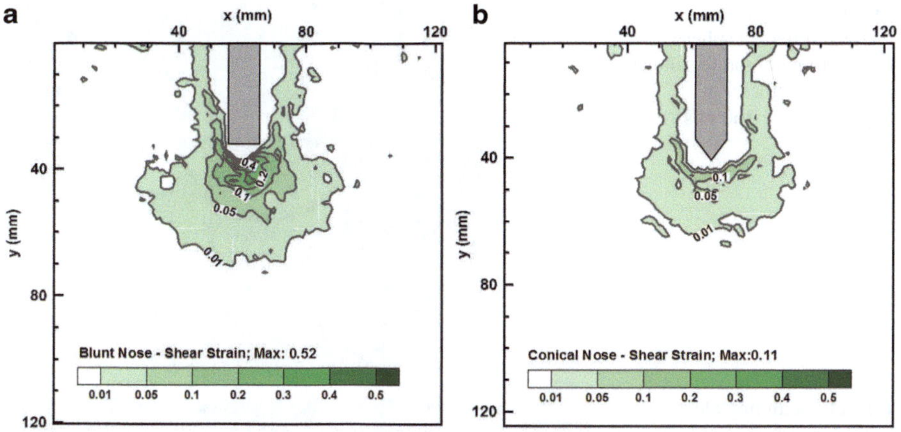

Fig. 9 Incremental shear strains at a penetration depth of approximately 4D in soil simulant: (**a**) blunt nose projectile; (**b**) conical nose projectile

strains were defined as the difference between the major and minor principal logarithmic strains. Further details can be found in [30].

Plots of incremental shear strain are shown in Fig. 9. It can be seen that large shear strains occur immediately ahead of the projectile. Shear strains are significantly larger in the blunt projectile, in excess of 0.5 for the increment of penetration shown at a depth of approximately 4D. Shear strains in the conical nose projectile are significantly smaller, with a maximum value of 0.11 for the increment of penetration shown.

Another observation is that in dynamic penetration into transparent soils there is a region of lost transparency around the penetrating projectile, which recovers its transparency after the projectile comes to rest. This observation may be a consequence of a loss of saturation in the medium, which is later recovered as liquid flows back into the affected area. Loss of saturation could have several causes. Considering the large shear strains encountered in the soil, the loss of transparency may be due to cavitation of the pore fluid. Cavitation can result from the combination of negative hoop stresses and shear-induced dilation in the granular media. In both cases, the granular media tend to expand at rates that exceed the time scale required for pore fluid to flow into the expanding voids. The result is a reduction in the pore fluid pressure, which can lead to cavitation.

In addition to the tests on quartz sand, a series of shots were performed with crushable round sand consisting of aragonite particles. A spherical projectile was shot into the densely packed aragonite target at an impact velocity of approximately 160 m/s [24]. Together with the transparent soil tests, the following qualitative observations can be made from post-test analyses:

- A conical shaped region of dense sand forms ahead of the projectile during the initial stages of penetration and travels with it during the later stages of

Fig. 10 False noses recovered from a sphere penetrating aragonite sand at an impact velocity of approximately 160 m/s

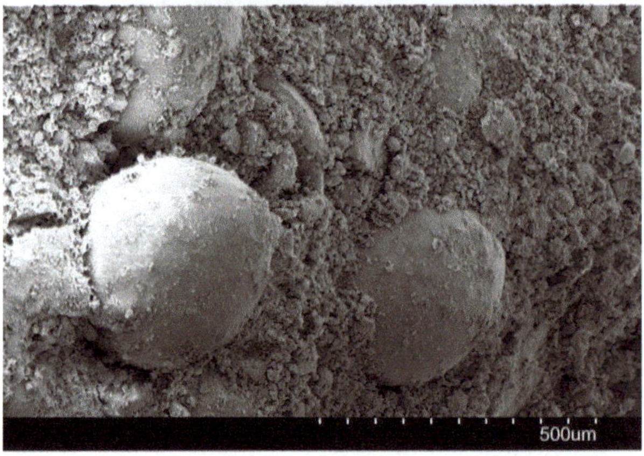

Fig. 11 Scanning electron microscope image of surface of false cone recovered from penetration test into aragonite sand

penetration. We term this false cone a "kernel", after Backofen [31]. There are several observations of kernels in sand, including Kotov et al. [32], Penumadu and Kim [33], and Glossner et al. [21]. The compacted material in this zone was recovered after a test on aragonite sand, shown in Fig. 10. This kernel may function as a *false nose,* which helps explain why the penetrator nose shape has relatively little effect on penetration depth, while influencing the soil kinematics. The false nose has only been observed in small-scale projectile penetration, with projectile diameters of up to 30 mm. The existence of the false cone has been reported in penetration of relatively blunt projectiles. It is not known whether the false cone affects penetration of sharp nose projectiles.

- An important feature of the composition of the kernel, as, shown in Fig. 10, is that there appears to be a bimodal distribution of particle sizes. The kernel consists of intact sand grains surrounded by pulverized particles, which are several orders of magnitude smaller than the original particles. A scanning electron microscope (SEM) image of the surface of the kernel is shown in Fig. 11. It was also recently reported by Parab et al. [34] that when two individual sand grains are forced into contact, one pulverizes while the other suffers minimal damage. As particles

Fig. 12 Evidence of the abrupt transition between crushed and non-crushed sand grains along the path of a penetrating rod [11]. This figure shows the plane of penetration after the material above the plane was excavated. Travel of projectile was left to right

pulverize, the resulting smaller particles surround the neighboring particles. This produces a cushioning effect whereby the surviving larger particles have much higher coordination numbers. (The coordination number is the number of particles in contact with a given particle.) This in turn distributes the stress on the surviving particles and reduces stress at individual contact points. The result is that some larger particles are preserved while other neighboring particles are pulverized, leading to a bimodal distribution of particle size in the kernel.

- Penetration in silica sand at speed above about 80 m/s involves pulverization of sand with a reduction of grain size of several orders of magnitude. Evidence for the transition from crushing to non-crushing includes a trail of comminuted *white particles* in the penetration path which comes to an abrupt end when the penetration speed drops below a particular value. The speed associated with the terminus of the comminuted particle trail is termed the critical penetration velocity, v_*. An example of this observation can be seen in Fig. 12. Similar observations have also been reported by others [35, 36].

The emerging picture is that there are distinct regimes or stages in the deceleration time history of projectiles penetrating granular media. Moreover, it appears that the transitions between these stages are marked by changes in mechanical response of the granular media in the path of the penetrator. For the case of silica sand, and probably for most granular materials, there can be four stages in penetration, as summarized in Table 2. There is a low velocity range, $v < v_t$, in which the penetration resistance is velocity independent; an intermediate region in which inertia dominates shear strength, but the grains are not damaged; a high velocity regime above a critical velocity v_* at which many of the grains are comminuted; lastly, there is a supersonic region in which shock compression dominates. Transitions between these stages are often gradual, and in the case of loosely packed sand, the transition between stages two and three seems to be especially gradual. In silica sand, a review of published data found that v_* was between 80 and 100 m/s [37]. This chapter does not cover stage four because there is little quantitative data for rigid projectiles at these high velocities, and when large deformation of the projectile occurs the response of the target medium cannot be decoupled from the failure of the projectile material. Regardless, accurate modeling of rapid penetration into granular media should take into account these different phenomena.

Table 2 Stages in penetration of a granular material by rigid projectiles

Stage	Dominate physical mechanism in target
1. Low velocity ($v < v_t$)	Shear failure associated with bearing resistance. Force chains may play a major role in stress transmission
2. Intermediate velocity ($v_t < v < v_*$)	Inertial drag associated with moving mainly intact particles out of the path of the penetrator. Particle crushing is minimal. Particle composition is relatively unimportant. Particles experience largely elastic collisions with penetrator nose
3. High velocity ($v > v_*$)	Inertial drag and strain hardening associated with filling in of porosity with crushed material. The crushing transition does not occur in loose soils, or is not significant. The existence of large pore space allows for movement of the particles, and prevents stress concentration. In dense sands, the crushing transition causes a nearly 50% increase in the Poncelet drag. Above the transition threshold crushing facilitates pore collapse
4. Shock	Impact transients, bow shock, and supersonic penetration

8 Engineering Formulas for Penetration of Dry Sand

This section develops approximate formulas for engineering applications in which the penetration of a projectile into sand needs to be predicted. We use the Poncelet framework, in which projectiles are decelerated by a force given by Eq. (4). The equations in this section are based on experiments reported in this study and have not been generalized to include other experimental data from the literature. For engineering purposes, we allow C to be a function of velocity and R to be a function of depth. As discussed in preceding sections, there is a great deal of evidence that in silica sand, and other sand as well, there is a change in the penetration resistance at the velocity associated with the onset of crushing of sand grains. This velocity is termed v_* in Table 2.

For the low velocity regime, the data in Omidvar et al. [23] for force measurements in sand provide a reasonable basis for empirical treatment of the Poncelet C coefficient. The data were analyzed by computing force divided by $\rho A v^2$ and plotting it as a function of time. At early times where drag dominates, the ordinate is equal to C. At late times, as the strength term comes to dominate, the ordinate becomes large and negative. Results are shown in Figs. 13 and 14 for both low velocity and high velocity penetration into loose and dense silica sand, respectively. As shown in Fig. 13, the Poncelet coefficient C for all loose materials is slightly lower than one, regardless of the impact velocity.

In Bless et al. [37], it is shown that a value near unity for C results from elastic collisions between the penetrator nose and sand particles. That this value is observed for several types of sand at low relative density can be explained if the particles have little dynamic interaction with one another (before contact with the penetrator.) Thus, for relatively low density granular materials, we take $C = 1$ for the region in which the collisions may be assumed elastic, i.e., the particles do not break. This simple approach fails to account for values of C in dense sand, as shown in Fig. 14. For this case, apparently particle interactions and fracture are more important.

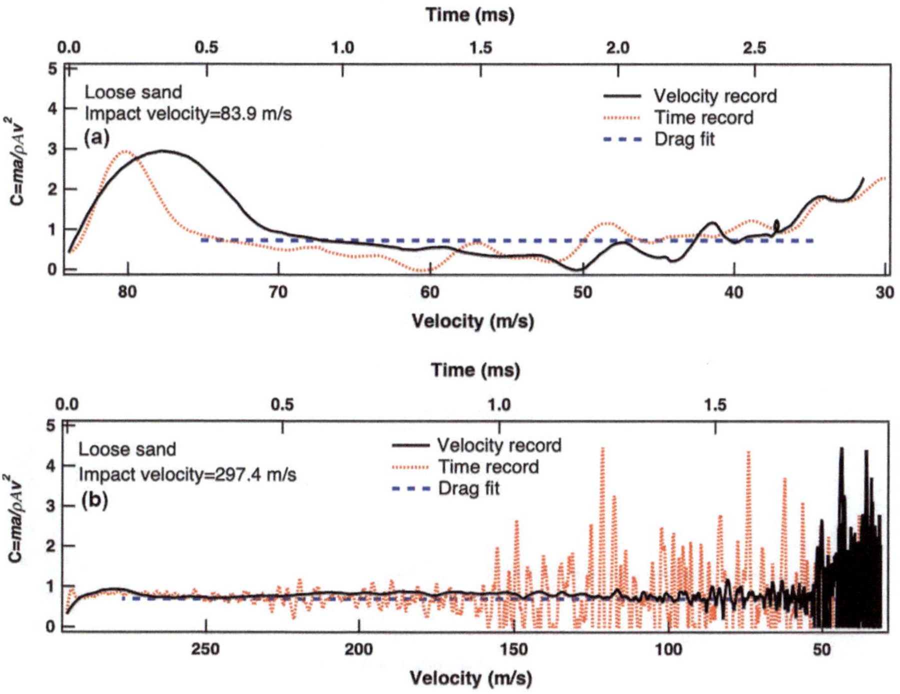

Fig. 13 Acceleration normalized by velocity squared as a function of time, for (**a**) low-speed penetration in loose sand; (**b**) high-speed penetration in loose sand

Analyzing the data for silica sand presented in previous sections, it is found that the Poncelet drag in the low velocity regime below the crushing threshold varies with density as

$$C = 3.94\rho – 5.85 \tag{14}$$

for density in g/cm^3. The velocity-independent resistance to penetration, R, is comparable to the bearing stress that has been analyzed for deep foundations in civil engineering applications. Accordingly, for cohesionless soils, the bearing stress arises from the stress level in the target soil at the penetrator nose, and the friction resistance of the target soil. Prediction of bearing stress requires knowledge of the soil properties and the stress level where the soil self-weight is predominant. Alternatively, cone penetrometers can be used to measure the bearing stress in situ. For the rod impact tests reported in previous sections, the bearing stress contribution, R, was independently measured in a quasi-static test. The result, shown in Fig. 15, was obtained by quasi-statically penetrating a rod into the soil target using a hydraulic jack. Resistance to penetration was measured using a load cell. R is seen to be a quadratic function of penetration depth over the penetration depths considered:

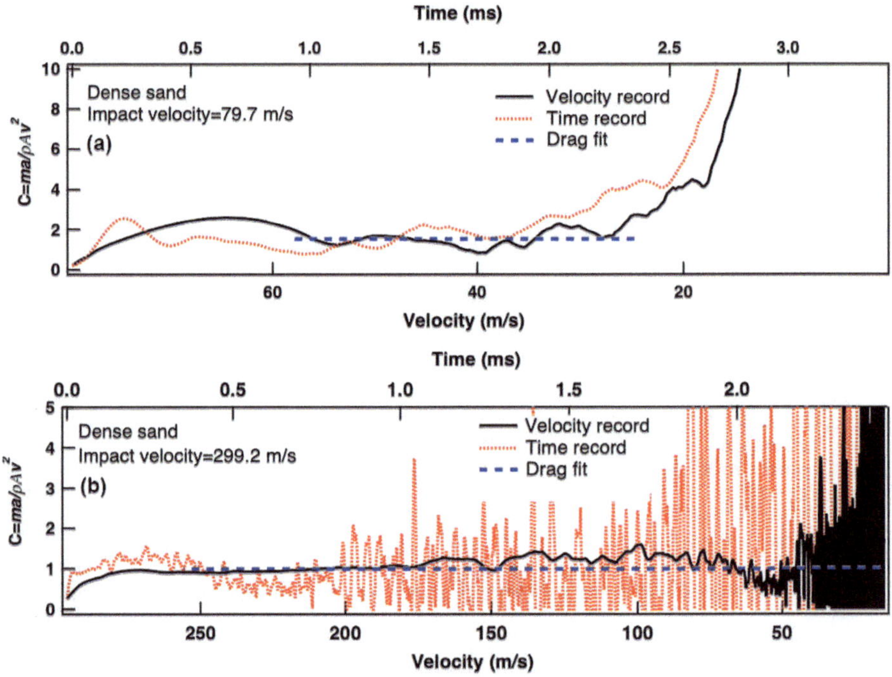

Fig. 14 Acceleration normalized by velocity squared as a function of time, for (**a**) low-speed penetration in dense sand; (**b**) high-speed penetration in dense sand

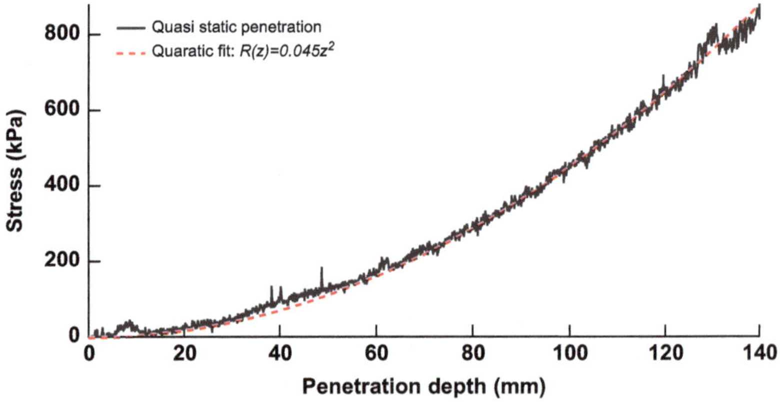

Fig. 15 Measurement of Poncelet R for silica sand in a quasi-static penetration

$$R = 0.045z^2 \tag{15}$$

for z in mm and R in kPa. It is important to note that the bearing stress is depth-dependent, and proper measurement of bearing stress requires in situ tests, or elevated gravity small-scale experiments. Nevertheless, the measurements reported herein provide a basis for future work. The quasi-static lab experiment in this study produced values on the order of 1 MPa for bearing stress, as shown in Fig. 15. Penetration depth in high velocity impacts is relatively insensitive to bearing stress, since most penetration takes place when the velocity-squared term dominates (See Fig. 1).

A projectile resembles a pile in a number of ways. The bearing stress resistance of piles can also be directly computed using equations developed for use in civil engineering applications, where the resistance is derived from two components: the tip resistance, and side friction. One such equation is due to Meyerhof [38] (see, for example, Das [39]). In this model, both tip and skin resistances are constant, despite increasing effective stress with depth, once a certain critical depth is exceeded. The critical depth would be exceeded in many applications in dynamic penetration. For demonstration purposes, a constant R term is adopted here for simplification, and to serve as an upper limit. In this analysis, the tip bearing resistance is written as $Q_p = Aq_{p,}$ where A is the cross-sectional contact area, and the end bearing resistance q_p is a function of friction angle, ϕ, when strength is described in a Mohr–Coulomb framework. The limiting value of q_p is $Q_L = (1/2)P_a N_{q\,*} \tan\phi$. For sand, we can take the friction angle as 30°, of which the tangent is 0.58. P_a is atmospheric pressure, approximately equal to 100 kN/m^2. Meyerhof theory gives $N_{q*} = 56.7$ for friction angle of 30°. Hence, the limiting end bearing stress turns out to be approximately 1.6 MPa. This is remarkably close to the value of about 1 MPa from the presented lab data. For deep penetration, one could also perform an integration for shaft friction. However, there is very little data for how this would be affected by the disturbed area around a high-speed penetration channel. Nor is it clear how this would be affected by shaft geometry. A thorough treatment of the contribution of bearing resistance to deep penetration requires realistic modeling of geostatic stresses. A predictive model for final penetration depth that includes bearing stress is beyond the scope of this paper.

A test application of incorporating the lab measurements of depth-dependent quasi-static resistance was applied to a low-speed penetration test into dense-packed quartz sand. The combination of the drag fit and the bearing resistance fit is shown in Fig. 16. The fit shows the sum of the constant drag fit described above, and the quasi-static resistance to penetration. Note the multiplier ξ was added to improve the fit; physical interpretation of this multiplier requires further testing of the parameter space. It can be seen that the Poncelet model adequately captures the salient features of penetration into the soil target beyond the initial transient region where the soil particles are set in motion.

For velocities above the comminution threshold, Bless et al. [37] analyzed data from several sources and conclude that $C = 0.8$ fits the available data for silica sand for intermediate relative densities. Reasonable values of R do not influence penetration resistance at these velocities. Therefore, for engineering estimates of kinetic energy driven penetration into sand we recommend using the Poncelet Eq. (3) to compute deceleration with the following values:

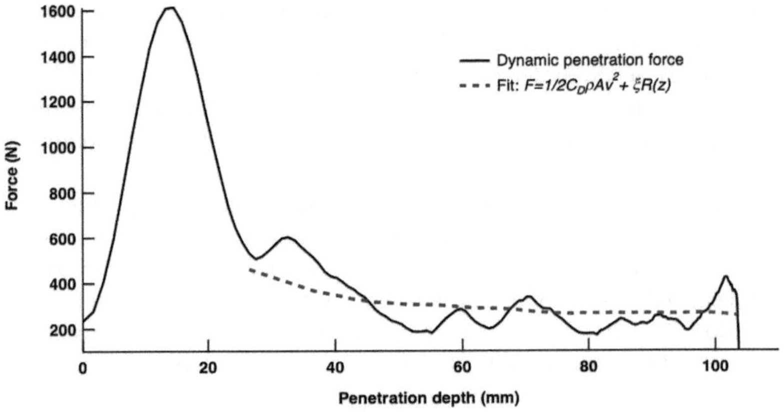

Fig. 16 Poncelet model for penetration into dense-packed sand using both values for C and R. Note, $R(z)$ in the fit is the best-fit function for measured resistance from the quasi-static penetration test. The second multiplier ξ was to improve the fit

$$\begin{aligned} &\text{For } \upsilon < 80\ \text{m/s}, C = 3.94\rho - 5.84, \\ &\text{For } \upsilon > 80\ \text{m/s } C = 0.8 \\ &R = \text{MIN } \left\{0.0045\ z^2, 1600\ \text{kPa}\right\} \end{aligned} \tag{16}$$

These equations fit small-scale laboratory data for dry silica sand up to speed of about 500 m/s. For applications that require safety factors, uncertainties of 30–50% in these parameters would be prudent. A factor of safety of 2 (100%) is common in civil engineering piling problems, since the problem is affected by many parameters, which are difficult to capture analytically. Possible variations of R with depth for larger scale penetrators are not yet determined. For low velocities that are influenced by gravity, the treatment by Clark and Behringer [40] is recommended. Also yet to be determined are modifications for effect of liquid saturation.

9 Concluding Remarks

Understanding of penetration resistance of soils and granular materials has been considerably advanced by newly developed experimental techniques. A series of penetration tests were reported herein on spherical projectiles and long rods impacted sand targets at impact speed in the range of 80–300 m/s. The following conclusions were made from analysis of the tests:

- Mesoscale phenomena that influence sand penetration are altogether different from those that dominate penetration of conventional engineering materials. This is because of the importance of pore collapse and friction in silica sand.

- At penetration velocities above approximately 30 m/s, deceleration of projectiles is drag dominated. Values of drag coefficient are near one for silica sand.
- There is a well-defined penetration velocity of 80–100 m/s above which sand is comminuted. Comminuted and intact particles form a false nose ahead of the projectile, contributing to projectile-soil interactions. This observation has been corroborated in other studies as well [41]. For loose sand, there is very little change in penetration resistance at this velocity. For dense sand, it has a measurable but modest effect.
- For low and intermediate velocities, prediction of final penetration depth requires a model for bearing stress. For lab-scale events, a bearing stress of 1 MPa will give reasonable results. The dependence of bearing stress on depth and velocity for larger scale penetrators can be estimated from quasi-static cone penetration tests.
- Pore saturation affects penetration dynamics in several ways. The role of pores saturation depends on the porosity of the soil target and on the penetration velocity, among other factors. At high penetration velocities, penetration is influenced by the drag coefficient of the pore fluid. At low penetration velocities, the role of pore fluid is to restrict shear-induced dilation. This effect is greater in dense sands compared to loose sands.
- In situ observations of soil-projectile interactions show that much larger shear strains occur at the penetration front in blunt projectiles compared to conical nose projectiles. Moreover, downward soil displacement along the penetration direction is significantly larger than lateral displacements.

References

1. Iskander M, Bathurst R, Omidvar M (2015) Past, present, and future of transparent soils. Geotech Test J 38:1–17. https://doi.org/10.1520/GTJ20150079
2. Iskander M, Bless S, Omidvar M (2015) Rapid penetration into granular media. Elsevier, Amsterdam
3. Peden R, Omidvar M, Bless S, Iskander M (2014) Photonic Doppler velocimetry for study of rapid penetration into sand. Geotech Test J 37:139–150
4. Omidvar M, Iskander M, Bless S (2012) Stress-strain behavior of sand at high strain rates. Int J Impact Eng 49:192–213
5. Clark AH, Kondic L, Behringer RP (2012) Particle scale dynamics in granular impact. Phys Rev Lett 109:238302
6. Borg J (2017) Projectile penetration into sand targets. In: American Physical Society topical conference on shock compression of condensed matter, St Louis, MO. 2017APS.SHK.D6003B
7. Kondic L, Goullet A, O'Hern CS, Kramar M, Mishaikow M, Behringer RP (2012) Topology of force networks in compressed granular media. Eourophys Lett 97:54001. https://doi.org/10.1209/0295-5075/97/54001
8. Myers MA (1994) Dynamic behavior of materials. Wiley, New York
9. Tate T (1969) Further results in the theory of long rod penetration. Journal of Mechanics and Physics of Solids 17:141

10. Flis W, Jann D, Shan K (2008) Supersonic penetration by Wedges and Cones into dry sand. In: 24th international symposium on ballistics, Sept. 22–26, Orlando, FL
11. Bless S, Berry D, Pedersen B, Lawhorn W (2009) Sand penetration by high-speed projectiles. In: 16th American Physical Society shock compression of condensed matter, June 28–July 3, 2009, Nashville, TN
12. Dwivedi SK, Teeter RD, Felice CW, Gupta UM (2008) Two dimensional mesoscale simulations of projectile instability during penetration of dry sand. J Appl Phys 104:083502
13. Savvateev AF, Budin AV, Kolikov VA, Rutberg PG (2001) High-speed penetration into sand. Int J Impact Eng 26:675
14. Schneider E, Stilp A (1984) Projectile penetration into low density media. In: 8th international symposium on ballistics, Orlando, FL
15. Satapathy S (2001) Cavity shape evolution during penetration of yawed long rods. In: 20th international symposium on ballistics. Interlaken, Switzerland, May 2001
16. Bless S, Peden B, Guzman I, Omidvar M (2013) Poncelet coefficients of granular media. In: Song B, Casem D, Kimberley J (eds) Dynamic behavior of materials, Conference proceedings of the Society for Experimental Mechanics series, vol 1. Springer, New York, p 528
17. Forrestal MJ, Luk VK (1992) Penetration into soil targets. Int J Impact Eng 12:427–444
18. Young CW (1997) Penetration equations. Report no. SAND97–2426. Sandia Laboratories, Albuquerque
19. Guzman I, Iskander M, Bless S, Qi C (2014) Terminal depth of penetration of spherical projectiles in transparent granular media. Granul Matter 16:829–884. https://doi.org/10.1007/s10035-014-0528-y
20. Thompson JB (1975) Low-velocity impact penetration of low-velocity soil deposits. Ph.D. dissertation. University of California, Berkeley
21. Glössner C, Moser S, Külls K, Hess S, Nau S, Penamadu D, Petrinic N (2017) Instrumented projectile penetration testing of granular materials. Exp Mech 57:271–272. https://doi.org/10.1007/s11340=016-0228-0
22. Collins AL, Addiss JW, Walley SM, Promratana K, Bobaru F, Proud WG (2011) The effect of nose shape on the internal flow fields during ballistic penetration of sand. Int J Impact Eng 38:951
23. Omidvar M, Doreau Malioche J, Bless S, Iskander M (2015) Phenomenology of rapid projectile penetration into granular soils. Int J Impact Eng 85:146–160
24. Omidvar M, Iskander M, Bless S (2016) Soil-projectile interactions during low velocity penetration. Int J Impact Eng 93:211–221. https://doi.org/10.1016/j.ijimpeng.2016.02.015
25. Chen Z, Omidvar M, Iskander M, Bless S (2014) Modelling of projectile penetration into transparent sand. Int J Phys Model Geotech 14:68–79. https://doi.org/10.1680/ijpmg.14.00003
26. Omidvar M, Malioche JD, Chen Z, Iskander M, Bless S (2015) Visualizing kinematics of dynamic penetration in granular media using transparent soils. Geotech Test J 38:18. https://doi.org/10.1520/GTJ20140206
27. Chapman DJ, Tsembalis T, Proud WG (2006) The behaviour of water saturated sand under shock loading. In: Proc 2006 Society of Engineering Mechanics annual conference and exposition on experimental and applied mechanics, vol 2. Society for Experimental Mechanics, Bethel
28. Taylor T, Fragaszy RJ, Ho CL (1991) Projectile penetration in granular soils. J Geotech Eng 117:658–672
29. Guzman I, Iskander M, Bless S (2015) Observations of projectile penetration into a transparent soil. Mech Res Commun 70:4–11
30. Omidvar M, Chen Z, Iskander M (2014) Image-based Lagrangian analysis of granular kinematics. J Comput Civ Eng 29:04014101
31. Backofen J (1989) Supersonic compressible modeling of shaped charge jets. Int Symp Ballist 2:395–406
32. Kotov VI, Balandin VV, Bragov AM (2013) Quasi-steady motion of a solid in a loose soil with developed cavitation. Dokl Phys 58:309–313

33. Penumadu D, Kim F (2015) Multimodal radiation based tomography and diffraction of granular materials using neutrons and photons and instrumented penetration mechanics. In: Iskander M, Bless S, Omidvar M (eds) Rapid penetration into granular media. Elsevier, New York
34. Parab ND, Claus B, Hudspeth MC, Black JT, Mondal A, Sun J, Fezzas K, Xiao X, Luo SN, Chen W (2014) Experimental assessment of fracture in individual sand particles at different loading rates. Int J Impact Eng 68:8–14
35. Allen WA, Mayfield EB, Morrison HL (1957) Dynamics of a projectile penetration sand. J Appl Phys 28:370–376
36. Cooper WL, Breaux BA (2010) Grain fracture in rapid particulate media deformation and a particulate media research roadmap from the PMEE workshops. Int J Fract 162:137–150
37. Bless S, Omidvar M, Iskander M (2017) Poncelet coefficients for dry sand. In: American Physical Society topical conference on shock compression of condensed matter, St Louis, MO
38. Meyerhof GG (1976) Bearing capacity and settlement of pile foundations. J Geotech Eng ASCE 102:195–228
39. Das BM (2007) Principles of foundation engineering. Chapter 11.7, 6th edn. Cengage Learning, Stamford
40. Clark AH, Behringer RP (2013) Granular impact model as an energy-depth relation. Europhys Lett 101:64001
41. Omidvar M, Iskander M, Bless S (2014) Response of granular media to rapid penetration. Int JImpact Eng 66:60–82. https://doi.org/10.1016/j.ijimpeng.2013.12.004

Applications of Reactive Materials in Munitions

Suhithi M. Peiris and Nydeia Bolden-Frazier

1 Introduction

In weapons, munition and ordnance communities, reactive materials, similar to explosives and propellants, are considered a class of energetic materials. Reactive materials react to produce energy, though, currently not at the rate of energy production usually seen in deflagrating or detonating explosives. Reactive materials are being investigated for use in ordnance to enhance the energy of munition systems that are usually volume or mass constrained. While liquid reactive materials do exist, this chapter is limited to the more common solid reactive materials, such as thermites (metal and metal-oxide mixtures), intermetallics (metal-metal mixtures), metals with oxidizers, and metals with polymeric 'binder' materials.

The applications of reactive materials in munitions fall in to two broad categories. One is their use in the energetic payload as powders or particles, and the other is their use in structural parts or weapon components, such as the weapon case. Therefore, this chapter is organized into sections, starting with Reactive Material Powders, their preparation methods, then Applications of Reactive Material Powders, and finally, Reactive Material Structures and Their Applications.

S. M. Peiris (✉) · N. Bolden-Frazier
Air Force Research Laboratory – Munitions Directorate, Eglin Air Force Base, FL, USA
e-mail: Suhithi.Peiris@us.af.mil

T. J. Vogler, D. A. Fredenburg (eds.), *Shock Phenomena in Granular and Porous Materials*, Shock Wave and High Pressure Phenomena,
https://doi.org/10.1007/978-3-030-23002-9_6

2 Reactive Material Powders

Initial research in this field in the 1990s studied micron-sized powders, similar in size to aluminum powders incorporated in typical aluminized explosive formulations. Early reactive materials consisted of Al powder in a polymer matrix, often a per-fluorinated polymer, like Teflon [1]. This century has focused on the preparation of nano-sized powders, with the hope that the smaller sized particles would burn or react more efficiently, could be mixed more intimately together to enhance reaction rate, etc. [2].

After 30 years of reactive material research, one of the most useful reactive metals is still aluminum. Aluminum burns easily to produce significant energy, as shown in Fig. 1 [3]. While it is highly attractive because of its high oxidation enthalpies and high combustion temperatures, its full potential have not yet been achieved in practical applications. One reason considered is ignition delay caused by the aluminum oxide (alumina) surface that passivates aluminum metal. According to the best known burning mechanism for larger particles (>100 micron), aluminum ignition is inhibited until the alumina shell melts at 2350 K [4]. For particles in the 10–100 μm range typically used in energetic applications, ignition is delayed until the particle temperatures rise to the 1000–2300 K range [5]. Therefore, other metals such as those listed in Fig. 1 continue to be explored in different applications.

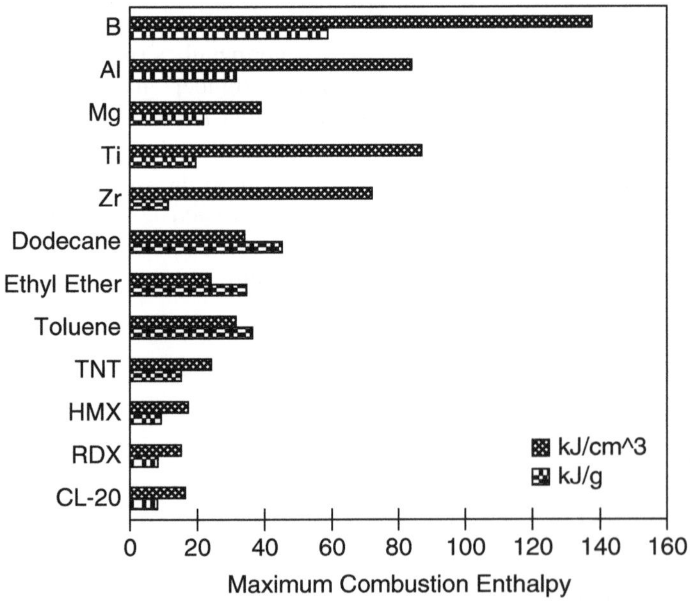

Fig. 1 Maximum combustion enthalpies (per volume and per mass) for select reactive materials, liquid fuels, and explosives [3]

As shown in Fig. 1, metals contain more combustion or reaction energy per unit volume or per unit mass than liquid fuels or explosives. However, the combustion or reaction rate of these metals is not fast enough to add to the energy from explosives in similar time scales of detonation. Therefore, since the early 2000s investigators, anticipating that smaller particles would ignite quicker (or ignite at lower temperature) and burn faster because of the intimate mixing of fuel and oxidizer at the nano-scale, focused their efforts on preparing nano-sized reactive material particles. The 2003 Materials Research Society Symposium on Synthesis, Characterization, and Properties of Energetic/Reactive Nanomaterials highlighted many of these early attempts [6]. They include sol-gel methods at Lawrence Livermore National Laboratory [7] an *in situ* nano-Al synthesis and passivation by U.S. Navy researchers [8].

While early work with these methods and the materials prepared were popular, other methods still being employed have proved more useful for scale up and testing. Some of these methods are lamination, electro-spraying, and ball milling, with lamination including cold rolled, radially forged, and electro-sprayed layered materials that are later broken/ground into powders. The next few pages will discuss each of these methods.

2.1 Lamination Methods

Lamination is a relatively straightforward method for producing reactive material powders that may be used as additives in energetic fills or consolidated for structural applications. Powders are produced by hand breaking and/or grinding bi-layered foils or sheets. The foils or sheets are fabricated by mechanical processes such as cold rolling, radial forging, etc. or by vapor deposition methods such as cold-spray, electro-spraying, etc. Irrespective of the method, reactivity is determined by the bilayer thickness, which controls diffusion distance of reactants, defining ignition thresholds and propagation velocities. The process used for producing laminates for munitions applications would be selected based on advantages such as energy gain weighed against disadvantages such as capital costs, throughput, and microstructural uniformity in mass production.

Cold rolling (CR) and radial forging (RF) have been extensively investigated by the Weihs group at Johns Hopkins University in Baltimore, Maryland [9]. CR uses a two-step process, as shown in Fig. 2, created initially by rolling stacks of alternating metals into bi-layers that are then cut, restacked, and rerolled until the desired thickness is achieved. Each pass reduces the thickness 20–50%, further refining the microstructure. Microstructures are much coarser and less uniform, than the finer, more structured microstructures achieved from other methods to be discussed later in this chapter [9].

Stover et al. [9] successfully developed Ni-Al powders ranging from 53 to 850 μm using this technique. Laminates with varying bilayer thicknesses were hand-broken and ground into final powders. Laminates were produced by cold

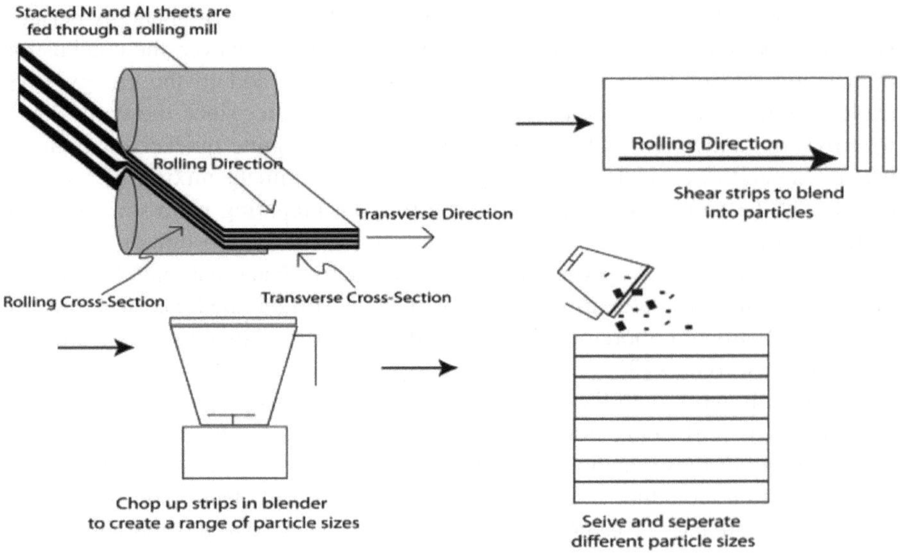

Fig. 2 Cold rolling is a method used to prepare laminates that are then broken by hand or ground in a blender to produce reactive material powders [9]

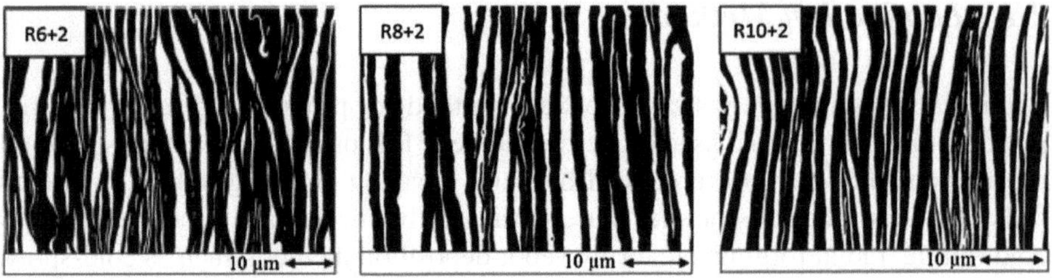

Fig. 3 Bi-layers of cold rolled Al:Ni composites shown to increase as number of rolls increase. The micrographs are thresholded to black and white with black representing Al and white representing Ni [9]

rolling T6 Al and fully annealed 201 nickel stacks at room temperature to achieve 50% thickness reduction per pass. Scanning electron micrographs (SEM) seen in Fig. 3 shows the changes in microstructures as the number of rolls increases from 6 to 10. The average bilayer thickness decreased from 2.9 to 2.0 μm as the number of rolls increased from 6 to 8, but minimally decreased to 1.8 μm with ten rolls. The fine microstructure of the bilayers thus obtained determines the density and heat of the resulting reactive material powder.

Radial forging, or as some communities call it "rotary swaging", is another option for producing reactive laminate composites. Here, severe plastic deformation is used to significantly change microstructures through grain refinement from micron to sub-micrometer or nanometer range sizes [10]. During swaging, the material (rod or

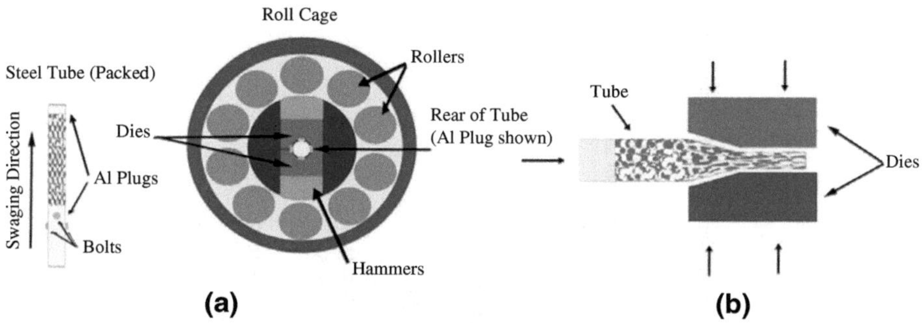

Fig. 4 Schematic of 3Fenn Rotary Swager, front view, and cross-sectional view of a tube being deformed to a smaller diameter by the dies shown in the middle of the front view [11]

tube) is rapidly compressed through a series of radial reductions during which large plastic strains shear the constituent powders, breaking the oxide surface and yielding enhanced bonding [11]. Figure 4 shows the process of the rapid compression of the roller bearings, reducing the rod/tube to specific outer diameters. Under this method, microstructures can be refined through rod reduction, particle size of constituent powders, and/or alloying of constituent powders.

Effects of these parameters can be seen in a systematic investigation conducted by Gibbins et al. [11] on Al:Ni based powders prepared from various powder chemistries and geometries. Table 1 displays the formulations and geometries of the starting materials, which included elemental powders, flakes, and laminate powders. Samples were prepared first by adding loose powders to stainless steel tubes that were compacted using an Instron 5582 load frame. Following the previously described procedure, each tube, starting from 15 mm outer diameter (OD) received a series of five radial reductions until the OD was 6.35 mm, and then five more reductions until the OD was 2.54 mm. Excess stainless steel was removed from the final samples. Near complete densification was expected as the diameters are reduced ~15% with each swaging pass.

Figure 5 shows cross-sectional images taken normal to the length of the rods, revealing the various microstructures achieved from the different starting materials detailed in Table 1. As expected, the reactant spacing consistently decreased in each of the rods as the swaging increased. However, the interfaces for the laminate powders were notably more uniform, compared to the discontinuous interfaces seen in the Powder/Powder and Powder/Flake samples. Ignition temperatures, measured using a hotplate, all decreased as the OD decreased from 6.35 to 2.54 mm, an indication of microstructural refinement. Reaction self-propagation was successfully achieved in most of the smaller samples with smaller bilayer distributions, thus confirming the potential of swaging, or radial forging, as a viable method of fabricating reactive materials for munition applications. Successful utilization of swaging to fabricate other reactive materials such as Al/NiO-Ni thermites has been demonstrated [12].

Table 1 Formulations, geometries of AlNi laminates and reactant spacing produced using radial forging starting at 15 mm outer diameter and reducing first to 6.35 mm and then to 2.54 mm, where P = particle, F = flake, and LP = laminate particle

	Composition %	Initial powder shape and size (mesh)	Average reactant spacing (μm) for 6.35 mm samples	Average reactant spacing (μm) for 2.54 mm samples
Al:Ni P/P	50 Al	Al: Sphere: −325	3.16 ± 0.42	2.01 ± 0.31
	50 Ni	Ni: Sphere: −325	Figure 5a	Figure 5b
Al:Ni P/F	50 Al	Al: Sphere: −325	6.17 ± 2.06	1.82 ± 0.20
	50 Ni	Ni: Flake: −325	Figure 5c	Figure 5d
Al-Mg:Ni P/P	46 Al	Al-Mg: Sphere: −325	4.11 ± 1.06	1.57 ± 0.31
	4 Mg	Ni: Sphere: −325	Figure 5e	Figure 5f
	50 Ni			
Al-Mg:Ni P/F	46 Al	Al-Mg: Sphere: −325	8.38 ± 2.71	1.99 ± 0.28
	4 Mg	Ni: Flake: −325	Figure 5g	Figure 5h
	50 Ni			
Al:Ni LP	48 Al	Al:Ni: Sphere:	0.35 ± 0.08	0.15 ± 0.02
	52 Ni	212–355 μm laminate structure	Figure 5i	Figure 5j

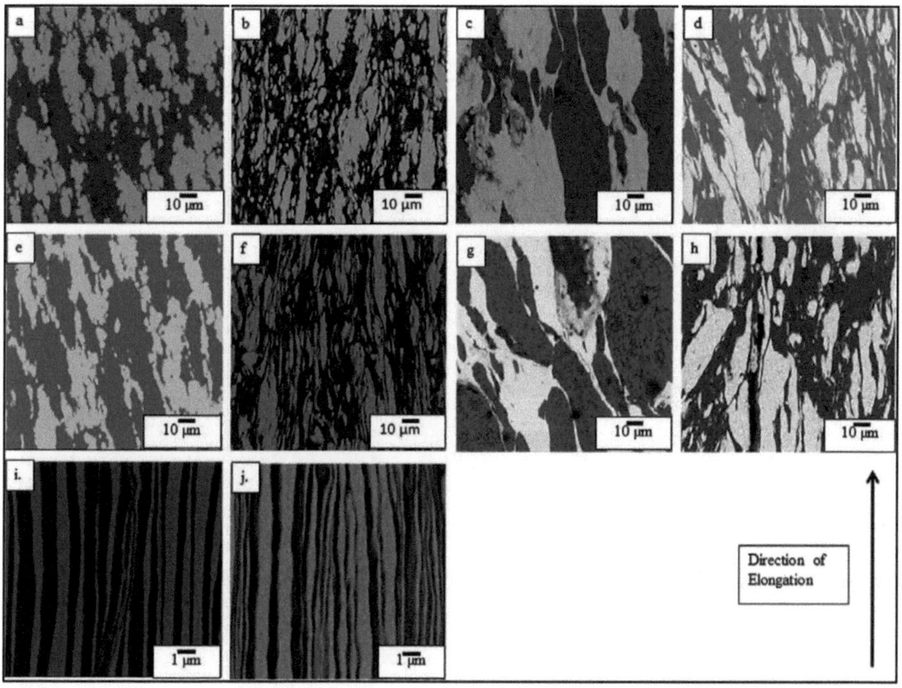

Fig. 5 Microstructural images taken normal to the length of the swaged rods, with compositions (**a**–**j**) as shown in Table 1

A third method for producing laminates is electro-spraying using deposition, instead of mechanical deformation, to create reactive bi-layer or tri-layer sheets or foils. The electrospray approach uses a high electric field to overcome surface tension and intermolecular forces within a solution to create smaller more-even-sized aerosol droplets [13]. Nanoparticles can simply be added to the precursor solution and dispersed with a narrow size distribution by tuning electro conductivity. This *in situ* nano-powder processing alleviates safety concerns with utilizing nano powders in commercial processing [14]. Because the droplets formed are electrically charged, they can be easily directed to create thin films with various nanoparticle loadings. This allows more control over layer thickness and final microstructure uniformity.

The study by Li et al. investigated the utilization of electro-spraying to produce layered materials with up to 60% nano particle (NP) loading [13]. Using layer-by-layer deposition, Al-NPs/CuO-NPs thermites were fabricated and found to be uniform and mechanically flexible. Similarly, alternate layers of polyvinylidene fluoride (PVDF an energetic binder), and Al/CuO nano-thermite were fabricated using a dual spray setup shown in Fig. 6. The particles, fed by a syringe pump, were ejected onto a rotating collector at a distance that enabled a wide spray pattern and sufficient time for solvent evaporation. Layer thicknesses were controlled by the duration of deposition. SEM images in Fig. 7 show a crack-free uniform PVDF

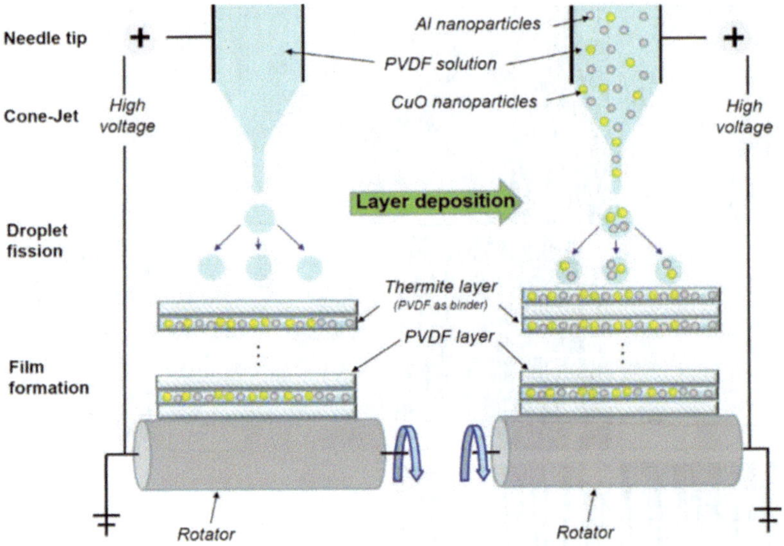

Fig. 6 Electro-spraying with dual spray set up for obtaining a bi-layer laminate [13]

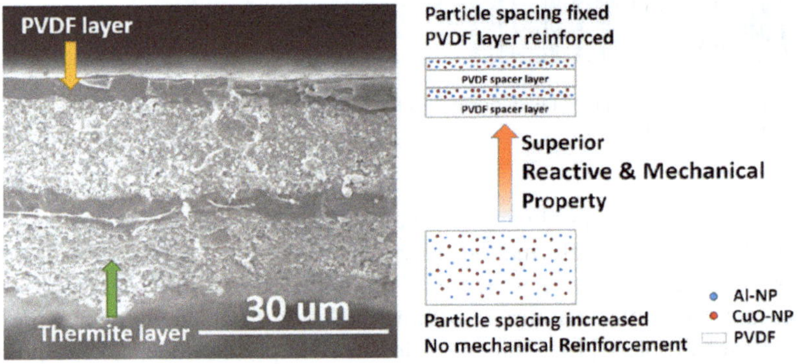

Fig. 7 Electro-sprayed laminate or layered material made of thermite composition and PVDF polymer [13]

layer as compared to the uniform, but more textured thermite layer. The alternating bilayers are well-defined, with lateral conformity and thickness and no significant differences in interface.

2.2 *Electro-spraying*

In addition to producing powders from grounded reactive thin films or laminates, electro-spraying can also be used in the direct fabrication of reactive meso-particles. The assembly of meso-particles using electro-spraying has been extensively inves-

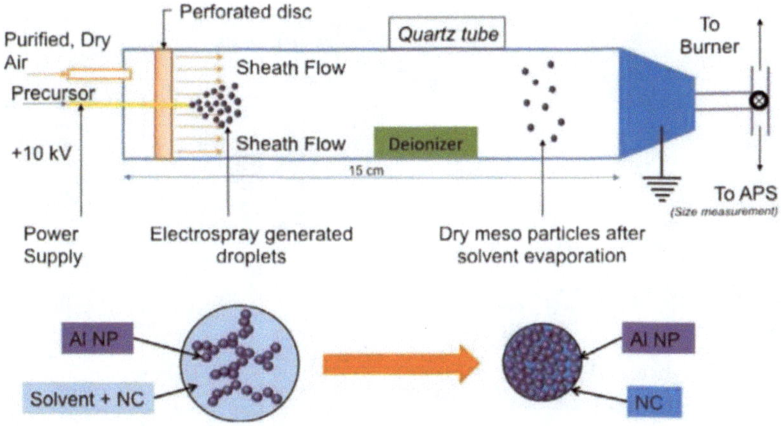

Fig. 8 Electrospray method to fabricate mesoscale particles [16]

tigated by the Zachariah group at the University of Maryland, College Park [15]. The principle is the same as previously described; it is based on liquid jet break-up under the influence of a strong electric force. Simple modifications of the precursor solution can be utilized to achieve microspheres composed of constituent nanoparticles embedded within a binder, as shown in Fig. 8. Wang et al. [15] used this method to create nano-aluminum (nAl) containing microspheres with enhanced burning behavior. These microspheres were produced by adding nitrocellulose (NC), an energetic binder, to nAl in an ethanol/ether solvent precursor solution. This solution containing the gelled fuel/binder mixture was sprayed at a speed and distance that allowed full solvent evaporation. As the solvent evaporates, aluminum nanoparticles aggregate and are left embedded in nitrocellulose [15]. Figure 9 shows the highly textured particle surfaces of the nAl/NC powders and the small size distribution achieved. The nitrocellulose matrix serves as a reactive binder with a low dissociation temperature and showed an order of a magnitude reduction in average burn times as that of the individual embedded nanoparticles.

2.3 *Ball Milling*

Ball milling or arrested reactive milling is one of the simplest methods of producing micron-sized particles that are mixed or layered at the nano-scale. This method of mechanical alloying has been accepted as a tailorable, scalable, and viable technique for producing various advanced materials [17]. The Dreizin group at New Jersey Institute of Technology has extensively studied ball milling and arrested reactive milling as a method of manufacturing reactive composite powders. Traditional ball milling uses mechanical impaction of hard balls falling and hitting material(s) while under rotation. Most commonly, cylindrical jars (Fig. 10) filled with balls and precursor materials are rotated at a speed that allows the balls to rise and

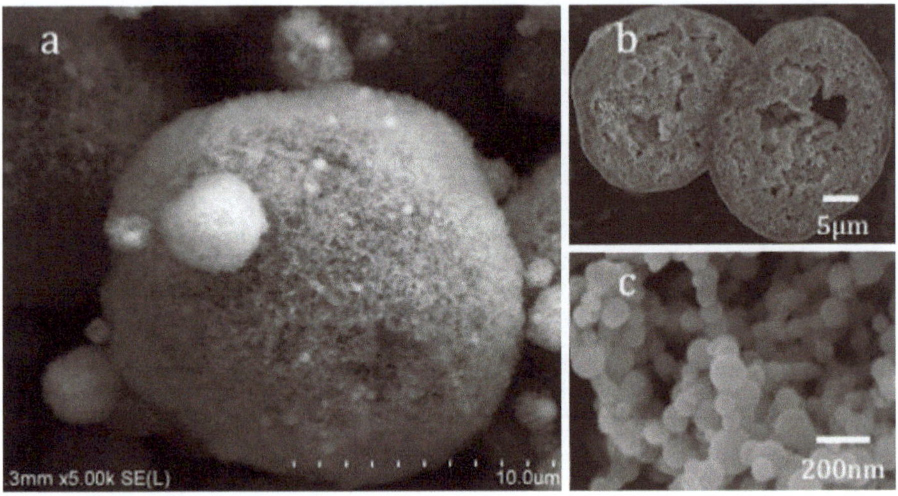

Fig. 9 Meso-particles or micron-sized particles composed of multiple nano-sized particles [15]

Fig. 10 Ball milling equipment and process [18]

fall, impacting the particles, decreasing the size, and/or mechanically alloying the materials. Optimal milling requires the ball diameters to be ~30×'s that of the constituent material(s) with the balls occupying ~50% of the total cylinder volume, and a precursor to be no more than 25% of the total volume [19]. Figure 10A, B demonstrates the milling process for a copper-doped $BiVO_4$ nanocomposite [18], and Fig. 11 shows a ball milled Al/Bi_2O_3 sample [20].

Aly et al. prepared Al·Mg particles via mechanical milling of elemental powders and compared combustion characteristics to cast-alloyed Al·Mg of the same exact composition with similar particle size distributions [21]. The goal was to show viability of the production method as well as the effects of microstructure on

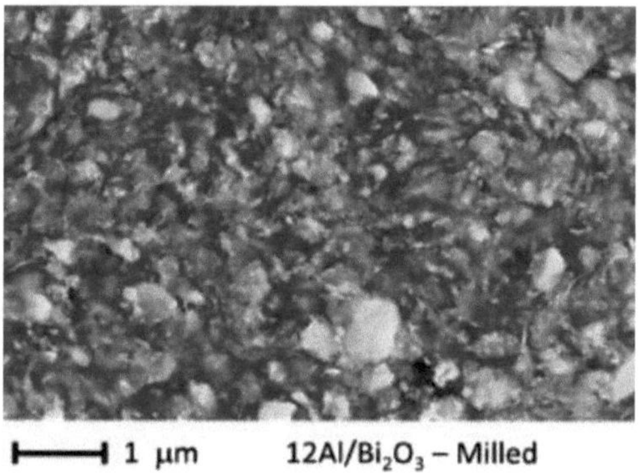

Fig. 11 Back-scattered scanning electron micrograph of a sample produced by ball milling an Al and Bi_2O_3 powder mixture. Note the light-colored micron- and nano-sized inclusions of Bi_2O_3 embedded into a dark-gray Al matrix [20]

combustion characteristics. Powders were loaded in 9.5 mm diameter steel milling vials under argon and rotated at 350 rpm for 120 min. Rotation direction was changed every 15 min. Ignition temperatures measured with an electrically heated filament as a function of heating rate showed ignition temperatures of the cast-alloyed (CA) powders ranging from 950 to 1175 K compared to the slightly lower ignition temperatures of 925–1025 K from the mechanically alloyed (MA) powders. MA powders outperformed CA powders in aerosol combustion tests (maximum pressure and rate of pressure rise) and ignited faster, resulting in a shorter ignition delay [22]. Other binary and ternary materials such as $Al{\cdot}I_2$, $Al{\cdot}B{\cdot}I_2$, and Mg·B·I2 with lower ignition temperatures and longer burn times have also been successfully prepared using the arrested reactive milling method [23].

3 Applications of Reactive Material Powders

Reactive material powders, prepared as described above and using other methods not detailed here, are micron-sized particles with nano-scale layers or mixtures of components. The laminates are usually metal/metal layers providing fuel that needs a separate oxidizer (or air) to react and produce energy. The electro-sprayed or ball-milled particles are more frequently fuel/binder or fuel/oxidizer compositions capable of producing energy with self-reaction.

No matter the fabrication method, a key feature of these materials is the internal layering or mixing of individual ingredients. Under shock compression, a chemical reaction is initiated. In a collaboration between Thadhani's group at Georgia Tech and Weihs' group at Johns Hopkins, the effects of impact velocity, layering orientation, and bilayer thickness on the evolution of reaction in bilayer materials have been investigated using numerical simulations [24, 25].

Specht's early work demonstrates that cold-rolled multilayer composites, which are not uniform in terms of their inter-layer characteristics, can result in complex wave interactions that will affect the overall shock compression response [24]. Real, heterogeneous microstructures, obtained from optical micrographs of the multilayered composite cross-section, were incorporated into the Eulerian, finite volume code CTH, and simulations were performed to understand shock dispersion and dissipation in cold-rolled materials. A simulation performed at an impact velocity of 1000 m/s, corresponding to a particle velocity of 597 m/s, was used to illustrate the effects of non-uniformities associated with the shock front propagating perpendicular to the composite layers. The pressure response under these conditions at 80 ns and 100 ns after impact is shown in Fig. 12. It can be seen that the multilayer composite generates localized areas of elevated pressure, which change over time. This suggests a quasi-steady response since the perturbations are not

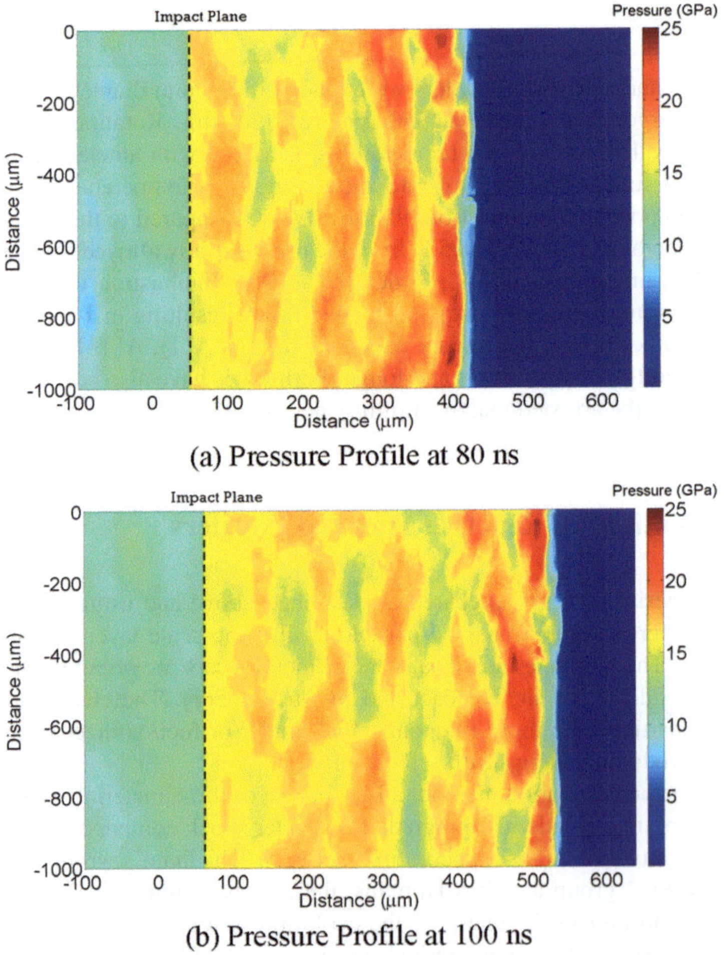

(a) Pressure Profile at 80 ns

(b) Pressure Profile at 100 ns

Fig. 12 Shock front propagation 80 ns (left) and 100 ns (right) after impact at 1000 m/s by a semi-infinite copper piston from the left of the composite [24]

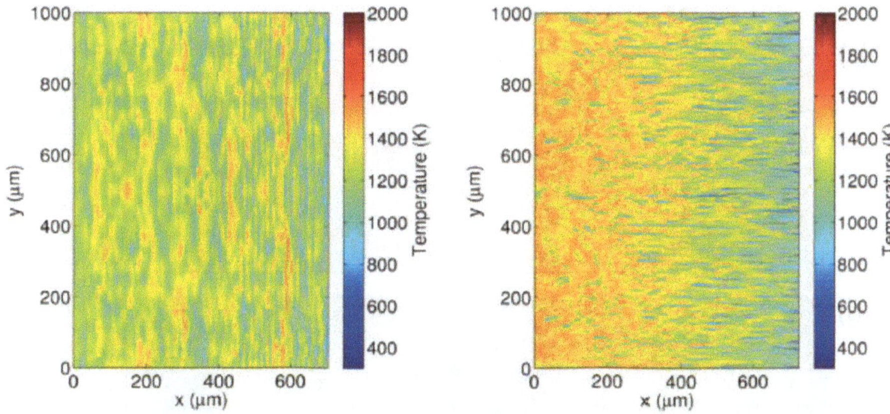

Fig. 13 Temperature distribution from numerical simulations of Ni/Al after shock compression. The direction of the shock is perpendicular (left) and parallel (right) relative to the direction of material layering [25]

consistent in the shock wave, in contrast to the steady pressure behavior (with periodic oscillations) seen for highly uniform layers [24].

Later, Sraj et al. demonstrated that the orientation of the layers with respect to the wave direction significantly affects energy dissipation and heating, as shown in Fig. 13 [25]. CTH predictions were used to obtain the initial temperature distribution after shock impact. They noted that the predicted peak temperature values are higher for the parallel shock orientation than for the perpendicular orientation. They applied a "Reduced Model" to compute the transient evolution of reaction [25]. The evolution of the reaction was seen to depend on the bilayer thickness of the composite and on the impact velocity. When the layers are thin (~250 nm) and initiated with a higher shock (3 km/s), a homogeneous reaction is observed. When the layers are thick (~2000 nm) and initiated with a lower shock (2 km/s), a self-propagating reaction front is observed.

Interfacial strength was also numerically shown to affect energy dissipation and thereby heating. When the interface was weak (when the bi layers were not bound with any strength), the two materials moved freely. When the interfacial binding was strong, interfacial shear caused increased energy dissipation and heating at the interfaces [26].

In collaboration between another group at Georgia Tech and UCSD, shock-initiated deformation and mixing of Ni/Al powders were studied using the Eulerian finite element code RAVEN in 2D [27]. The powders shown in Fig. 14 were spherical with 5 μm Ni particles and 20 μm Al particles mixed in a 2.6Ni:1Al composition and packed to 55% packing density. The simulations were performed on powders rendered in two dimensions (under plane strain) with all material interfaces being perfectly bonded. They report that the deviatoric stress–strain responses of the particles were described using a physically based macroscopic model of high strain-rate viscoplasticity. The powder mixtures were impacted with

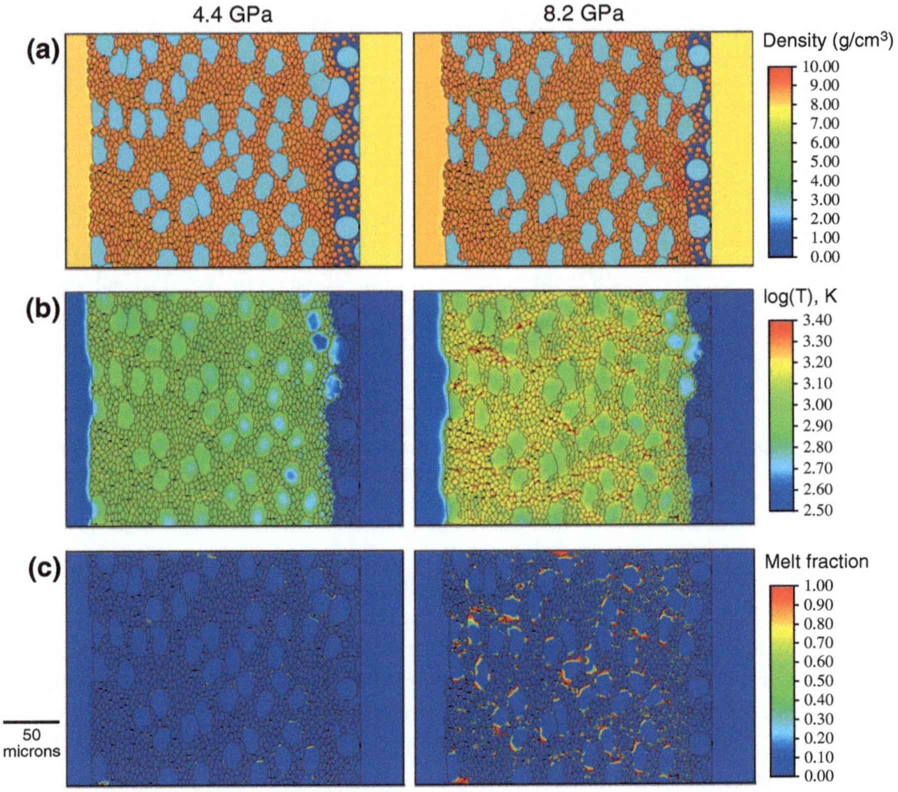

Fig. 14 5 μm Ni particles and 20 μm Al particles mixed in a 2.6Ni:1Al composition and packed to 55% packing density, impacted at 4.4 GPa (left column) and 8.2 GPa (right column): (**a**) mass density fields, (**b**) logarithmic temperature fields, and (**c**) melt fraction fields [27]

flyer plates assigned initial velocities in the range of 0.6–1.2 km/s, which for the powder in Fig. 14 resulted in initial shock stresses of 4.4 GPa and 8.2 GPa. At 4.4 GPa, the stress wave was strong enough to crush out all mixture porosity. At 8.2 GPa, the Al particles are flattened and look more ellipsoidal. Note the "hotspots" or red Ni particles in the second row, column (b) of Fig. 14 showing temperatures, indicating the beginnings of reaction. Thus numerical studies highlight the shock impact conditions necessary for initiating reaction in reactive materials, particularly how particle or layer morphology, and interfacial properties affect reactions [27].

3.1 Agent-Defeat Applications

One utility of reactive material powders is shock or detonation induced late-time burning, important for applications in weapons designed to burn the target or burn everything in the target. For example, targets containing biological warfare agents and/or chemical warfare agents, if attacked by a weapon that is detonated, may just further spread these agents with the blast. If the weapon contained reactive materials that continued to burn after the detonation, then active warfare agent within the target would be burned before being released to the environment. Further, if these reactive materials had low ignition temperatures and long burn times, they would start burning with lower energy localization and continue to burn over long times. Such specialized weapons, known as agent-defeat weapons, could utilize reactive material powders. The following paragraphs describe these applications.

One example of reactive materials that continue to burn over extended burn times are intermetallics. Metal/metal mixes in laminates, with effective choice of metals, can produce heat from initial intermetallic reactions, followed by burning or reaction with an oxide to release significantly more heat than typical metal/oxide reactions [28]. The layered microstructures facilitate early time mixing, resulting in self-propagating reactions started with low ignition energy. Continued combustion in air allows the formation of nitrides and oxides, releasing additional heat over an extended time. This effect was demonstrated by Overdeep et al. in Al-Zr-based laminates prepared using vapor deposition, with substitution of increasing quantities of Mg for Al [28]. Al-Zr, Al-8Mg-Zr, and Al-38Mg-Zr laminates were prepared and evaluated in a bomb calorimeter in four different environments of oxygen, air, nitrogen, and argon. The extent of oxidation, calculated as the ratio of experimental heats to theoretical heats (Fig. 15), was determined to be highest for Al-8Mg-Zr in all environments except inert Ar. This is believed to be a direct result of lowered ignition temperatures from the added magnesium combined with the specific metal composition which yielded the most efficient intermixing and self-propagating

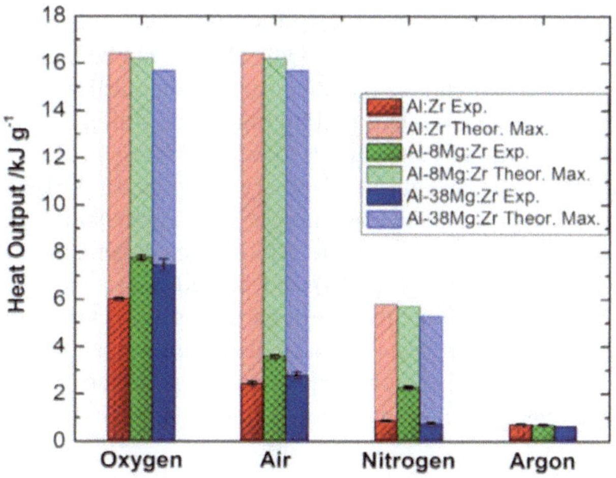

Fig. 15 Combustion efficiency or extent of combustion of a series of compositions of laminates in four environments [28]

Table 2 Compositions of ball-milled ternaries with biocidal Iodine [23]

Sample	Mass percentage (%)			
	Mg	Al	B	I_2
A	0	30	50	20
B	0	40	40	20
C	33	0	47	20
D	50	0	30	20

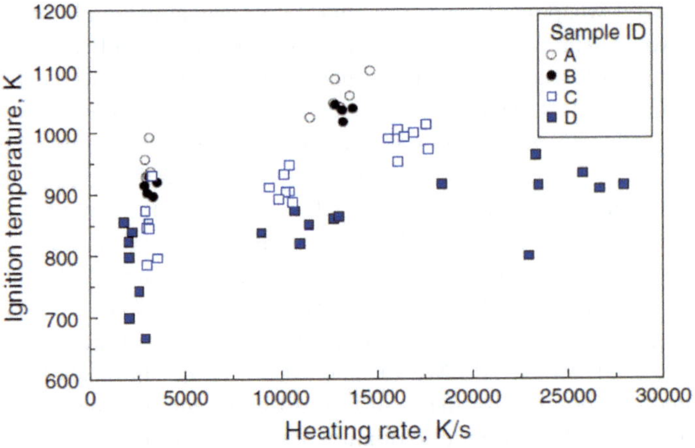

Fig. 16 Ball-milled, iodine-containing powders ignite at lower temperature than pure ball-milled Al or Mg, which ignite at 1750 K and 1050 K, respectively [23]

reactions to extend burn times. Such materials with extended burn times and lower ignition temperatures could be utilized in agent-defeat weapons applications.

An alternative approach for agent-defeat weapons is to use powders composed of metals and biocidal or anti-bacterial chemicals that can be harmful to biological warfare agents or that can change the chemistry of chemical warfare agents. Such mixtures can be produced from ball milling processes, with the simplest one being Al-I compositions, because of our familiarity with Al burning. The Dreizin group at NJIT prepared varying compositions of ternary compounds, $Al{\bullet}B{\bullet}I_2$ and $Mg{\bullet}B{\bullet}I_2$, as shown in Table 2, by mechanical ball milling [23]. Combustion of the powders (as aerosols) was tested in a constant volume explosion chamber. All of the composites ignited at lower temperatures (Fig. 16) and burned consistently longer than the pure Al and Mg powders with the same particle sizes. Other powders such as $Al{\cdot}I_2$, $Al{\cdot}CHI_3$, $Al{\cdot}B{\cdot}I_2$, and $Mg{\cdot}B{\cdot}I_2$ have also been prepared using ball milling techniques [23]. The lower ignition temperatures, longer burn times, and resultant extended iodine release make such compositions suitable for bio agent defeat applications.

A third variation of reactive material powders used in agent-defeat applications is known as shock-dispersed fuels, where particles of reactive materials are packed as a loose powder or in liquid or in a soft binder. An advantage of this method

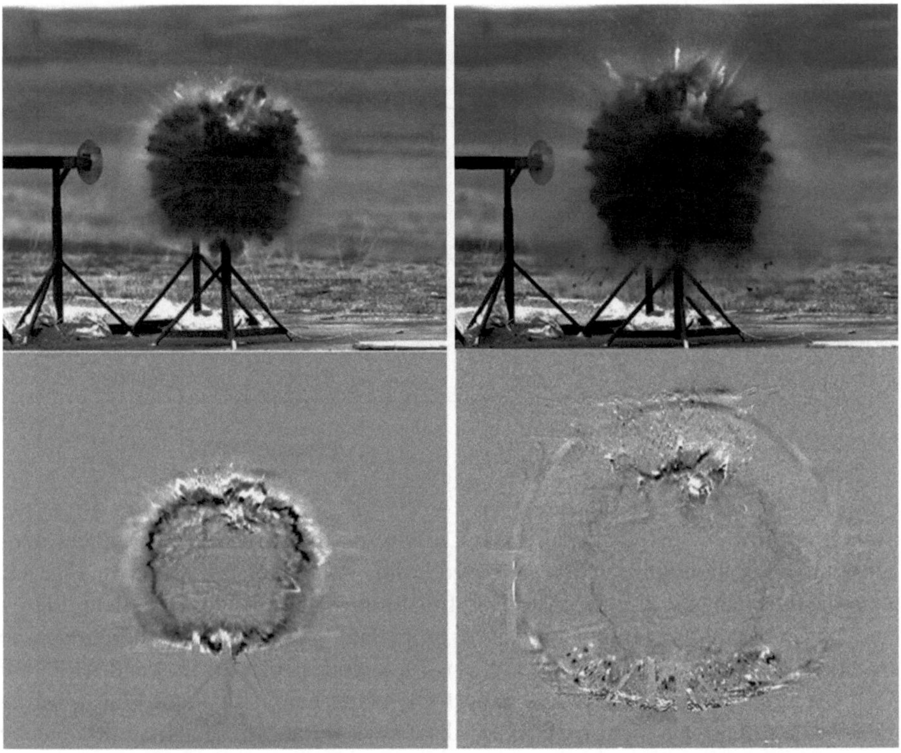

Fig. 17 Explosive dispersal of Ni powder at 0.8 ms and 2.6 ms after detonation, where the shockwave is seen breaking out and preceding the powder cloud. The lower images were generated by subtracting the previous video frame leaving only the last 100 μs showing [29]

is that reactive metal particles are first dispersed, and then ignited, reducing shock-initiation and mixing challenges encountered with solid composite reactive materials, as shown in Fig. 17. Such configurations are being investigated at Defense Research and Development Canada (DRDC). Basic research on metal particle dust clouds performed by the Frost group at McGill University is providing understanding of the necessary metal particle concentrations, burning velocities, and scaling effects [29].

3.2 *Propellant Applications*

A second utility of reactive material powders is in propellant applications. An example is research in Steve Son's group at Purdue, where they investigate nano-scale metal fuels in composite solid rocket propellants to improve performance. Their recent publication discusses the effect of Al/PTFE nano foils, Ni-Al nano

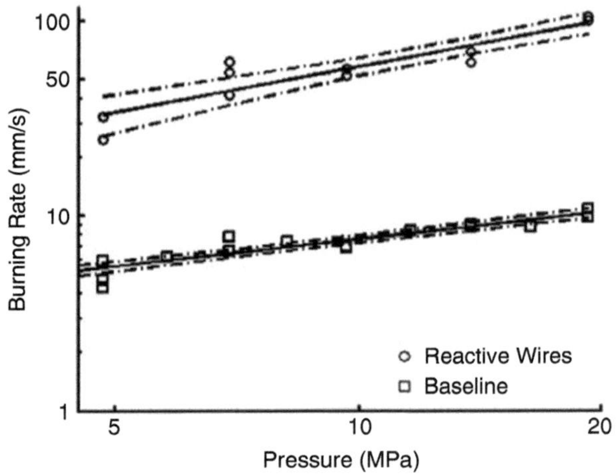

Fig. 18 Comparison of the burn rate of propellant near Al/PTFE foils and baseline propellant burn rate, as a function of pressure [30]

foils, Pyorfuze® wires, and traditional Cu wire on the burn rate of a typical non-aluminized ammonium perchlorate-based solid propellant. As shown in Fig. 18, their results indicate that the Al/PTFE foils contributed to the propellant energy and gas production yielding increased propellant efficiencies in comparison to traditional wired propellants [30]. Other work includes B/PTFE [31] and Al/PVDF [32]. As seen in these papers, the nano foils, being nano-structured with layering (similar to materials that have been discussed in above sections on reactive powders), burns faster than singular micron materials. Therefore, these reactive nano foils could be utilized in future propellant applications.

3.3 Explosive Formulation Applications

A third application of reactive powders is their use directly in explosive formulations. An explosive formulation is usually a polymer-bound composite containing pure explosive particles. Some 'metalized explosives' also contain micron-sized aluminum particles. These Al particles burn during the detonation of an explosive, and, therefore, can be considered a reactive material. Since the beginning of this century, researchers have substituted micron-Al with nano-sized Al powders and shown that the smaller sized particles would burn more efficiently in certain specific explosive formulations [33]. There are multiple theories about why nano-Al in explosives formulations is not as great as it was hypothesized to be. One theory highlights the effect of the amorphous alumina (γ-Al_2O_3) coating commonly found to passivate the surfaces of Al. For large Al particles this coating of 2–5 nm is irrelevant. However, for nano-Al this coating could be a significant portion of the particles mass, reducing the percent Al available for reaction when nano-Al is equally substituted for micron-Al in formulations. Another theory focuses on the diffusion limitation of the Al oxidation (or combustion) reaction. The

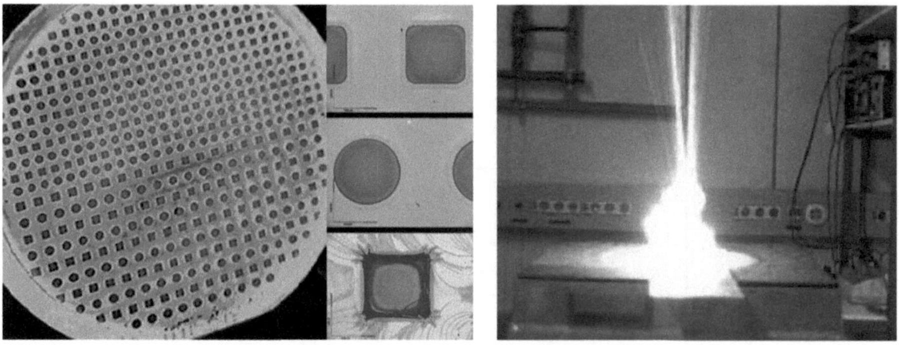

Fig. 19 Nano-porous Si wafers, then used in explosive formulations and ignited [35]

resulting alumina (α-Al_2O_3) forms a cap around the Al, and further reaction is limited by the ability of oxygen to diffuse through this oxide cap, resulting in incomplete burning of all the Al within particles. A very recent understanding of the limitation of nano-Al is based on the fact that most nanoparticles are agglomerated together (not isolated) as they are processed in to a formulation. Agglomerated particles as they are rapidly heated by the explosion start sintering or aggregating together, dramatically increasing their size and therefore burning more like regular micron-Al [34].

These hindrances and challenges with using nano-Al in explosive formulations initiated searches for other reactive powders that could be similarly used in metalized explosives, such as porous Si. Bulk silicon has not traditionally been considered for explosive applications, in part due to the slow diffusion of oxygen through the silica layer on the Si surface. However, the potential energy yield is much higher than from most common carbon-based explosives. Therefore, Clement et al. investigated the potential of explosive composites that included porous silicon (PSi) filled with perchlorate oxidizers, as shown in Fig. 19 [35]. The energy yield determined from calorimetric bomb-test was found to be 7.3 kJ/g, almost double that of TNT (~4 kJ/g). Their work resulted in a proto-type airbag ignitor application.

3.4 Munition Liner Applications

A fourth application for reactive material powders in munitions is their incorporation into weapon 'liners' or flexible sheets used in various warhead designs. Reactive liners are commonly used in munition applications as annular liners (lining the inside of the case or the outside of the explosive fill) to improve blast and fragmentation and as impedance mismatch layers between the explosive and warhead to decrease sympathetic reaction. These liners are typically metal sheets or metal-binder mixtures that are expected to ignite with the explosive shock wave. Crouse et al. reported the preparation of a liner material from functionalized nano-Al particles chemically integrated into a fluorinated methacrylate polymer, termed AlFa

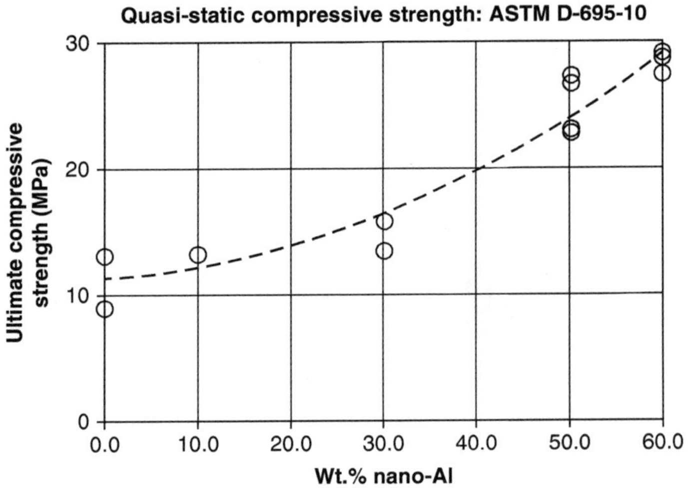

Fig. 20 Quasi-static compressive strength as a function of nano-Al content in AlFA composites. The dashed curve is an aid to the eye [36]

[36]. AlFA was prepared with 0, 10, 30, 50, and 60% Al loadings, with compressive strength increasing from 11 MPa in neat polymer to 25–28 MPa in AlFA-50, as shown in Fig. 20. All formulations up to 60% Al maintained thermoplastic behavior and could be prepared with little to no agglomeration and voids. Self-sustaining deflagration was seen in compositions with 30% or more Al content, with AlFA-50 producing the most intense and rapid reaction [36].

The preceding pages have described several applications of reactive material powders in munition. Many other groups not mentioned here are doing basic or fundamental research on reactive material powders. When funded by defense organizations and agencies, the materials being researched are bound to have conceivable applications. Future research on reactive powders with munition applications would continue to investigate microstructures and other particle geometries, to tailor burn rates and to achieve rates closer to deflagrating explosives. Such a significant gain in burn rate would open more application possibilities for reactive material powders in future munitions.

4 Reactive Material Structures and Their Applications

In addition to reactive material powders, reactive material structures or structural reactive materials are also being investigated for use in ordnance to increase energy or add energy to the munition system. This work has focused on consolidating micron-sized powders to make structures with reactive materials, or on incorporating reactive materials in composite structural components. For instance, many munition parts are made from steel. While steel has highly desirable strength properties, it does not contribute to the energy produced by the weapon. If we make a reactive material with the density and strength of steel, we can swap out the steel in a munition with a steel-like reactive material that would burn or react and produce additional energy when the weapon was detonated.

There are several methods being researched for preparing structural reactive materials or reactive material structures, herein abbreviated to SRM. Since these are to be SRM, the method to process the material has to be scalable, so that large pieces or components can be made without reduction in material quality and material strength. The later requirement is a challenge. One of the more successful approaches used is based on powder metallurgy where metal powders are heated and sometimes pressed to achieve high strength structures. Another approach is fiber-reinforced SRM. The next paragraphs will describe the structural materials made by these processes.

4.1 Powder Metallurgy

Powder metallurgy (PM) is the process of converting metallic powders into structures using heat and sometimes compaction or pressure. Powders, like those produced from previously described techniques, can be elemental, premixed, or pre-alloyed. Initially, the powders may be cold compacted in a die, punch, or mold can with or without mandrels. Then the compacts are heat and pressure treated, depending on the quality and strength desired of the structure and the optimal temperature and pressure conditions for the chosen powders. Irrespective of the process cycle, the strength of the final material is determined by final density and the extent of porosity present in the structure [19].

One PM method used to produce SRM is hot isostatic pressing (HIP), which uses the simultaneous application of heat and isostatic pressure (gas pressures) to achieve low porosity structures. The shape of the structure is held by a mold or HIP can, as shown in Fig. 21. The HIP can filled with powder is placed in a vessel heated and pressurized with gas pressure. Powder volume reduction is measured throughout the process, and the HIP process is stopped when the volume appears to be minimized. Zahrah et al. and others have successfully used this technique to produce structural reactive components [38, 39].

Another PM method used to produce SRM is sintering, which uses pressure to compact the powders and then heat to cure the structure. Ren et al. prepared tungsten-zirconium SRM using this method [40]. Three compositions of W-Zr were successfully sintered into pellets with final densities ranging from 8.01 to 9.15 g/cm^3 and percent theoretical maximum densities (TMD) of 87.5–99.2%. Quasi-static tests indicated brittle materials with 1022–1880 MPa comprehensive strength and linear stress–strain curves. Dynamic tests with strain rates from 200 to 1200 s^{-1} were performed using a modified split Hopkinson pressure bar system with bars made of maraging steel. The elastic wave velocity of the sample was about 4800 m/s,

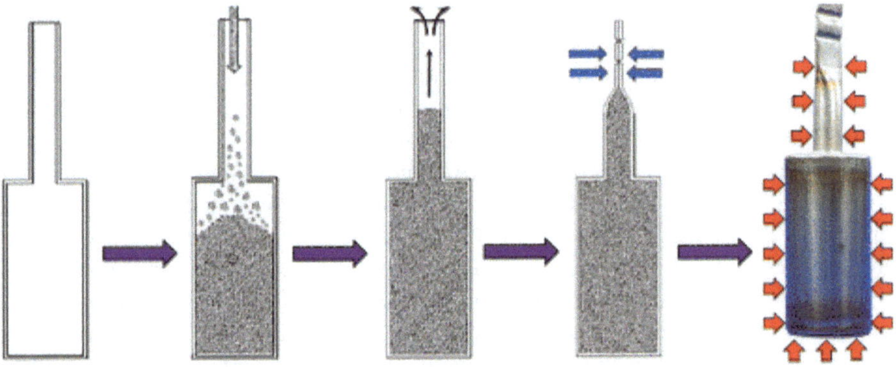

Fig. 21 Hot-isostatic pressing (HIP) process [37]

Fig. 22 Difference in reaction of two different batches of W-Zr SRM subjected to strong impact. The brighter reaction is from the batch sintered to 99.2% TMD [40]

and the samples are reported to be in nearly dynamic stress equilibrium over the test duration. When subjected to strong loads, the samples failed and reacted with varying energy depending on the %TMD, as shown in Fig. 22 [40]. These materials have potential as high-density reactive materials due to their mechanical and energetic properties.

Munitions applications of SRM produced by PM are being evaluated in defense labs. Considering these are intermetallic materials, the intermetallic reactions followed by combustion reactions in aerobic environments can enhance the energy obtained from traditional explosives. Therefore, one popular application of SRM in munition is to enhance blast. For such applications, the SRM must react during the detonation or soon after, producing its energy even as the detonation energy from the explosive is dying off, as shown in Fig. 23. Aluminum is one metal proved to react with this criteria, but since the strength properties of Al makes it weaker than steel, other metals or alloys that react with similar burn rates and ignite at temperatures within the detonation fireball, with densities and strength closer to that of steel, must be investigated.

A second munition application of SRM produced by PM is their use as fragments. In some munitions, the weapon is designed to maximize the production of fragments, rather than blast. The fragments from such weapons puncture targets,

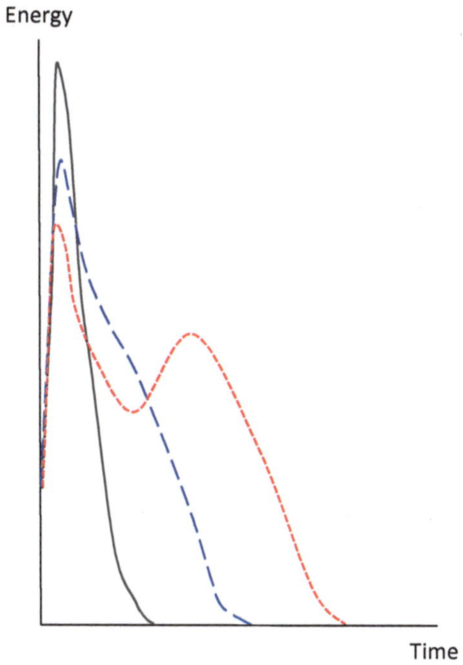

Fig. 23 Structural reactive materials (SRM) in enhanced blast applications. The solid line represents the detonation of a high explosive, the dashed line an aluminized explosive, and the dotted line an enhanced blast, where a second set of reactions from SRM continue to add energy to the explosive event

damaging critical components to impede the target or destroy its function. For this application, the SRM must break into chunks or fragments during the weapons detonation, reach the target with sufficient mass to puncture or pierce the target and then react or burn. The metals included in such SRM must survive the detonation and react to produce energy only on impact with the target. Finding such reactive metals that are insensitive to ignition from the detonation and have strength properties to produce SRM with strength closer to that of steel remains a research challenge.

A third application for SRM is their use in penetration-based ordnance. These could range from small-caliber projectiles like bullets, to large-caliber projectiles that are gun launched, to very large weapons like air-dropped penetrators. The SRM used in such applications must be sufficiently insensitive to survive launch conditions and the trajectory conditions such as the acceleration and high Gs experienced by the projectile or penetrator. Similar to fragments, these SRM would react on impact, or during penetration, or after penetration of the target, and, considering the extreme shock conditions these reactive materials need to survive, these too remain a research challenge.

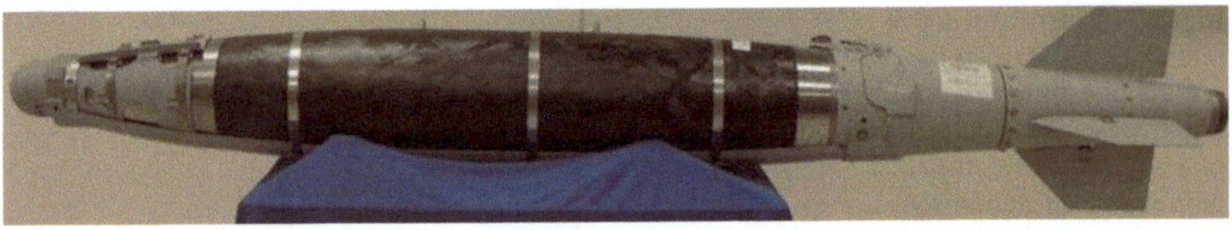

Fig. 24 Fiber-wrapped warhead case that disintegrates with minimal fragmentation during warhead detonation, thereby minimizing collateral damage [41]

4.2 Fiber Winding

Another method for making SRM is using fiber-reinforced polymer matrix composites manufacturing techniques herein incorporating reactive materials. Manufacturing fiber composites can be carried out in a number of ways, depending on the desired final shape. Fiber winding is particularly useful in producing cylindrical shapes, as shown in Fig. 24 [41]. These cylindrical structures are prepared by winding the fibers around a mandrel and impregnating with a low viscosity polymer binder.

While the main goal of fiber winding is to combine the fibers and binders to provide structural strength, the weave could also be embedded or functionalized with reactive metal powders. These SRM could be utilized in munitions applications where fragmenting of metal is not desired, while adding energy or enhancing blast is desired. Similar to the enhanced blast applications detailed above, such SRM must ignite and react with the explosive shock wave or soon after.

4.3 Other Methods

There are other methods for making SRM, such as bulk metallic glass production methods, other melt casting methods, or sintering methods. Producing large pieces of SRM with sufficient strength to be a structural component of a weapon continues to be a challenge. This is mainly due to the inherent properties of reactive materials, which ignite and react before melting or soon after melting. Most processes that produce strong metal parts like steel require very high temperature processing close to the metals melting point or higher temperatures. Perhaps, future spray methods and highly localized laser-melt methods as seen in the burgeoning 3D printing industry or additive manufacturing industry could prove useful. For structural reactive materials, the criteria for application in munitions remains the reliable reproduction of components with high strength, with reaction ignition only under the shock conditions suitable for that application.

5 Conclusion

This chapter described research on reactive materials for conventional weapon applications. One application is the use of particulate reactive materials, where recent research has focused on tuning the burn rate or reaction rate as needed for various applications. For agent-defeat applications, lower ignition and longer burn times are more desirable. For propellant and explosive formulation applications, a very high burn rate or reaction rate is desired. Since reactions takes place at the interfaces of two ingredients, successful approaches attempt to increase interfaces by mixing ingredients at the nano-scale. As reviewed in this chapter, nano-scale mixing has been achieved by processes such as lamination, electro-spraying, and ball milling. Of these methods, while lamination produces excellent quality and burn rates, increasing production throughput and scaling up could prove challenging. Similarly, electro-spraying being a layer-by-layer deposition scheme could prove costly to scale up. Therefore, ball milling, might succeed as the most utilitarian method, being easy to scale up to produce kilogram quantities. Further research on the ball-milling parameters needed to decrease ignition temperature and tune the burn rate could result in particulate reactive materials that are incorporated into future munitions.

Another application of reactive materials is in structural parts or weapon components, such as the weapon case. For this application, the density and strength of the produced reactive material structure is important, and methods such as hot isostatic pressing and fiber reinforcing methods are a good beginning. Again, the scalability of the method, so that large pieces or components can be made without reduction in material quality and material strength, will prove a challenge to overcome.

References

1. Woody DL, Davis JJ, Miller P (1996) Metal/metal exothermic reactions induced by low velocity impact. In: MRS Proceedings, vol 418. Materials Research Society, Boston, pp 445–449
2. Johnson CE, Higa KT (1999) USA patent no. 5,885,321
3. Dreizin EL (2009) Metal-based reactive nanomaterials. Prog Energy Combust Sci 35:141–167
4. Beckstead M (2005) Correlating Aluminum burning times. Combust Explos Shock Waves 41:533–546
5. Trunov MA, Schenitz M, Dreizin EI (2005) Ignition of aluminum powders under different experimental conditions. Propellants Explos Pyrotech 30:36–43
6. Armstrong R, Thadhani NN, Wilson W, Gilman J (2004) Synthesis, characterization and properties of energetic/reactive nanomaterials. In: MRS Proceedings, vol 800. Materials Research Society, Boston, pp 3–392
7. Gash AE, Satcher JH, Simpson RL (2003) Strong akaganeite aerogel monoliths using epoxides: synthesis and characterization. Chem Mater 15:3268–3275
8. Jouet R, Warren AD, Rosenberg DM, Bellito VJ, Park K, Zachariah MR (2005) Surface passivation of bare aluminum nanoparticles using perfluoroalkyl carboxylic acids. Chem Mater 17:2987–2996

9. Stover AK, Krywopusk NM, Gibbins JD, Weihs TP (2014) Mechanical fabrication of reactive metal laminate powders. J Mater Sci 49:5821–5830
10. Abdulstaar M, El-Danaf E, Waluyo N, Wagner L (2013) Severe plastic deformation of commercial purity aluminum by rotary swaging: microstructure evolution and mechanical properties. Mater Sci Eng A 565:351–358
11. Gibbins JD, Stover AK, Krywopusk NM, Woll K, Weihs TP (2015) Properties of reactive Al:Ni compacts fabricated by radial forging of elemental and alloy powders. Combust Flame 162:4408–4416
12. Woll K, Gibbins JS, Kinsey A, Weihs T (2016) The utilization of metal/metal oxide core-shell powders to enhance the reactivity of diluted thermite mixtures. Combust Flame 167:259–267
13. Li X, Guerieri P, Zhou W, Huang C, Zachariah MR (2015) Direct deposit laminate nanocomposites with enhanced propellent properties. Appl Mater Interfaces 7:9103–9109
14. Guerieri P, DeCarlo S, Eichhorn B, Connell T, Yetter R, Tang X, Bowen KZ (2015) Molecular aluminum additive for burn enhancement of hydrocarbon fuels. J Phys Chem 119:11084–11093
15. Wang H, Jian G, Yan S, DeLisio J, Huang C, Zachariah M (2013) Electrospray formation of gelled nano-aluminum microspheres with superior reactivity. ACS Appl Mater Interfaces 5:6797–6801
16. Jacob R, Wei B, Zachariah M (2016) Quantifying the enhanced combustion characteristics of electrospray assembled aluminum mesoparticles. Combust Flame 167:472–480
17. Abraham A, Zhong Z, Liu R, Grinshpun S, Yermakov M, Indugula R, Schoenitz DE (2016) Preparation, ignition, and combustion of Mg-S reactive nanocomposites. Combust Sci Technol 188:1345–1364
18. Merupo V, Velumani S, Ordon K, Errien N, Szade J, Kassiba A (2015) Structural and optical characterization of ball-milled copper-doped bismuth vanadium oxide (BiVO4). CrystEngComm 17:3366–3375
19. German RM (1998) Powder metallurgy of iron and steel. Wiley, New York
20. Abraham A, Schoenitz M, Dreizin EL (2016) Energy storage materials with oxide-encapsulated inclusions of low-melting metal. Acta Mater 107:254–260
21. Aly Y, Dreizin E (2015) Ignition and combustion of AlMg alloy powders prepared by different techniques. Combust Flame 162:1440–1447
22. Abraham A, Obamedo J, Schoenitz M, Dreizin E (2015) Effect of composition on properties of reactive AlBI2 powders prepared by mechanical milling. J Phys Chem Solids 83:1–7
23. Wang S, Abraham A, Zhong Z, Schoenitz M, Dreizin E (2016) Ignition and combustion of boron-based AlBI2 and MgBI2 composites. Chem Eng J 293:112–117
24. Specht PE, Thadhani NN, Weihs TP (2012) Configurational effects on shock wave propagation in Ni-Al multilayer composites. J Appl Phys 111:073527–073521
25. Sraj I, Specht PE, Thadhani NN, Weihs TP, Knio OM (2014) Numerical simulation of shock initiation of Ni/Al multilayer composites. J Appl Phys 115:023515
26. Specht EP, Weihs PT, Thadhani NN (2016) Interfacial effects on the dispersion and dissipation of shock waves in Ni/Al multilayer composites. J Dyn Behav Mater 2:500–510
27. Austin AR, McDowell LD, Benson JD (2014) The deformation and mixing of several Ni/Al powders under shock wave loading: effects of initial configuration. Model Simul Mater Sci Eng 22:025018
28. Overdeep KR, Livi KJ, Allen DJ, Glumac NG, Weihs TP (2015) Using magnesium to maximize heat generated by reactive Al/Zr nanolaminates. Combust Flame 162:2855–2864
29. Goroshin S, Frost DL, Ripley R, Zhang F (2016) Measurement of particle density during explosive particle dispersal. Propellants Explos Pyrotech 41:245–253
30. Isert S, Lane CD, Gunduz IE, Son FS (2016) Tailoring burn rate using reactive wires in composite solid rocket propellants. Proc Combust Inst 6:141–149
31. Connell TL, Risha GA, Yetter RA (2015) Boron and polytetrafluoroethylene as a fuel composition in hybrid rocket applications. J Propuls Power 31:373–385

32. Wu T, Li X, Hu X, Delisio JB, Zhou W, Zachariah MR (2015) Direct-deposition to create high particle loading propellants with controlled architecture: combustion and mechanical properties. Am Inst Aeronaut Astronaut 688:1–9
33. Brousseau P, Anderson CJ (2002) Nanometric aluminum in explosives. Propellants Explos Pyrotech 27:300–306
34. Sullivan KT, Piekiel NW, Wu C, Chowdhury S, Kelly ST, Hufnagel TC, Zachariah MR (2011) Reactive sintering: an important component in the combustion of nanocomposite. Combust Flame 159:2–15
35. Clement D, Diener J, Gross E, Kunzmer N, Timosh VY (2005) Highly explosive nanosilicon-based composite materials. Phys Status Solidi 202:1357–1364
36. Crouse CA, Pierce CJ, Spowart JE (2012) Synthesis and reactivity of aluminized fluorinated acrylic (AlFA) nanocomposites. Combust Flame 159:3199–3207
37. Nguyen CV (2015) HIP simulation. http://www.iwm.rwthaachen.de/indes.php?id=549. Accessed Dec 2016
38. Zahrah T, Kecskes L, Rowland R (2008) Smart processing of tungsten-bulk metallic glass composites. In: International conference on tungsten, refractory & hard materials VII, Washington, DC
39. Biancaniello F, Zahrah T, Jiggetts R, Kecskes L, Rowland R, Maters S, Ridder S (2003) Structure and properties of consolidated amorphous metal powder. In: Powder materials: current research and industrial practices III. Wiley, Hoboken, pp 265–272
40. Ren J, Liu XN (2016) Microstructure and mechanical properties of W-Zr reactive materials. Mater Sci Eng A 660:205–212
41. Bloomfield S (2012) Precision lethality responds to urgent operational need. http://www.wpafb.af.mil/News/Article-Display/Article/399665/precision-lethality-responds-to-urgent-operational-need. Accessed Dec 2016

Part III
Emerging Areas

X-Ray Phase Contrast Imaging of Granular Systems

B. J. Jensen, D. S. Montgomery, A. J. Iverson, C. A. Carlson, B. Clements, M. Short, and D. A. Fredenburg

1 Introduction

Dynamic compression experiments have proven useful for decades in examining material response at high pressures and providing equation-of-state and other information on numerous phenomena including phase transitions [1–3], strength [4], and kinetics [5]. Although there has been much success in relating the shock-wave profile to the material response, these methods provide indirect information about the microscopic level mechanisms responsible and have difficulty when the processes and/or materials are heterogeneous. More advanced diagnostics are required to study complex materials such as foams, powders, explosives, as well as more complicated processes such as hot-spot formation in explosives and jet formation in metals. Over the years, significant effort has focused on the development of experimental methods that couple shock wave or dynamic compression platforms to more advanced diagnostics including X-ray imaging and diffraction [6–12] and proton radiography [10]. More recently, efforts have focused on the use of X-rays from third and fourth generation light sources (synchrotrons, X-ray free-electron lasers) which are ideal for dynamic experiments since they have extremely high spatial coherence (collimated, low divergence), are bright ($>10^{10}$ photons/pulse), and are tunable [9, 12–15].

Developments in synchrotron facilities and diagnostics (e.g., X-ray diffraction and phase contrast imaging or PCI) are providing unique opportunities for high-resolution, spatially resolved, in situ measurements of materials subjected to

B. J. Jensen (✉) · D. S. Montgomery · B. Clements · M. Short · D. A. Fredenburg
Los Alamos National Laboratory, Los Alamos, NM, USA
e-mail: bjjensen@lanl.gov

A. J. Iverson · C. A. Carlson
Mission Support & Test Services, LLC, Los Alamos, NM, USA

T. J. Vogler, D. A. Fredenburg (eds.), *Shock Phenomena in Granular and Porous Materials*, Shock Wave and High Pressure Phenomena,
https://doi.org/10.1007/978-3-030-23002-9_7

dynamic loading conditions [16–20]. One of the first capabilities to use dynamic phase contrast imaging at a synchrotron on the ns-μs timescale with μm spatial resolution was developed by Los Alamos researchers and collaborators at the Advanced Photon Source (Argonne, IL) using the IMPULSE (IMPact System for ULtrafast Synchrotron Experiments) single-stage gas-gun with multiple gated frames [13–15, 21, 22]. These experiments demonstrated the usefulness of multi-frame X-ray PCI (MPCI) for studying a wide variety of phenomena including crack propagation in materials [23, 24], jet formation in metals [25, 26], dynamic extrusion tests on polymers [27], response of additively manufactured (AM) materials to shock loading [28], dynamic response of detonator initiator systems [29], and spall and damage.

One area in which phase contrast imaging (PCI) techniques have the potential to offer significant near-term advances is in the theoretical development of compaction models for initially distended materials, specifically, those under dynamic loading conditions. Historically, high-strain-rate compaction modeling at the bulk, or continuum, level was performed using phenomenological frameworks, for example, P-α [30, 31] and P-λ [32] models. While these models offer a convenient means of capturing the high-strain-rate densification behavior for a given material, their largely empirical nature requires model calibration for each distinct initial condition, e.g., initial density. PCI offers a unique means of characterizing the deformation and/or failure properties of granular materials under intense dynamic loading, such that a paradigm shift in the way granular materials are understood and modeled can now be realized. Specifically, PCI can be used to validate models at the grain-scale that capture localized mechanical phenomena, which can in turn be used to develop physics-based bulk level compaction models.

Studying the mechanical properties of idealized granular systems in the form of spherical particles has occurred for many years; however, previous diagnostics have been limited. Low-strain-rate indentation tests [33], as well as higher strain-rate impact tests [34, 35], on single and multi-particle systems have often been limited to optical techniques, thereby imposing practical lower limits on particle size as well as the size of any observable features. Directly imaging the deformation and fracture response of spheres using the dynamic impact capabilities at the APS coupled with PCI over the range of available beam energies at APS allows features at the 1–3 μm level to be easily observed in real-time. As shown in Fig. 1, X-ray PCI has been used to examine the impact response of an idealized linear system of spheres to understand the initial deformation and subsequent failure of the spheres subjected to dynamic loading. By studying variations in particle size (mm to μm) and/or increasing the complexity of the configuration (pyramidal, sheets, and random) of the distribution of particles toward more realistic powder or granular samples. PCI can be used to investigate a wide array of relevant compaction-scale phenomena.

This chapter is organized in the following way. First, in Sect. 2 the basic concept and theoretical framework for X-ray phase contrast imaging are discussed including the optimal target-to-detector distances to optimize contrast and a discussion on density retrieval for single composition materials. This is followed by details of the multi-frame X-ray PCI (MPCI) system and its operation in Sect. 3 including

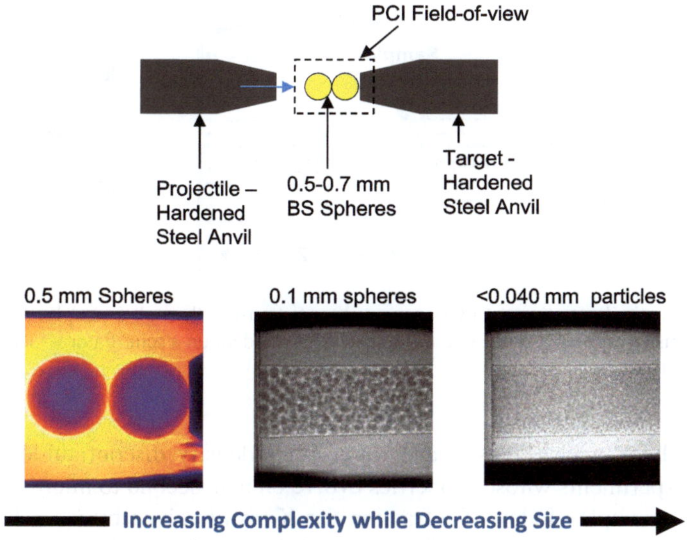

Fig. 1 Schematic of an impact experiment using a linear stack of ideal borosilicate spheres. In this work, the particle size is systematically reduced to approach the compaction of more realistic systems including sand and powder materials

recent improvements that include dual-imaging of the scintillator to obtain greater light efficiency and allow for multiple, simultaneous magnifications. In Sect. 3.2, experimental results and analysis of images obtained using the MPCI system to examine the compression of an idealized system of 0.6-mm borosilicate spheres are shown. The data were compared to theoretical simulations from the Lagrangian finite element code ABAQUS [36] and a damage model for brittle materials [37]. This chapter is concluded with a discussion and summary.

2 X-Ray Phase Contrast Imaging

For over a century, X-ray imaging techniques have relied on absorption, as the X-rays propagate through an object, to produce contrast in the resultant image. For traditional absorption-based imaging methods, the spatial coherence of the X-ray source plays little role in the image formation process and can be understood using simple geometric optics. However, spatially coherent sources, such as from a micro-focus X-ray tube, synchrotron radiation, or X-ray lasers, have enabled phase-based imaging methods over the past two decades, and require a more sophisticated theoretical treatment to understand the image formation process. Phase sensitive X-ray techniques rely on overlap and interference of the wave fields as they propagate due to spatial gradients in the transmitted phase [38, 39]. As first suggested by Montgomery et al. [40], X-ray phase contrast imaging can be used

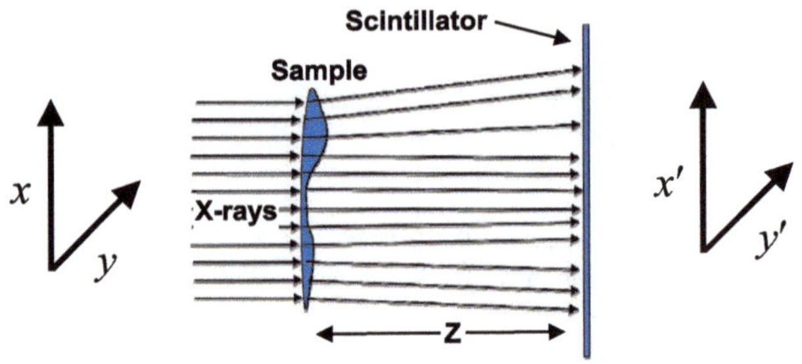

Fig. 2 Schematic illustrating the PCI concept. X-rays transmit through a sample, diffract according to the density (or index of refraction), and are then incident on a scintillator which converts the X-rays into visible light

to image shock waves, blast waves, and other density discontinuities arising in dynamic experiments whose properties evolve on nanosecond to microsecond timescales. The complex index of refraction for X-rays is given by $n = 1 - \delta + i\beta$, where the δ term accounts for refraction, and β accounts for absorption. The phase shifts imposed on the X-ray wave front transmitted through an object are given by $\phi(x, y) = (2\pi/\lambda) \int \delta(x, y, z)dz$, where λ is the X-ray wavelength, and z is the propagation direction. These spatial variations in the phase produce local curvature in the transmitted wave front, causing overlap and interference of the waves as they propagate, modulating the intensity. Thus, even for a purely transparent object (β = 0), phase gradients modulate the intensity after propagating some finite distance, and can be used to image an otherwise transparent object. A major requirement for realizing this imaging technique is for the X-ray source to have a high degree of spatial coherence, thus a sufficiently distant point source or highly collimated beam of X-rays is needed. X-ray phase contrast imaging is described in detail in [39].

It is instructive to examine a simple derivation in the geometric-optics limit to understand the image formation process for X-ray phase contrast imaging. Following Fig. 2, consider a uniform, collimated beam of X-rays propagating in the z direction, illuminating a transparent object at the plane $z = 0$. Immediately past the object, the rays deviate by some small angle due to refraction by the object. If the ray deviates by an angle θ at $z = 0$, then its transverse coordinates at (x, y) are shifted to the transverse coordinates (x', y') at a plane located some distance z, and the positions can be simply written as

$$x' = x + z\tan\theta \tag{1}$$

or since the ray deviation is related to the gradient of the phase front

$$x' = x - \frac{\lambda z}{2\pi} \frac{\partial \phi}{\partial x} \tag{2}$$

$$y' = y - \frac{\lambda z}{2\pi} \frac{\partial \phi}{\partial y} \tag{3}$$

which maps x to x', y to y'. An area element $(\partial x \partial y)$ now has area $(\partial x' \partial y')$ at the plane z. Assuming uniform intensity at the object plane $(x, y, z = 0)$, the intensity at the plane z will be proportional to the ratio of the area elements $\left(\frac{\partial x'}{\partial x}\frac{\partial y'}{\partial y}\right)^{-1}$, whose elements are written as

$$\frac{\partial x'}{\partial x} = 1 - \frac{\lambda z}{2\pi}\frac{\partial^2 \phi}{\partial x^2} \tag{4}$$

$$\frac{\partial y'}{\partial y} = 1 - \frac{\lambda z}{2\pi}\frac{\partial^2 \phi}{\partial y^2} \tag{5}$$

which is the Jacobian of the transformation. The intensity at some plane $z = z$ is given by

$$I(x', y') = I_0\left(1 - \frac{\lambda z}{2\pi}\nabla_\perp^2 \phi\right)^{-1} \tag{6}$$

valid in the geometric-optics approximation only when $\frac{\lambda z}{2\pi}\nabla_\perp^2 \phi \neq 1$, otherwise a caustic is formed at the singularity. In the limit of small argument $\frac{\lambda z}{2\pi}\nabla_\perp^2 \phi << 1$, then

$$I(x', y') \approx I_0\left(1 + \frac{\lambda z}{2\pi}\nabla_\perp^2 \phi\right) \tag{7}$$

as also derived in Pogany et al. using a first-order simplification of diffraction theory [39]. In these equations, the Laplacian of the phase $\nabla_\perp^2 \phi$ is the second partial derivative with respect to (x, y) only, i.e., only the components perpendicular to the propagation direction z. From Eq. (7), it can be easily seen that the intensity is unperturbed at $z = 0$, and the intensity perturbation due to phase curvature (non-zero second derivative) increases with increasing z. To first order, X-ray phase contrast imaging can be viewed as an object consisting of small "lenslets" that either slightly focus or defocus the X-rays by refraction after the rays propagate a finite distance. While this geometric optics simplification offers insight, it is often necessary to consider the full diffraction theory when applying the formalism to experimental data as discussed next.

Using scalar diffraction theory, consider now a monochromatic plane wave $E_0 e^{-ikz}$, where $k = 2\pi/\lambda$, propagating along the z-axis through a "thin object" such that the small angle refraction within the object can be ignored. Then the phase of the transmitted electric field just past the object (defined to be the plane $z = 0$) is simply given by $\phi(x, y) = 2\pi/\lambda \int \delta(x, y, z)dz$. Similarly, the absorption coefficient at the plane $z = 0$ is given by $\mu(x, y) = 4\pi/\lambda \int \beta(x, y, z)dz$, so that the transmitted complex electric field at the plane $z = 0$ may be written as $E(x_0, y_0) = \left(E_0 e^{-ikz}\right) \exp[i\phi(x_0, y_0) - \mu(x_0, y_0)/2]$. The transmitted wave then propagates in free space for $z > 0$ according to the Helmholtz wave equation, whose solutions in the paraxial limit (E varies slowly with z) yield the scalar diffraction integral [39]

$$E(x, y, z) = \int\int_{-\infty}^{\infty} E(x_0, y_0)h(x, y)dx_0dy_0 \tag{8}$$

where

$$h(x, y) = \frac{1}{i\lambda z}\exp\left[-\frac{i\pi}{\lambda z}(x^2 + y^2)\right] \tag{9}$$

As a physical picture, the resulting electric field at some plane $z > 0$ is simply a convolution of the field $E(x_0, y_0)$ at $z = 0$ with an optical transfer function $h(x, y)$, which is the paraxial approximation of a spherical wave in the limit $z \gg x, y$.

Since the scalar diffraction integral represents a convolution, Fourier analysis can be used to solve Eq. (8). Let $H(u, v)$ represent the 2D Fourier transform of $h(x, y)$, so that $H(u, v) = \exp[i\pi\lambda z(u^2 + v^2)]$, where (u, v) are the spatial transform coordinates of (x, y). To simplify for illustrative purposes and following Ref. [39], consider the 1D case, $H(u) = \exp[i\pi\lambda zu^2]$, and assume $\chi = \pi\lambda zu^2$. This can be expanded as $H(i\chi) = \cos\chi + i\sin\chi$, where the real part (cosine term) can be viewed as responsible for absorption contrast, i.e., when $z = 0, \chi = 0$ the real part is maximized. Alternatively, the imaginary part (sine term) can be viewed as responsible for phase contrast, and reaches its first maximum when $\chi = \pi/2$ or when $\lambda zu^2 = 1/2$. The function $H(u, v)$ can be viewed as an optical transfer function that can optimize image contrast by choosing the appropriate propagation distance z given the X-ray wavelength λ and some limiting feature size a. Imagine that the highest spatial frequency u is set by the smallest feature size a that can be resolved, $u = \sqrt{2}/a$. For the case of plane wave illumination, the propagation distance that optimizes phase contrast for this feature size is $z_{opt} = a^2/\lambda$. For example, assuming X-ray wavelength $\lambda = 0.5$ Å and feature size $a = 5\,\mu$m, the optimum propagation distance (distance from object to detector) is $z_{opt} \sim 0.5$ m. While this simple estimate is useful, other factors such as absorption, spatial coherence (beam divergence), diffraction broadening, detector resolution, pixel size, and noise are considered when choosing a propagation distance that optimizes both image contrast and feature resolution, and will be discussed later.

Equation (8) can be generalized for the case of a point source [39], with source to object distance R_1, and object to detector distance R_2, such that the geometric magnification can be written as $M = (R_1 + R_2)/R_1$, and the effective propagation distance becomes $z_{eff} = R_2/M$. The coordinates at the detector plane are $(x', y') = (Mx, My)$. The case for plane wave propagation is recovered when $R_1 \gg R_2$ such that $M = 1$.

It is useful to define the dimensionless Fresnel number $N_F = a^2/\lambda z$ to distinguish between various regimes of diffraction. In the case of large Fresnel number $N_F \gg 1$, this regime is accessible for large feature sizes a, short wavelength λ, and small propagation distances z. Here, diffraction plays a small role in image formation, and this regime is also known as the near-field regime or the geometric-

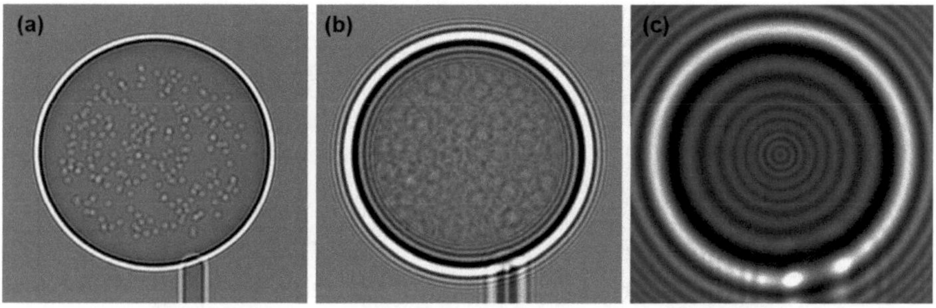

Fig. 3 Simulations of image formation from a polystyrene sphere illuminated by 20-keV X-rays from a simulated synchrotron source. (**a**) Near-field geometric-optics limit with $z = 0.15$ m. (**b**) Intermediate-field or Fresnel diffraction limit with $z = 0.4$ m. (**c**) Far field or Fraunhofer diffraction limit with $z = 4.0$ m. The object shape and areal density can be retrieved in all imaging regimes

optics limit, where absorptive and refractive effects dominate the image formation. In the intermediate case of $N_F \sim 1$, diffraction effects become large, but many of the image features observable in the near-field regime are still visible and also include diffraction around sharp edges and features. This is known as the Fresnel regime or intermediate-field regime. Finally, in the limit $N_F \ll 1$, accessible for small feature sizes a, long wavelength λ, and large propagation distances z, the image is dominated by diffraction, and any features discernible in the near-field regime are nearly unrecognizable due to multiple overlapping diffraction patterns.

These three regimes are illustrated in Fig. 3 by simulating the image formation from a polystyrene sphere illuminated by 20-keV X-rays from a simulated synchrotron source. The simulated sphere, comprised of C_8H_8 at $\rho_0 = 1.05$ g/cm^3, is 120 μm in diameter, and is affixed to a 12-μm diameter SiO_2 stalk with density 2.2 g/cm^3. Up to 200 spherical voids with 2 μm diameter are randomly distributed within the C_8H_8 sphere. Propagation distances of 0.15 m, 0.4 m, and 4.0 m are simulated using algorithms described in Ref. [40–42] to illustrate regimes where $N_F \gg 1$, $N_F \sim 1$, and $N_F \ll 1$, and the pixel size is simulated as 0.8 μm with a 1-μm resolution blur included. In the near-field regime, refraction-enhanced edges of the sphere, stalk, and voids are visible, but no diffraction is observed. In the intermediate-field regime, the image is very similar to the near-field image, except diffraction is observed around all large features, and diffraction broadening is seen for the small voids. The far-field image is dominated by diffraction.

An illustrative example for X-ray phase contrast is given by Cloetens et al. [43], and the images for increasing propagation distances are shown in Fig. 4. In this work, 18-keV X-rays from a synchrotron ($\lambda = 0.7$ Å) were produced using a double crystal monochromator and were incident on a 1-mm thick polystyrene foam sample. Images from a detector placed various distances from the object were recorded. At the closest distance (1 cm), there is practically no image contrast since the object is mostly transparent at 18 keV. As the detector is placed further away from the sample, up to 91 cm, image contrast is observed to increase due to phase contrast, and the small-scale features become wider due to diffraction broadening.

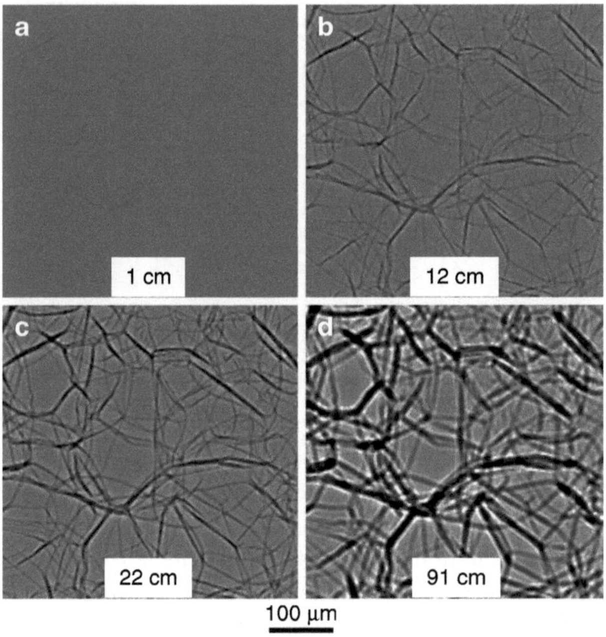

Fig. 4 Example X-ray PCI images of a polystyrene foam for four detector distances [43]

2.1 *Optimizing Phase Contrast Imaging for Synchrotron Experiments*

The simple result for the optimum propagation distance between object and detector derived in the previous section, $z_{opt} \approx a^2/\lambda$, is a rough guide and useful for illustrative purposes, but it ignores other factors such as absorption, spatial coherence (beam divergence), diffraction broadening, detector resolution, pixel size, and noise. To optimize both phase contrast and resolution, simulations of synthetic objects were employed using scalar diffraction theory and including the effects of beam divergence, detector pixel size and resolution, synchrotron source spectrum, and detector spectral response, as well as considerations of noise. We further discuss and derive simple estimates for optimum imaging conditions at the end of Sect. 2.1.

Fast-Fourier-transform (FFT) based methods were used to solve the scalar diffraction integral, Eq. (8), for a range of X-ray wavelengths by binning the source spectrum into 10–20 spectral bins weighted by integrated flux and detector spectral response [40–42]. The angular spectrum method is used for most cases where the propagation distance is less than some critical distance $z_{crit} < (L_x \Delta x)/\lambda$, where L_x is the image width and Δx is the pixel size [42]. For propagation distances larger than the critical distance z_{crit}, the optical transfer function used in the angular spectrum method, $H(u, v) = \exp[i\pi\lambda z(u^2 + v^2)]$, is insufficiently sampled for large z and becomes aliased, and a direct method is used to solve the diffraction integral [42].

Synthetic objects are simulated by constructing a thickness map $T(x, y)$ consisting of simple geometric objects (spherical solids, spherical voids, cylindrical rods, cones), or smoothly varying analytic functions such as Gaussian bumps and error functions. A material and its density is chosen (e.g., polystyrene, vitreous carbon, SiO_2, magnesium) and maps of the phase $\phi(x, y)$ and absorption $\mu(x, y)$ are computed using the thickness map $T(x, y)$ and the complex index of refraction $n(\lambda) = 1 - \delta + i\beta$ for each wavelength. The input electric field at the object plane $z = 0$ is written as $E(x, y, 0) = e^{i\phi(x,y)-\mu(x,y)/2}$. For the typical case using the angular spectrum method, the electric field at the plane z is computed in Fourier-space by $\hat{E}(u, v, z) = \hat{E}(u, v, 0)H(u, v)$, where $\hat{E}(u, v, 0) = \mathcal{F}\{E(x, y, 0)\}$, $\hat{E}(u, v, z) = \mathcal{F}\{E(x, y, z)\}$, and the transform spatial frequencies have a step size $du = 1/L_x$, $dv = 1/L_y$ (for non-square images $L_x \neq L_y$). The relative intensity at the detector plane z is $I(x, y, z) = |E(x, y, z)|^2$, and is computed for each wavelength bin. The intensity for each wavelength bin is then weighted according to spectral flux within the bin and detector spectral response, and summed over wavelength. The resulting total intensity is normalized so that an unattenuated X-ray beam corresponds to $I(x, y) = 1$.

For undulator radiation source characteristics, XOP [44] was used to model the source spectrum and flux for a given undulator and electron beam parameters, with input measurements of the spot size and divergence. The source to object distance, R_1, and object to detector distance, R_2, are used to model the blur due to beam divergence for a source size σ_x,σ_y by $(\alpha_x, \alpha_y) = (\sigma_x, \sigma_y)R_2/(R_1 + R_2)$. For most experiments performed at APS, $\alpha_x \sim 10\alpha_y$, and the blurring due to beam divergence is modeled as an elliptically shaped Gaussian of width (α_x, α_y). The blurring due to both beam divergence and detector resolution is treated as Gaussian point-spread-functions and convolved with the normalized total intensity. For many simulated objects, it is necessary to compute the diffraction integral with a pixel size smaller than the pixel size of the detector to avoid aliasing of the input electric field. For those cases, the resulting totaling intensity, after accounting for blurring, is resampled to the proper detector pixel size. This simulation method is accurate, and has been compared to measurements of laser-fusion capsules [41], as well as various simple test objects measured at the APS.

For a given set of synthetic objects, a series of simulations with increasing propagation distance are performed. A contrast metric is calculated for an object, where contrast is defined as $(I_{max} - I_{min})/(I_{max} + I_{min})$, with I_{max} and I_{min} being the maximum and minimum local intensities associated with a given object or feature. The propagation distance is chosen that provides the largest contrast for a given feature of interest without substantial broadening due to diffraction or blurring from beam divergence. Figure 5 shows a series of calculations for a 10-μm horizontal and vertical void in PMMA for propagation distance R_2 increasing from 200 to 3200 mm. The top row shows the simulated images with no noise added, and the bottom row shows the same images with 1.4% noise added, corresponding to an average of 5000 photons detected per pixel. For the case with $R_2 = 200$ mm,

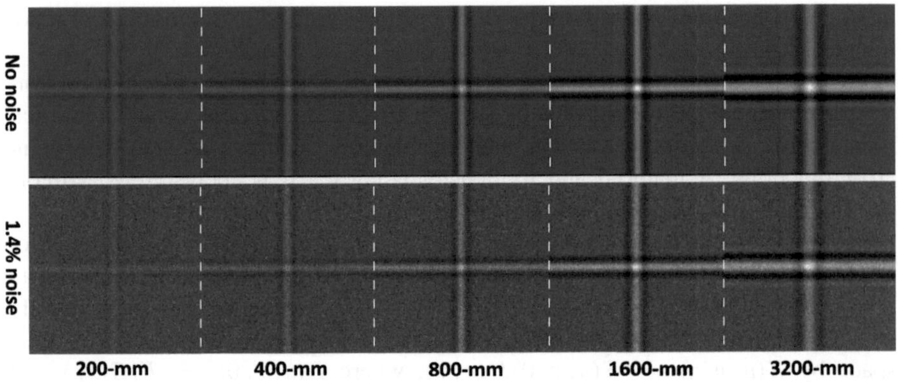

Fig. 5 Series of calculations for 10-μm cylindrical voids (both horizontal and vertical) in PMMA for a range of propagation distances R2. (top) Simulations with no noise added. (bottom) Simulations with 1.4% noise added

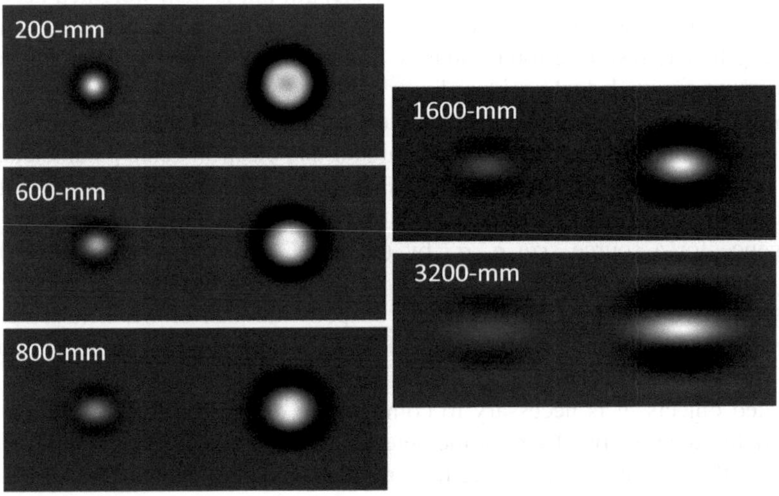

Fig. 6 Calculations for 10-μm and 20-μm spherical voids in a 2-mm thick piece of magnesium. Significant blurring is observed for propagation distances greater than $R_2 = 800\,\text{mm}$

the feature contrast is 1.8%, of order the noise level of 1.4%, and the features are just barely discernable. For $R_2 = 400$ mm, the feature contrast increases to 3.6%, about 2.5× the noise level, and is sufficiently discernable. As a rule of thumb, the limit of detectability for a feature is considered to be when the image contrast is 2–3× the noise level. Figure 6 shows a similar series of calculations for 10-μm and 20-μm spherical voids in a 2-mm thick sample of magnesium. In this case, the horizontal beam divergence is about twice as large as the case for Fig. 5. For propagation distances $R_2 > 800$ mm, substantial blurring in the horizontal direction is observed due to the larger beam divergence.

Optimizing the propagation distance z from object to the detector to get the maximum phase contrast for the smallest discernable features while avoiding resolution degradation due to diffraction and source divergence depends on the X-ray energy E_x (wavelength λ), the spatial resolution of the camera system σ_r, and the X-ray beam divergence α_x, where α_x is the ratio of the source spot size σ_x and the distance from source to object, $\alpha_x = \sigma_x / R_1$ in radians. While no simple analytic theory exists to express this, various parameters can be estimated based on a range of simulations from scalar diffraction. It is found that when $N_F = a^2/\lambda z \approx 2.5$, diffraction broadening of the first bright (or dark) bands surrounding a small feature, of size a, is only 20% broader than those same bands when $N_F \gg 1$. This choice is somewhat arbitrary, but causes a noticeable loss of resolution for features near the Nyquist resolution limit $a = 2\sigma_r$. The optimum distance to propagate from object to detector before the onset of noticeable diffraction broadening can be expressed as

$$z_{opt} \approx \frac{8}{5} \frac{\sigma_r^2}{\lambda} \tag{10}$$

The beam divergence α_x also limits the propagation distance before noticeable degradation of resolution occurs. The spatial blur due solely to beam divergence is $\sigma_{blur} = (\alpha_x R_1 R_2)/(R_1 + R_2)$, where R_1 and R_2 are the source to object and object to detector distances, respectively. Taking a similar 20% decrease in spatial resolution as the metric, and assuming the detector resolution σ_r adds in quadrature with the divergence blur σ_{blur}, then an acceptable divergence blur is limited to be such that $\sqrt{\sigma_r^2 + \sigma_{blur}^2} \approx 1.2\sigma_r$, or $\sigma_{blur} \leq \frac{2}{3}\sigma_r$. Combining and solving for $z_{lim} = R_2$, the limiting propagation distance to avoid blurring due to beam divergence is

$$z_{lim} \leq \frac{2R_1\sigma_r}{3\alpha_x R_1 - 2\sigma_r}, \text{ where } 3\alpha_x R_1 \neq 2\sigma_r \tag{11}$$

Plotted in Fig. 7 is the optimum propagation distance z_{opt} versus X-ray photon energy E_x for camera resolution σ_r of 2.5, 5.0, and 10.0 μm. The X-ray photon energy can be found from $E_x = hc/\lambda$. As an example, for X-ray energy $E_x = 12.5$ keV and camera resolution $\sigma_r = 5$ μm (pixel size Δx_0 is smaller to over-sample σ_r), the optimum propagation distance before noticeable diffraction broadening occurs is ≈400 mm.

Also shown in Fig. 7 is a plot of the limiting propagation distance z_{lim} versus beam divergence α_x for camera resolution σ_r of 2.5, 5.0, and 10.0 μm. For a beam divergence of $\alpha_x = 8$ μrad, typical of the synchrotron source at APS, and a camera resolution $\sigma_r = 5$ μm, the limiting distance is found to be ≈400 mm before blurring due to source divergence is noticeable. For more complex situations involving substantial absorption contrast, or complex features one wishes to resolve and optimize, simulations using scalar diffraction theory, as described earlier in this section, are required.

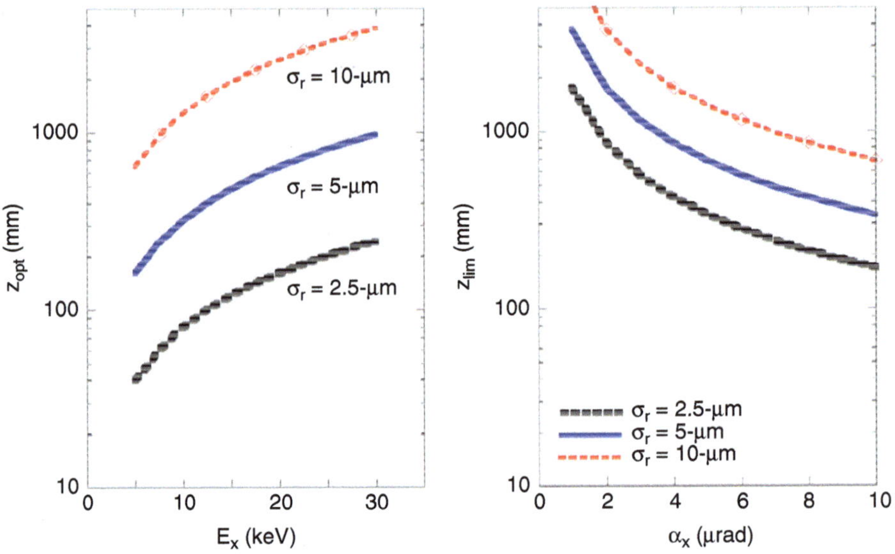

Fig. 7 (Left) Nomograph for optimum propagation distance versus X-ray energy and detector resolution σ_r (right) Nomograph for limiting propagation distance due to X-ray beam divergence α_x and camera resolution σ_r

2.1.1 Phase Contrast Imaging Data Analysis

Various data are required to calibrate the system and analyze dynamic PCI images. Measurements are typically made with a resolution target or grid to determine the pixel size referenced to the object plane, and well as the detector resolution. In addition, a test object, consisting of a transparent sample such as PMMA with a void or defect of known size, is imaged at several different propagation distances to validate the beam divergence and resolution parameters in the simulation model. The object to scintillator distance is measured accurately to determine R_2, and the undulator parameters are recorded. For phase retrieval or areal density retrieval, the following are required:

1. A dark-field or background image to measure the detector background without X-rays.
2. A bright-field or flat-field image to measure the X-ray source without a sample to calibrate 100% transmission levels and any slowly varying X-ray beam shape. This is effectively a measure of $I_0(x, y)$ at the plane $z = 0$ just before the object.
3. A static image with the object in place, which is used to position the object, and visualize object features prior to the dynamic shot, or persisting during frames from a dynamic shot.
4. Dynamic images

Prior to analyzing the dynamic data with a phase retrieval or areal density retrieval algorithm, the normalized intensity is defined as

$$\tilde{I}(x, y) = \frac{I(x, y) - BKG(x, y)}{I_0(x, y) - BKG(x, y)} \tag{12}$$

where $I(x, y)$ is the dynamic image, $I_0(x, y)$ is the bright-field or flat-field image, and $BKG(x, y)$ is the dark-field or background image, and $\tilde{I}(x, y)$ is the normalized intensity such that 100% transmission corresponds to 1.0. One may choose to smooth or fit a low-order polynomial to the denominator and ensure that all values are greater than zero or some limiting value to avoid division by small numbers. It is important that the bright-field image is taken prior to each experiment in case the detector system is moved with respect to the X-ray beam. For the case of the intensity of the dynamic image being of order the intensity of the bright-field image, $\tilde{I}(x, y) \sim 1$, the noise level in the normalized image is about $\sqrt{2}$ times the noise level of a single image, and is important to consider when doing error analysis for the retrieval algorithms.

2.2 Phase Retrieval and Areal Density Retrieval

For a highly transparent object, the assumption of a pure phase object can be made, Eq. (6) can be written as a Poisson equation:

$$\nabla_{\perp}^2 \Phi(x, y) = \frac{2\pi}{\lambda z} \left(\frac{\tilde{I}(x, y) - 1}{\tilde{I}(x, y)} \right) \tag{13}$$

where $\tilde{I}(x, y)$ is the measured normalized intensity from Eq. (12). Recalling that the Fourier transform of a derivative is

$$\mathcal{F}\left[\frac{\partial f(x)}{\partial x}\right] = ik\hat{f}(k) \tag{14}$$

where $\hat{f}(k)$ is the discrete Fourier transform of $f(x)$, the discrete 2D Fourier transform of both sides of Eq. (13) can be taken, which gives

$$-\mathbf{k}_{\perp}^2 \phi(k_{\perp}^2) = \mathcal{F}[RHS] \tag{15}$$

where $\phi(k_{\perp}^2)$ is the discrete Fourier transform of $\phi(x, y)$, $\mathbf{k}_{\perp}$ is a vector of the transform coordinates of (x, y) with components (k_x, k_y), and *RHS* is the right-hand side of Eq. (13) The phase $\phi(x, y)$ is then

$$\phi(x, y) = \mathcal{F}^{-1}\left[-\frac{1}{\mathbf{k}_{\perp}^2}\mathcal{F}\left[\frac{2\pi}{\lambda z}\frac{\tilde{I}(x, y) - 1}{\tilde{I}(x, y)}\right]\right] \tag{16}$$

for $\mathbf{k}_\perp^2 \neq 0$, where $\mathcal{F}\{\}$,$\mathcal{F}^{-1}\{\ \}$ are the discrete Fourier transform (FFT) and inverse transform. The discrete Fourier solution method assumes periodic boundary conditions, which can be enforced by zero-padding the right-hand side of Eq. (13) in a sufficiently large 2D array to satisfy periodicity prior to taking its FFT in Eq. (16). This recovers the phase within some additive constant due to the fact that the DC component is not allowed to be zero in the denominator, i.e., $\mathbf{k}_\perp \neq 0$. This can be ignored if one knows a priori that $\phi(x, y) \approx 0$ in the boundary regions and force that condition to be satisfied after solving Eq. (16). This solution method is only valid for highly transparent objects that are in the near-field (geometric-optics) limit of propagation $N_F \gg 1$, but can be used as an initial guess for higher-order iterative methods that include diffraction, such as Gerchberg–Saxton or Fienup iterative algorithms [45, 46].

For objects composed of a single material, both the attenuation (absorption) and the phase are related to the line-integrated areal density $\rho_0 T(x, y)$ [g/cm^2], where ρ_0 is the initial (static) mass density of the object in [g/cm^3]. In the geometric-optics limit, the areal density can be reconstructed from a single dynamic PCI image by solving the transport of intensity equation [47]. The reconstructed areal density is accurate to first order only since the transport of intensity equation is only a geometric-optics approximation to the scalar diffraction integral Eq. (8).

The transport of intensity (TIE) equation describes changes to the intensity of a scalar wave as it propagates [47]:

$$\nabla_\perp \cdot [I(\mathbf{r}_\perp, z)\nabla_\perp \phi(\mathbf{r}_\perp, z)] = -\frac{2\pi}{\lambda}\frac{\partial}{\partial z} I(\mathbf{r}_\perp, z) \tag{17}$$

where $\mathbf{r}_\perp$ is the position vector perpendicular to the propagation axis z, with components (x, y). The absorption per unit length $\mu = 4\pi\beta/\lambda$ so that the intensity just past the object is attenuated according to Beer's law of absorption, $I_0 e^{-\mu T(\mathbf{r}_\perp)}$, where I_0 is the uniform intensity of the incident X-rays, or is slowly varying in space $I_0(x, y)$. The phase delay can be written as $\phi(r_\perp) = (2\pi/\lambda)\delta T(\mathbf{r}_\perp)$, where $T(\mathbf{r}_\perp) = T(x, y)$ is the thickness map and δ is the refractive decrement, so that the TIE can be rewritten as [48]

$$\left(-\frac{z\delta}{\mu}\nabla_\perp^2 + 1\right) e^{-\mu T(\mathbf{r}_\perp)} = \frac{I(\mathbf{r}_\perp, z)}{I_0} \tag{18}$$

where the derivative $\partial I/\partial z$ is approximated by the measured intensity $I(\mathbf{r}_\perp) = I(x, y)$ at the detector plane situated a distance z from the object, and the incident intensity $I_0(x, y)$ at $z = 0$. As discussed by Paganin and co-authors [48], this form of equation can be solved for the projected thickness using discrete Fourier transforms

$$T(\mathbf{r}_\perp) = -\frac{1}{\mu} ln \left(\mathcal{F}^{-1}\left\{\mu \frac{\mathcal{F}\{I(\mathbf{r}_\perp, z)/I_0\}}{z\delta|\mathbf{k}_\perp|^2 + \mu}\right\}\right) \tag{19}$$

where $k_\perp$ is the vector of spatial frequencies (k_x, k_y) corresponding to transform coordinates of $\mathbf{r}_\perp$ with components (x, y). The areal density is simply $\rho_0 T(x, y)$. This solution is only valid in the geometric-optics limit for propagation $N_F \gg 1$. Further refinement of the areal density solution to include diffractive effects can be obtained by using the result from Eq. (19) as an initial guess in an iterative algorithm using forward and backward Fresnel propagation [49, 50].

3 Dynamic Multi-Frame X-Ray Phase Contrast Imaging

For decades, significant effort has focused on the development of X-ray diagnostics that are compatible with shock wave experiments so that scientists could study the microscopic level response of matter at extremes. Much of the work has focused on the use of X-ray diffraction techniques that utilized flash X-ray technology which provided divergent X-ray beams to study elastic–plastic deformation and phase transitions [6, 7, 51–54]. In contrast, X-rays from third and fourth generation light sources (synchrotrons, X-ray free-electron lasers) are ideal for dynamic phase contrast imaging since they have extremely high spatial coherence (collimated, low divergence), are bright ($>10^{10}$ photons/pulse), and are tunable. Although there are several examples of research coupling gun systems to synchrotrons in the past decade [8, 9], routine measurements have not been possible until recently [3, 13, 15].

One of the first capabilities that was designed specifically for conducting routine plate impact experiments at synchrotron sources to take advantage of diagnostics such as phase contrast imaging (PCI) and Laue X-ray diffraction (XRD) is the IMPULSE system (IMPact System for ULtrafast Synchrotron Experiments) [3, 15] located at the Advanced Photon Source (Argonne, IL). IMPULSE was designed specifically to couple with the synchrotron X-ray beam and consisting of a 12.6-mm bore gas-gun that can accelerate projectiles down a barrel to impact samples at velocities up to 1 km/s. A photo of the IMPULSE capability is shown in Fig. 8. The system is shown on the right where the breech, barrel, and target chamber are located on a support structure that is mobile and remotely operable with four degrees of freedom to allow for precise alignment of the target within the X-ray beam. IMPULSE included a standard suite of traditional diagnostics including PDV, impact pins, and timing/synchronization diagnostics. The success of IMPULSE relied on the ability to perform X-ray imaging experiments during dynamic loading and the development of a robust, optically multiplexed detector system that could capture images from single 80-ps width X-ray bunches from the synchrotron.

3.1 Multi-Frame X-Ray PCI System (MPCI)

The multi-frame X-ray phase contrast imaging (MPCI) system is described in detail in this section. The system, based on a single detector system previously

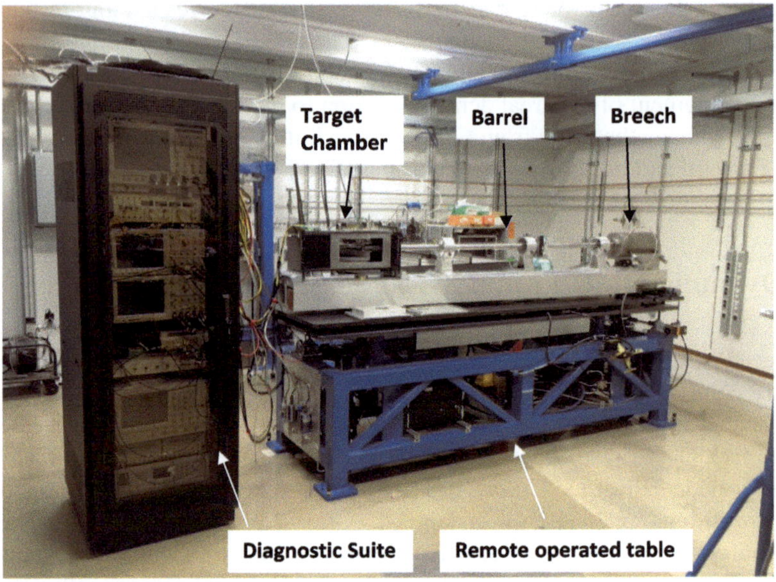

Fig. 8 Photo of the IMPULSE system which includes the gun system located on support structure (blue) along with the diagnostics rack. The gun structure is mobile and remotely operable allowing for four degrees of freedom to position targets within the beam. The diagnostics rack includes PDV, digitizers for impact pins, and the timing system. The multi-frame X-ray PCI (MPCI) system (not visible) was pioneered on this platform at Sector 32 ID-B of the APS where the first X-ray imaging "shock-movies" were obtained for dynamic compression experiments

reported [13, 26], now allows for multiple images per experiment (up to 8 frames), remote optimization for camera and scintillator alignment, and synchronization with the radio frequency (RF) pulse provided by the synchrotron to locate the 80 ps width X-ray bunch. A schematic of a standard experiment configuration is shown in Fig. 9. During the experiment, X-rays are transmitted through the sample and incident on the scintillator, typically LSO or LuAg single crystals, converting them to visible light. A turning mirror was used to direct the visible light into the detection system and away from the X-ray beam to prevent damage to the optics and detectors. Objective lenses with typical 5×, 7.5×, or 10× magnification were used to image the scintillator light onto the detectors. The 7.5× objective resulted in a fields-of-view of 1.42 mm, for example. Three non-polarizing beam splitters (Newport) and other optics hardware were used to optically couple the scintillator light with the PI-MAX intensified charge-coupled device (ICCD) detectors (Princeton Instruments). Because these ICCDs were independently controlled, they could be triggered in sequence to capture images from a series of X-ray bunches.

Synchronization of the ICCD detectors with the dynamic event and the desired X-ray bunches was accomplished using a radio frequency signal provided by the synchrotron (known as the bunch clock), a delay generator (DG), and a PZT impact pin (lead-zirconium-titanate) (see Fig. 10a). The RF signal is coincident with each

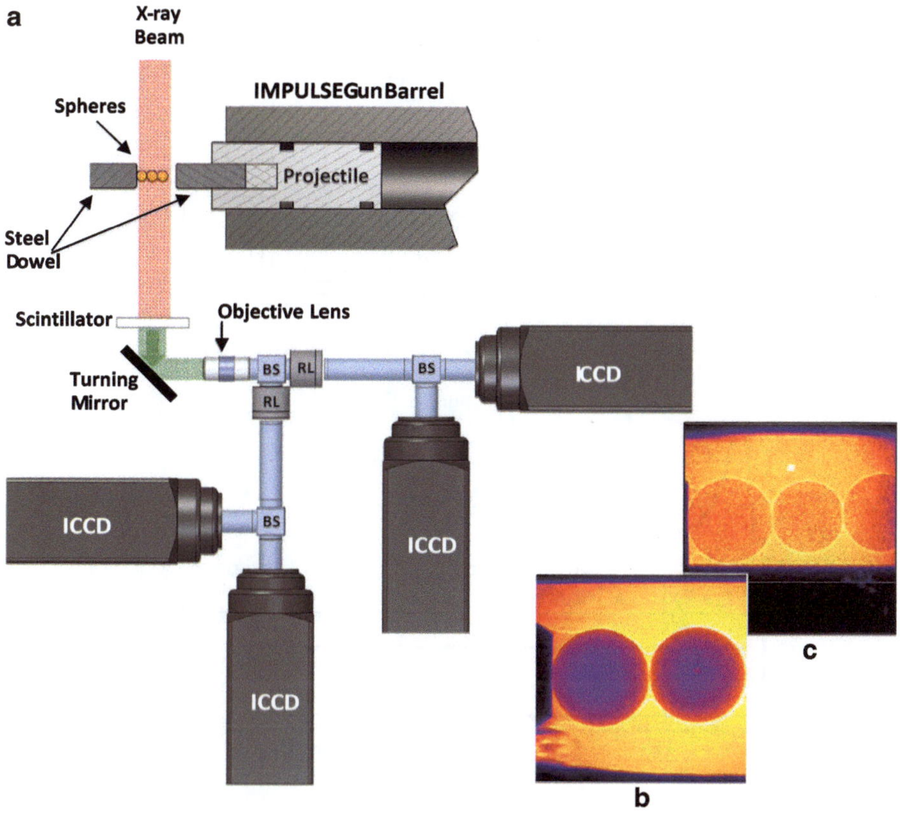

Fig. 9 (**a**) Schematic of the experimental configuration for dynamic MPCI experiments. The X-ray beam interacts with a sample subjected to projectile impact from the IMPULSE system (not shown). The transmitted X-rays are incident upon a scintillator converting them to visible light which is then optically coupled to the four ICCD detectors via a series of beam splitters (BS) and relay lenses (RL). The scintillator-to-sample distance is variable to optimize the image contrast and resolution. Static X-ray PCI images of 0.5-mm diameter borosilicate spheres using (**a**) Schematic of the experimental (**b**) 12 KeV X-rays, and (**c**) 23 KeV X-rays

80-ps width X-ray bunch arriving every 153.4 ns in the standard operating mode. This continuous stream of RF signals was connected to the external trigger input of the DG. To synchronize with the dynamic event, the PZT pin supplied a signal at impact that was sent to the inhibit input of the DG. Once the PZT impact signal was received the next RF signal triggered the DG to send four independent TTL signals to each ICCD. These TTL signals were delayed appropriately to center the camera frame on the first available X-ray bunch (fixed insertion delay) and an additional delay placed the image near the dynamic event in time.

Once a PZT signal is received, there is a minimum time before the cameras can acquire the first image caused by insertion delays in the DG, the cameras, and cable delays. This insertion delay time was determined in the following way.

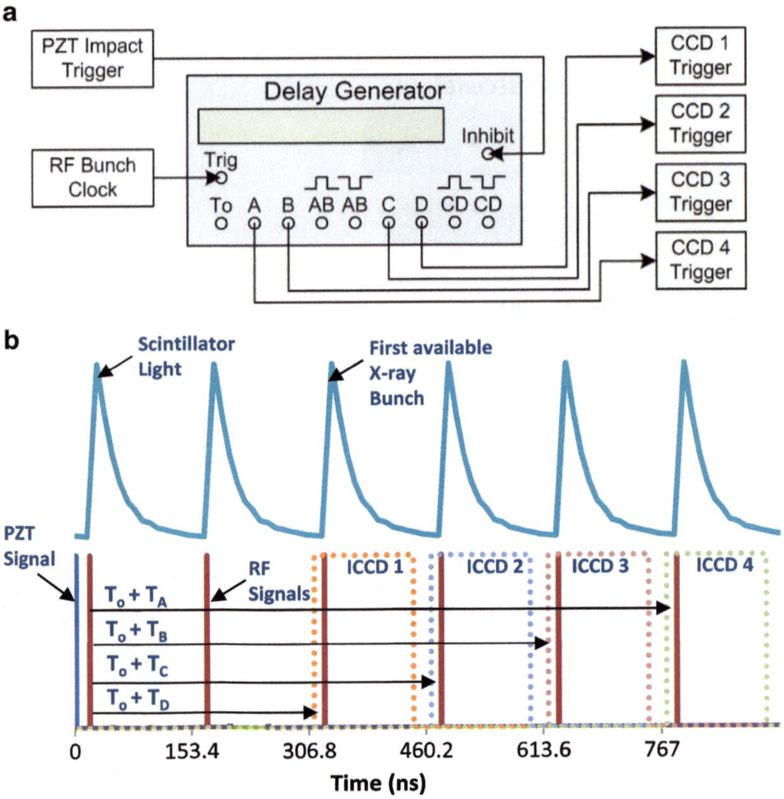

Fig. 10 (**a**) Detector trigger system which consists of a delay generator (DG) which is triggered using an impact PZT pin and a signal from the RF bunch clock. (**b**) A schematic time line showing the various signals used to synchronize the ICCD detectors with the X-ray bunch. The blue curve represents the scintillator light where the peak is coincident with the X-ray bunch arrival. Each ICCD is delayed by an amount T_o (minimum insertion delay) plus an additional delay T_n (where n = 1, 2, 3, or 4 corresponding to each camera) in 153.4 ns increments to position the image near the dynamic event in time

First, the integration time in the camera software was set to 20 ns and the DG was set at zero delay. Sequential images were acquired while changing the DG delay from 0 to 200 ns in increments of 10–20 ns. The mean intensity of each image was calculated and plotted to determine the intensity profile of the scintillator material as shown in Fig. 10 (blue curve). The peak intensity corresponds to the arrival of the X-ray pulse on the scintillator, and the corresponding time step was taken as the minimum delay required to center the camera frame on the first available X-ray beam. Each DG channel was then set to this minimum insertion delay plus an additional delay (in 153.4 ns increments) to position each frame in time near the dynamic event of interest. Note that as much as 153.4 ns of jitter is expected because of the asynchronous nature of the dynamic event with respect to the X-ray

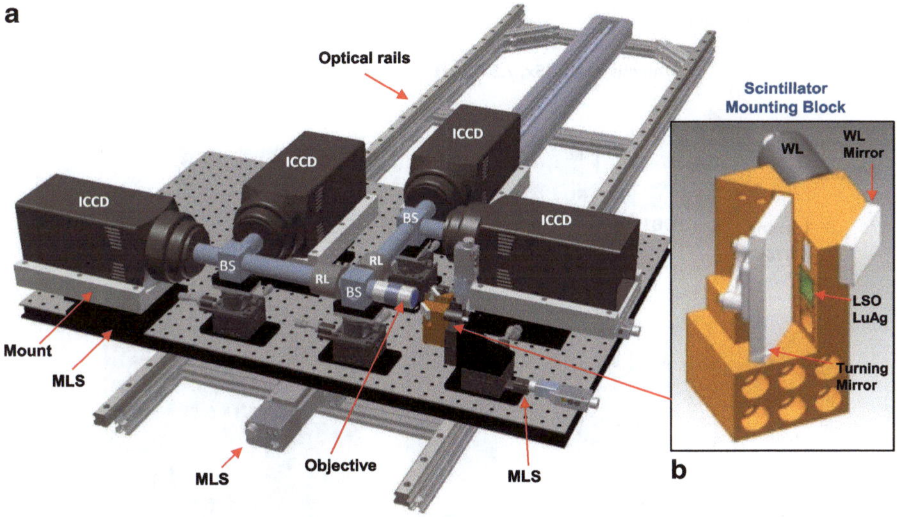

Fig. 11 (**a**) Model of the MPCI setup showing details of the ICCD mounts, the relay lenses (RL), beam splitters (BS), the motorized linear stages (MLS), objectives, and optical rails all used to remotely optimize the detectors. (**b**) Close-up view of the scintillator mounting block which includes a white-light source (WL) capability to align the system without X-rays, multiple scintillating material mounts, and a turning mirror to direct the light into the detector optics

pulse train. The exact timing of the images with respect to impact was determined using the recorded PZT impact signals and monitor outputs of each camera.

The entire MPCI system (Fig. 11a) incorporated motion control stages so that it could be optimized remotely while inside the closed experimental hutch with the X-ray beam on. Each ICCD was mounted on stages to adjust the alignment and focus of the image on the detector independently. A custom scintillator mount (Fig. 11b) was designed to three separate scintillators and two fiber-optic back-lit optical targets: (1) a USAF negative resolution test target to verify spatial resolution and monitor camera-to-camera alignment, and (2) a 200 lp/mm Ronchi ruling to align the object plane perpendicular to the optical axis and obtain an optimized full-field spatial focus. To optimize the contrast of certain target features, the entire system was secured to precision rails that allowed the oscillator position to be varied within a range of 200–1700 mm. The image contrast and resolution were optimized using the motorized linear stages shown in Fig. 11. The optimal distances for phase contrast and resolution were $z = 700$ mm for the U18 undulator (Sector 35 ID-B) with first harmonic at 23 keV, and $z = 400$ mm for the U33 undulator (Sector 32-B) with first harmonic at 12 keV. Determining the optimal distance is discussed in a subsequent section.

Recent efforts in the continued development of the IMPULSE MPCI system have focused on increasing the total number of frames from 4 to 8 to make longer shock-movies, and on optimizing the optical coupling from the scintillator to the cameras to increase light efficiency and allow for the use of multiple objectives

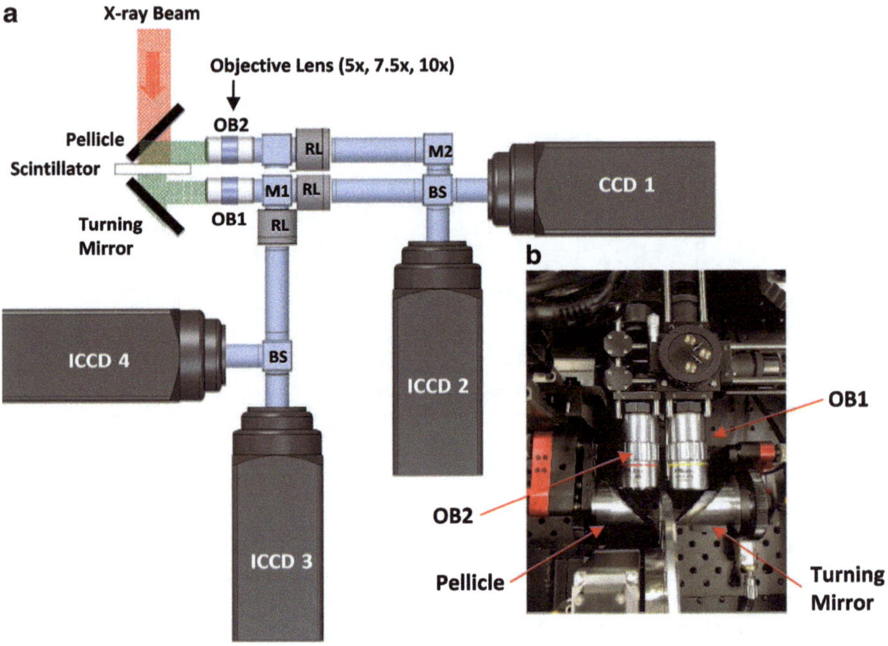

Fig. 12 (**a**) Schematic of the experimental configuration for the 8-frame, dual magnification MPCI system. (**b**) Photo of the imaging portion of the system. The two Mitutoyo objectives are shown focused onto a scintillator with the two angled pellicle and mirrors directing visible light into the objectives

simultaneously. A schematic of the next generation MPCI system is shown in Fig. 12. The increase in frames from 4 to 8 was accomplished by replacing the original PI-MAX II ICCD cameras (Princeton Instruments) with the PI-MAX 4 ICCD cameras that included the dual image feature (DIF). These cameras can provide two images separated by a minimum time duration of approximately 500 ns. Camera testing has shown that good results are obtained when there is a 614 ns delay between the first and second frame for a given camera. Thus, four cameras can be interleaved to accommodate this time delay and obtain a total of eight images for sequential X-ray bunches at the APS in standard mode (bunch every 153.4 ns).

The system design shown in Fig. 12 takes advantage of the ability to image both sides of the scintillator to increase the amount of light coupled into the system. In the figure, X-ray light is transmitted through an aluminum plated pellicle and incident upon the LSO scintillator. The X-rays are converted to visible light which reflects off a $\lambda/20$ aluminum mirror (downstream) and is imaged onto detectors ICCD-3 and ICCD-4 using a Mitutoyo objective (OB1), a mirror (M1), a relay lens (RL), and a beam splitter (BS). At the same time, the visible light on the upstream side of the scintillator reflects off the pellicle and is coupled to ICCD-1 and ICCD-2 using a second Mitutoyo objective (OB2), a mirror (M2), and a beam splitter.

Both the pellicle and the turning mirror were mounted to Zaber translation stages to position them within the beam and optimized remotely using picomotors piezo actuators (Newport). Commercially available 30-mm cage system components from Thorlabs Inc. were used which allowed for a rigid and robust optical setup for the beam splitters, mirror mounts, in-line rotation and tilt stages, relay lenses, and kinematic objective mounts. The system used 5×, 7.5×, and 10× Mitutoyo Plan Apo Infinity-Corrected Long WD Objective attached to kinetic mounts. Example images obtained using the dual-imaging MPCI system are shown in Fig. 13. The top set of images show eight consecutive images taken every 153.4 ns using a 7.5× Mitutoyo objective in the OB1 location (OB2 was not used in this example) showing the evolution of a metal jet formed when a shock wave interacted with a groove in a disc of cerium metal [26]. The remaining four images in the figure show results using the dual magnification feature to obtain simultaneous images at 5× and 10× magnification of a resolution target (middle) and a foam material (bottom) used to optimize phase contrast.

3.2 *Example Experiment: Compression of Idealized Spheres*

Impact experiments were performed on the IMPULSE system, described previously, at Sector 32 and 35 of the APS. The experimental configuration is shown in Figs. 1 and 9. Aluminum and Lexan projectiles, fitted with hardened steel dowel pins (ground flat), were accelerated down the barrel to impact a series of 0.5-mm diameter borosilicate (BS) spheres (purchased from Mo-Sci Specialty Products) mounted on a target using small amounts of Angstrom Bond epoxy. The borosilicate spheres used in this work had a chemical composition by weight of: silica (SiO_2) of 70–85%, boron oxide (B_2O_3) of 10–15%, sodium oxide (Na_2O) of 5–10%, and aluminum oxide (Al_2O_3), of 2–5%. The reported density is approximately 2.2 g/cc with an average sphere diameter of 0.65 mm resulting in a calculated areal density of 0.143 g/cm^2 at the center of the sphere. A photonic Doppler velocimetry (PDV) system was used to measure the projectile velocity up to impact.

The results for three representative experiments are shown in Figs. 14, 15 and 16. In the first experiment (Experiment 1), 12-keV X-rays from Sector 32 of the APS were used to image the impact of two BS spheres by a steel impactor traveling at a measured projectile velocity of 0.275 km/s. Four sequential images from the MPCI system are shown in Fig. 14a. Both the steel impactor and target are shown. In the first frame (T = 307 ns from impact), the steel impactor is observed to enter the field-of-view (FOV) as it impacts the first sphere (S1). In subsequent images, S1 begins to undergo significant change in shape, from circular to elliptical, and cracks begin to appear mostly along the direction of impact. By the third frame (T = 1226 ns), S2 begins to deform and complete fracturing of S1 occurs by the last frame. The results for Experiments 2 and 3 are shown in Figs. 15 and 16, respectively, where 23-keV X-rays from Sector 35 of the APS were used to image the compaction of three BS spheres stacked on the target. The additional sphere increases the time duration

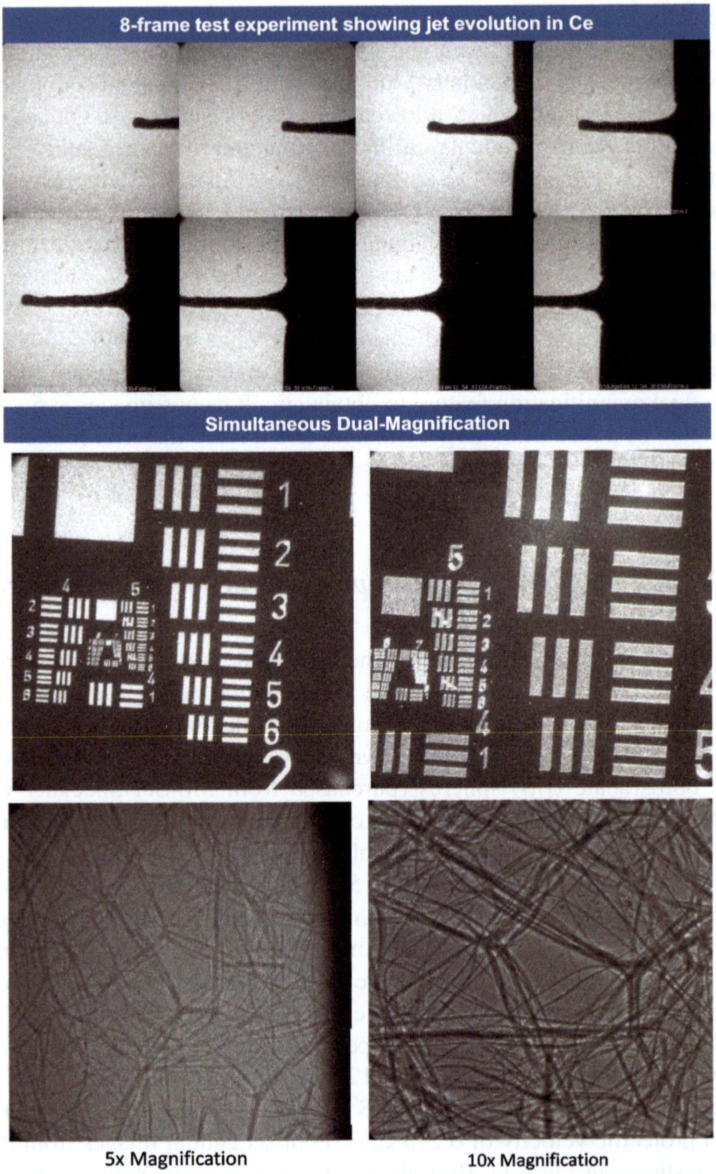

Fig. 13 (Top) images obtained using the PI-MAX 4 detectors with the dual image feature (DIF). The four detectors were interleaved in time to obtain eight images showing jet formation in cerium metal with a 7.5× magnification objective. Images taken simultaneously from ICCD-2 and ICCD-4 using the configuration shown in Fig. 9 using 5× and 10× objectives, respectively, for a resolution grid (middle images) and a foam material (bottom images) demonstrating the dual magnification feature

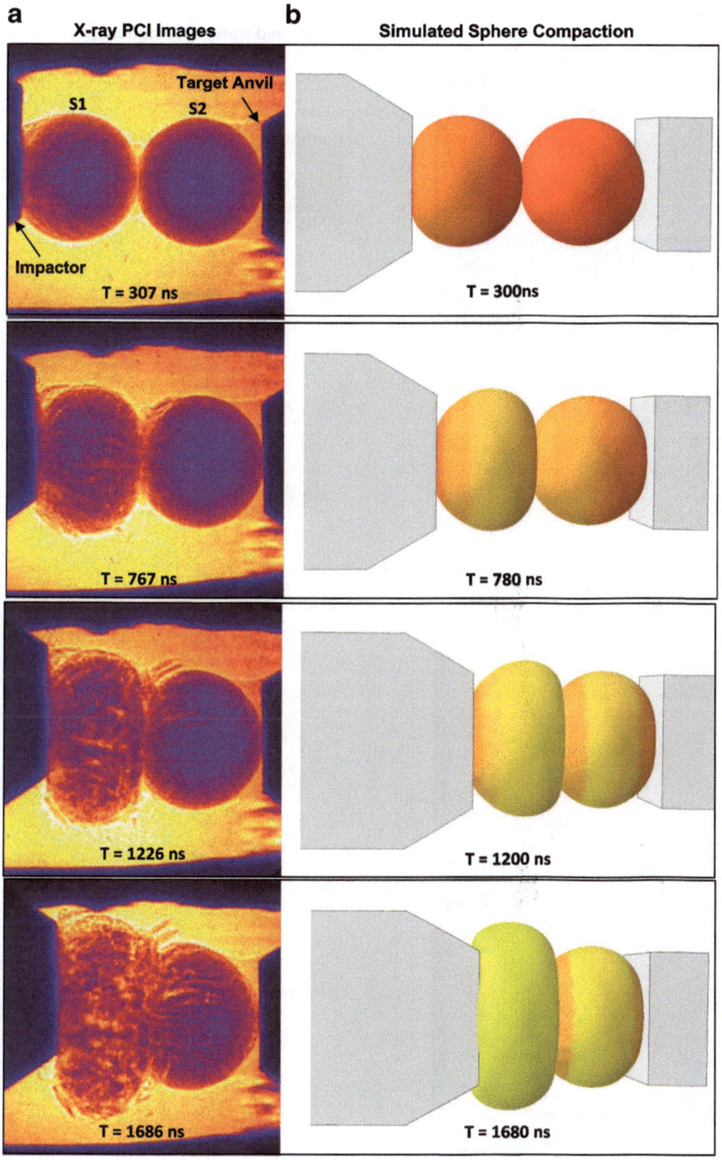

Fig. 14 (**a**) Dynamic MPCI images for Experiment 1 consisting of a steel projectile impacting two BS spheres at a velocity of 0.275 km/s. The times relative to impact are shown. The images are shown in false color where black represents complete absorption of the X-rays. (**b**) Results from three-dimensional simulations obtained using the Lagrangian finite element code ABAQUS are shown at comparable times relative to impact

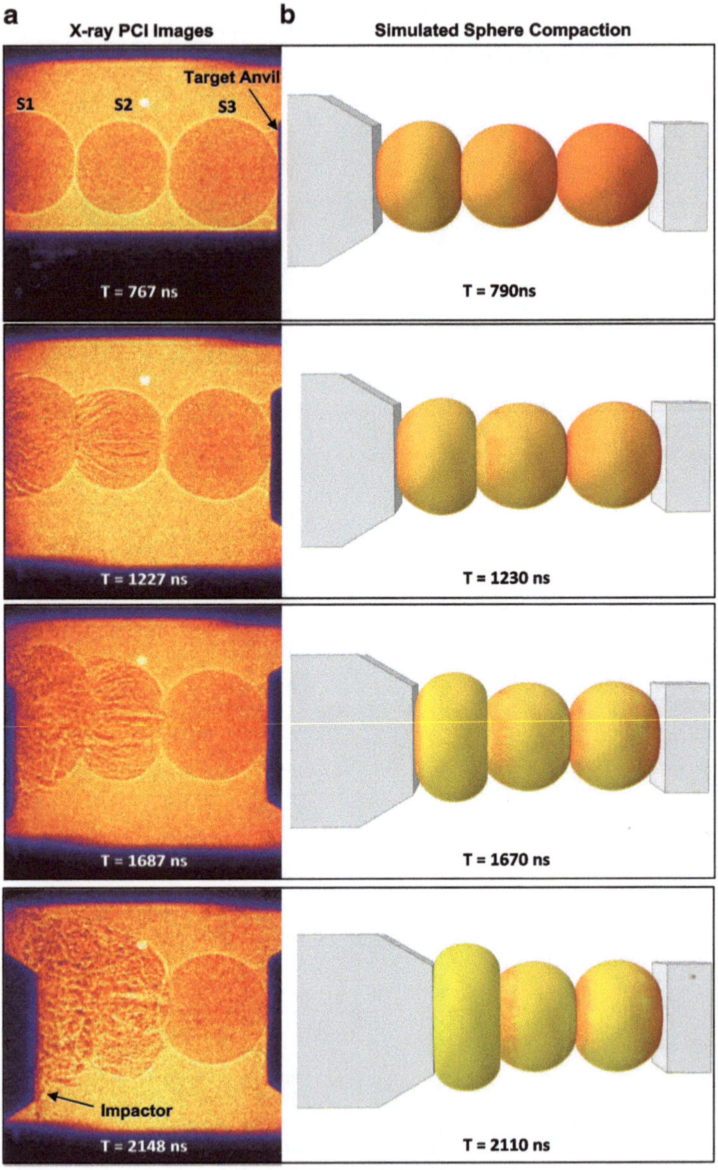

Fig. 15 (**a**) Dynamic MPCI images for Experiment 2 consisting of a steel projectile impacting three BS spheres at a velocity of 0.237 km/s. The times relative to impact are shown. The images are shown in false color where black represents complete absorption of the X-rays. (**b**) Results from three-dimensional simulations obtained using the Lagrangian finite element code ABAQUS are shown at comparable times relative to impact

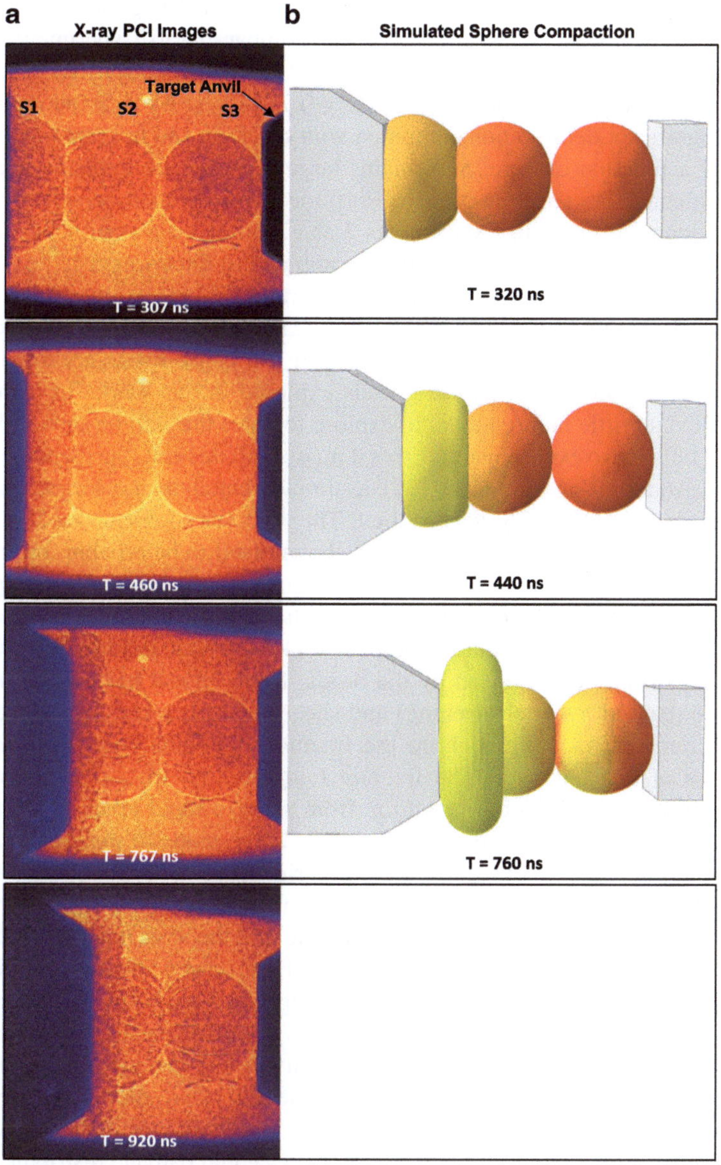

Fig. 16 (**a**) Dynamic MPCI images for Experiment 3 consisting of a steel projectile impacting three BS spheres at a velocity of 0.711 km/s. The times relative to impact are shown. The images are shown in false color where black represents complete absorption of the X-rays. (**b**) Results from three-dimensional simulations obtained using the Lagrangian finite element code ABAQUS are shown at comparable times relative to impact

of the experiment and the higher energy X-rays provided greater transmission. The projectile velocity of Experiment 2 was comparable to Experiment 1 with a measured velocity of 0.237 km/s. The measured projectile velocity for Experiment 3 was considerably higher at approximately 0.711 km/s. As in Experiment 1, we observe significant deformation combined with cracking and fracture of spheres S1 and S2 as a function of time. S3 remains largely unaffected until the final frame of Experiment 3. Because the X-ray absorption was less, many cracks or fracture interfaces were observed in Experiments 2 and 3 though it was difficult to discern actual voids as the spheres came apart, as compared to Experiment 1.

The high-resolution images obtained by X-ray PCI lead to immediate benefits for increasing our understanding of the complicated processes involved in dynamic fracture of brittle materials as well as increasing our capability to accurately model dynamic fracture in the borosilicate beads explored in this work. To demonstrate this, we used a brittle failure model to explore features that can be captured well by the model but also to expose features that it misses. To this end, the dynamic fracture model of Addessio and Johnson [37] as implemented in the Lagrangian finite element code ABAQUS [36] was applied. The Addessio–Johnson model has wide usage in the materials community and has shown itself to be reliable for predicting average damage properties in brittle materials including ceramics and explosives. For the present study, the model uses the hypothesis that small micron-sized cracks in un-deformed borosilicate glass will grow under dynamic deformation according to the local stresses experienced by the beads. In the model the evolution of both volumetric (caused by crack opening) and shear cracks is taken into account. Two critical parameters in the model are the fracture toughness corresponding to the two modes of crack growth. The fracture toughness controls the state of stress (for a given crack) at which a transition from slow to rapid crack growth occurs. By altering the fracture toughness, the model can be made to match the observed instantaneous beads sizes observed in the experiment.

A series of calculations were performed using ABAQUS with appropriate equation-of-state parameters for the borosilicate and the steel materials. The simulation configuration is shown in Fig. 17a. A steel impactor is shown impacting three identical 0.5-mm diameter BS spheres stacked up on a fixed boundary at a projectile velocity V. A more highly resolved mesh was used for the first sphere because it experiences higher degrees of deformation during the simulation. Example calculations are shown in Fig. 17b for Experiment 1 where the fracture toughness was varied with all other variables held constant. From these simulations, we can see that a large fracture toughness value, k_o, produces equal damage distributed among the three beads with a corresponding equal deformation in them. A small fracture toughness value produces early onset of damage growth in the impacted bead, with an observed enhanced deformation in that bead. The impact bead fractures first, then the second, and so on.

The value for the fracture toughness, K_o, was estimated by comparing simulated images for Experiment 1 with the experimental data shown in Fig. 14a. Specifically, we compared the shape of each sphere at comparable times from impact. The sim-

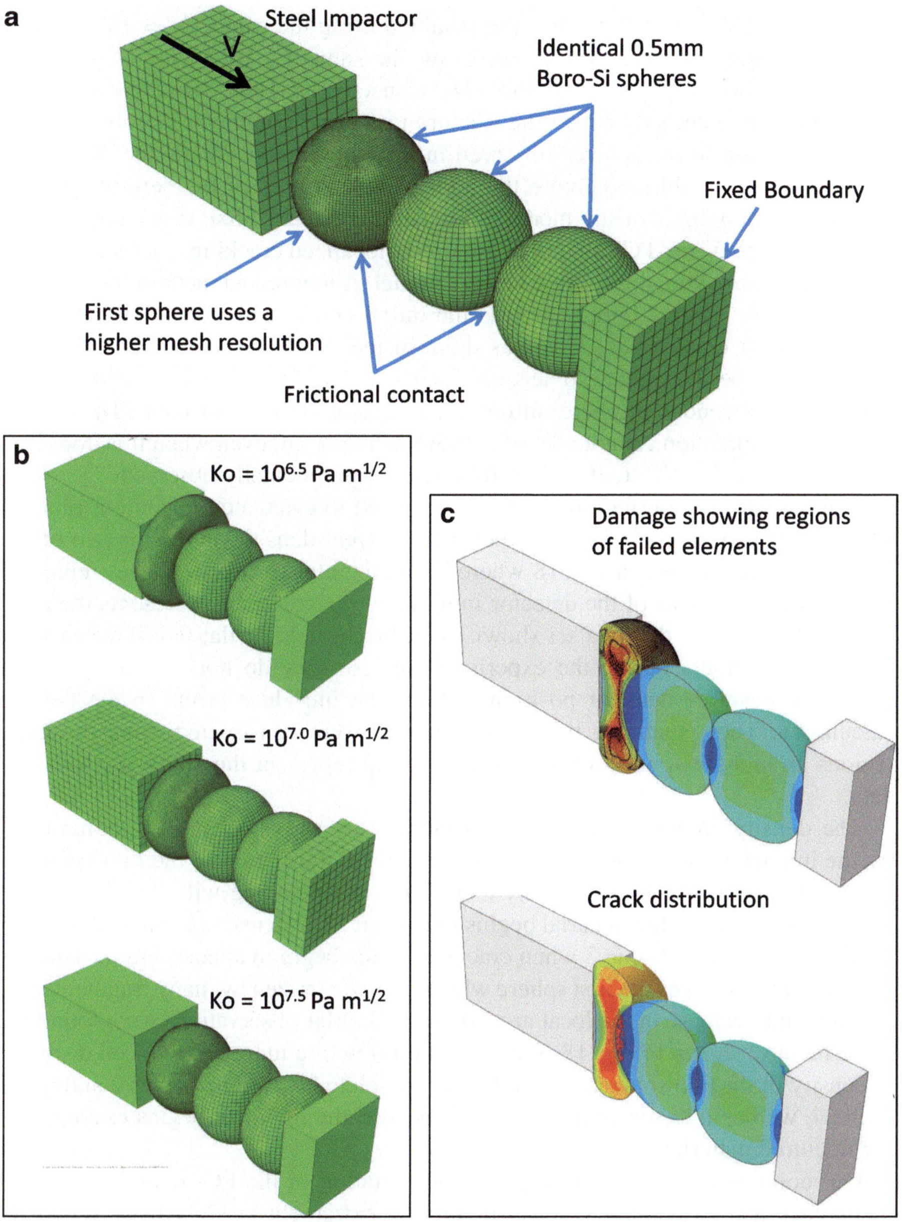

Fig. 17 (**a**) Model configuration for simulations using ABAQUS. (**b**)Three simulations for different values of the fracture toughness, K_o. (**c**) Example simulations illustrating the crack distribution for $K_o = 10^{6.5}\,\mathrm{Pa\,m}^{\frac{1}{2}}$

ulated images shown in Fig. 14b were obtained using the value $K_o = 10^{6.5}\,\mathrm{Pa\,m}^{\frac{1}{2}}$. Similar simulations were performed, using the same K_o value, for Experiments 2 and 3 as shown in Figs. 15b and 16b, respectively. Thus, for a given model, one can rather accurately adjust the fracture toughness parameter to correspond to the deformation of the spheres observed in the experiments. While the Addessio–Johnson model is able to capture the approximate shapes of the deforming glass spheres, shortcomings of the model are equally apparent. First, it is clear that the high-resolution X-ray PCI images show strong localized cracking, even in the early stages of deformation, not captured by the model. A theoretical method for inducing localization in brittle failure is to seed the initial cracks with a small non-uniform distribution of crack sizes. This was done in the displayed simulations, but was still found to be inadequate to account for the strong localization observed in the experiment. Second, the spheres ultimately fragment in the experiment. The current model and simulation cannot allow for fragmentation, but even when it is does (not shown) it can not replicate the sliver fragments experimentally observed.

The formalism described in Sect. 2.2 was used to calculate the contour plots of the areal density (g/cm^2) for the experiments. Areal densities plots for two of the experiments are shown in Fig. 18 where the horizontal and vertical axes represent the spatial dimensions of the detector in mm and the colormap represents the areal density in g/cc. For the data set shown here, bright-field or flat-field images were taken several hours before the experiment and so they do not accurately reflect the X-ray beam or detector positions which can introduce errors in the density calculation. Thus, synthetic bright-field images were constructed using the large regions around the spheres in the static image to represent the 100% transmission data.

The density contour plots for Experiment 1 are shown in Fig. 18 (column 1). As the impactor moves into the field-of-view, the deformation of the first sphere is observed, which is characterized by a change in shape along with visible areas of decreased density, as the material begins to fracture. The density of sphere 2 remains nearly constant until frame 3 when cracks or voids begin to appear. Frame 4 shows complete fracturing of the first sphere which is characterized by many fragments and a significant decrease in the local areal density. Similar observations were found for Experiment 2 shown in Fig. 18 (second column) with a more pronounced decrease in density of the two spheres caused by the complete fracture in frames 2 and 3. By frame 4, we begin to see an increase in density as the impactor begins to compress the fractured material.

Our comments above regarding the interpretation of the PCI images are based on the fact that X-ray phase contrast imaging is extremely sensitive to areal density variations, allowing one to easily detect cracks, voids, density variations, and defects in a sample. While the exact sensitivity will depend on the material, X-ray photon energy, propagation distance, beam divergence, detector resolution, and photon counting statistics, example calculations are shown in Fig. 19 for a 2-mm thick borosilicate sample, 2.23 g/cm^3, with 24.5 keV PCI imaging. For this example, a propagation distance of 700 mm from the sample to the scintillator was used along

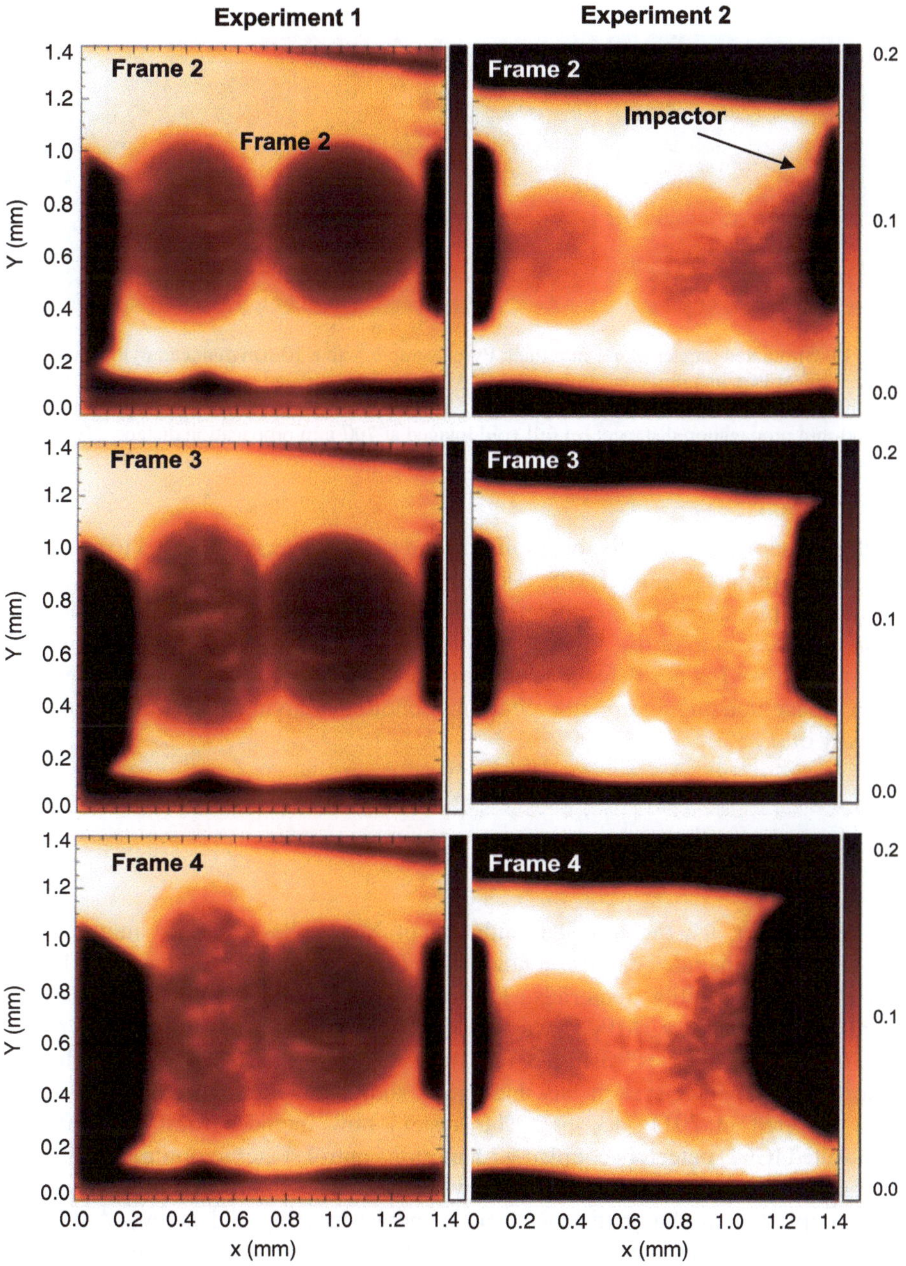

Fig. 18 Areal density retrieved from the images shown in Fig. 14 (column 1) for Experiment 1 and Fig. 15 for Experiment 2. Only frames 2 through 4 are shown and the direction of the impactor entering the field-of-view is indicated for both experiments. The areal density ranges from 0 to 0.2 g/cm^2

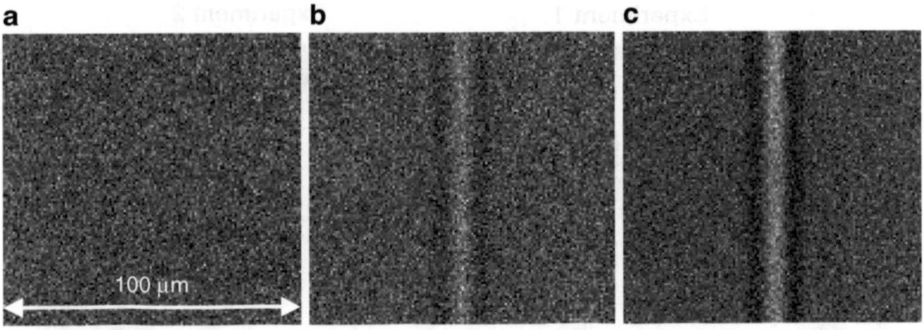

Fig. 19 Simulated images of a 2-mm thick Pyrex sample with a 10-μm diameter cylindrical void, under typical PCI imaging conditions used at APS. (**a**) Void imaged with a propagation distance of 1 mm, producing an absorption-only image. No void is discernible; (**b**) 1/2 density void imaged at 700-mm propagation distance; (**c**) full void imaged at 700-mm propagation distance. Estimated areal density sensitivity is 0.25% in (**b**)

with an assumed 5-μm detector resolution and a beam divergence of 3.3 μrad typical of our PCI imaging setup at APS. A cylindrical void, 10 μm in diameter was placed in the sample, either as a complete void, (a) and (c), or as a void at 1/2 normal density, 1.12 g/cm^3 in (b). Figure 19(a) is an absorption-only image, showing that a complete void is not detectable. In (b), the 1/2 density void is easily detectable under these imaging and photon counting conditions, giving a sensitivity for areal density of 0.25%.

4 Discussion and Summary

In this chapter, the theoretical foundation of X-ray phase contrast imaging was presented that included details on how to use PCI images to determine areal density for single composition materials. The recently developed dynamic, multi-frame X-ray PCI capability (MPCI) was also described that has been used to obtain the first "shock-movies" using X-rays to examine matter-at-extreme conditions. The dynamic MPCI system has undergone significant improvement since it was first reported in 2011 [13, 14]. The system now includes four independently controlled ICCD detectors coupled with motion control stages to optimize important parameters including the focus, field-of-view, spatial resolution, contrast by varying the sample-to-scintillator distance, and camera triggering and gating. Additional upgrades were presented for the first time that allow for simultaneous, dual-imaging of the scintillator resulting in a significant improvement in the optical efficiency of the system while allowing for the ability to use multiple objectives for dual magnification. Efforts are underway to further develop this diagnostic, which includes the use of two scintillators to allow for phase retrieval, and thus density retrieval, for heterogeneous or complex materials.

The MPCI system and theory were applied to begin a systematic study of granular systems focusing initially on an idealized system of 0.65-mm diameter borosilicate spheres in a linear geometry. These spheres were impacted using the IMPULSE gun system, and the MPCI system was used to diagnose the response from the initial compression to the complete failure and fragmentation of the spheres. The data revealed striking time-resolved images of the spheres showing significant changes in shape combined with localized cracking followed by complete fracture. The data were compared with calculations using ABAQUS [36] that included an equation-of-state for the borosilicate material and the Addessio–Johnson failure model [37]. The results have led to an increased understanding of the complex process involved in dynamic fracture while helping to identify the shortcomings of these models.

In particular, the details of the sphere deformation, both shape and timing, provide a way to estimate the fracture toughness for a given model, but the data content of the images is far richer than this simple analysis suggests. Because the images represent two-dimensional convolutions of the three-dimensional deformation that occurs throughout the bulk and at the interfaces, a more involved analysis is required to understand and interpret the images. The complexity of the process is shown in Fig. 17c where both the damaged regions and crack distributions are shown for a given simulation. Future work must utilize these simulated data to perform forward calculations of both the areal density and the detector images for direct comparison with experimental data. In this way, the data can inform the development and validation of failure models and equations-of-state for materials subjected to dynamic loading.

Density retrieval, using the PCI images as described in Sect. 2.2, has provided important information about the dynamic process from the initial material deformation, to failure, and then compaction. First, the analysis illustrates the ability to use X-ray PCI to obtain density snapshots essential for understanding the compaction process and for model development and validation. Second, the analysis shows that the decrease in density observed during failure is correlated with an increase in transmission likely caused by voids as the material fractures. The density also decreases as the fragments, which are unconfined, move out of the path of the impactor. Lastly, as the impact velocity increases the spatial resolution of the density contours decreases as the particle size of the fragments also decreases. To retain resolution when the particle size decreases, diffraction effects must be taken into account, which is beyond the scope of this work but is in development for future experiments.

To begin the transition from the more idealized linear stack of spheres, shown in the previous example, toward systems with increased complexity (e.g., random configurations, powders, sand, soils), new experiment configurations are needed that allow for confining the sample while clearly separating the effects caused by the compaction wave in the sample, shock propagation through the confining material, and edge waves from the sample/impactor edges. An example configuration designed to study the compression of a random distribution of particles is shown in Fig. 20. The configuration consists of a flyer plate (10 mm diameter and 2 mm thick)

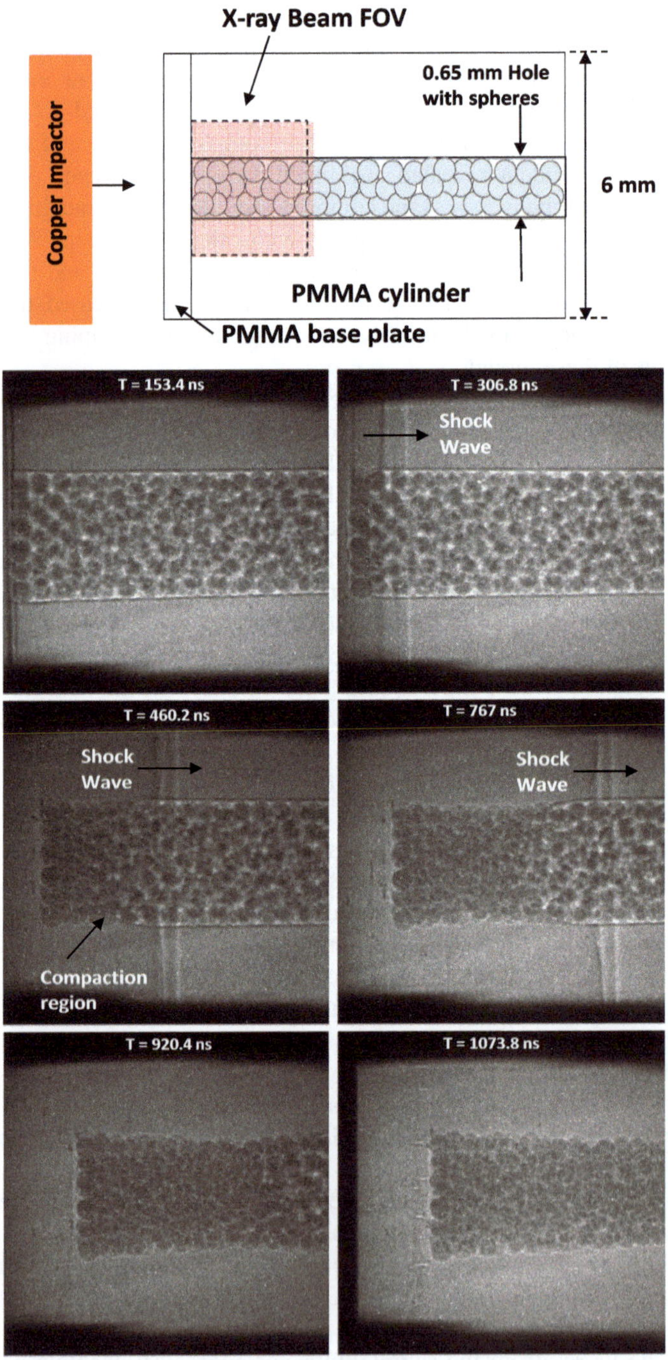

Fig. 20 (Top) Possible experiment configuration (not to scale) for using X-ray PCI to examine shock compaction of granular systems. In this example, 104-μm borosilicate spheres are loaded

impacting a PMMA cell (6 mm diameter) that contained 0.104-mm borosilicate spheres packed (tap density) into a 0.65-mm diameter hole. As shown in Fig. 20, shock waves generated at impact propagated into the cell leading to a shock wave in the PMMA cell followed by a slower moving compaction wave propagating through the spheres.

In summary, this work represents an important advancement in the development of capabilities that are available to examine the dynamic response of granular systems. Future efforts include incorporating diffraction effects in the analysis to improve spatial resolution coupled with the ability to use multiple scintillator distances to retrieve the phase and determine the volumetric density for heterogeneous materials with multiple compositions. We expect that this capability will continue to provide invaluable insight into the dynamic compression of matter at extremes and help to define the path and ensure success of future capabilities including the new Dynamic Compression Sector (DCS) at the APS and Los Alamos National Laboratory's proposed signature facility known as MaRIE [55]. In particular, MaRIE will provide more photons/pulse and well-characterized higher energy X-rays making possible the study of higher density materials while remaining in the transmission limit that simplifies the PCI analysis. In addition, the timing structure of the X-ray pulse will provide enhanced time resolution needed to resolve these complicated processes as they occur.

Acknowledgements This work was performed at Los Alamos National Laboratory (LANL) and at Argonne National Laboratory's (ANL) Advanced Photon Source (APS). Charles T. Owens (LANL), the lead technician for IMPULSE, is gratefully acknowledged for technical assistance with target and projectile fabrication, gun setup, system maintenance, and shot execution. The MPCI system was developed as a collaborative effort between LANL and National Security Technologies (NSTec). A. Deriy and K. Fezzaa (ANL) are thanked for technical support at Sector 32 ID-B of the Advanced Photon Source (APS) where the initial dynamic experiments were performed. One of the authors, B.J.J, would like to thank and acknowledge the other two IMPULSE co-founders, Sheng Lou and Dan Hooks, for their roles in the initial PCI detector setup and early project direction, respectively, which led to the first experiments on IMPULSE in 2011. Additional experiments were performed later at the Dynamic Compression Sector (Sector 35 of the APS) during the early commissioning phase of its development. Tim Graber and N. Sinclair are thanked for technical support for experiments conducted at the DCS. This work was supported by LANL's MaRIE concept, Science Campaign programs, Joint Munition Programs (JMP), and the NSTech (Los Alamos Office) Shock Wave Related Diagnostics program. LANL is operated by Los Alamos National Security, LLC for the U.S. Department of Energy (DOE) under Contract No. DE-AC52-06NA25396. Use of the Advanced Photon Source, an Office of Science User Facility operated by Argonne National Laboratory, was supported by the U.S. DOE under Contract No. DE-AC02-06CH11357. The Dynamic Compression Sector was supported by the Department of Energy, National Nuclear Security Administration, under Award Number DE-NA0002442 and operated by Washington State University.

Fig. 20 (continued) into a 6mm diameter cell which is impacted by an OFHC copper flyer plate. Six X-ray PCI images are shown (sample centered on the detector) taken during an experiment clearly showing the shock wave in the cell and the compaction wave in the sample. A thin 0.588-mm thick base plate was required to confine the sample and provide a sustained shock in the sample

References

1. Dolan DH, Knudson M, Hall C, Deeney C (2007) A metastable limit for compressed liquid water. Nat Phys Lett 3:339–347
2. Jensen BJ, Cherne FJ, Cooley JC, Zhernokletov M, Kovalev A (2010) Shock melting of cerium. Phys Rev B 81:214109
3. Jensen BJ, Cherne FJ (2012) Dynamic compression of cerium in the low-pressure $\gamma - \alpha$ region of the phase diagram. J Appl Phys 112:013515
4. Asay JR, Fowles GR, Duvall GE, Miles MH, Tinder RF (1972) Effects of point defects on elastic precursor decay in LiF. J Appl. Phys 43:2132
5. Jensen BJ, Gray GT, Hixson RS (2009) Direct measurement of the alpha-epsilon transition kinetics and stress for shocked iron. J Appl Phys 105:103502
6. Jensen BJ, Gupta YM (2008) Time-resolved X-ray diffraction experiments to examine the elastic-plastic transition in shocked magnesium-doped LiF. J Appl Phys 104:013510
7. d'Almeida T, Gupta YM (1999) Real-time X-ray diffraction measurements of the phase transition in KCl shocked along [100]. Phys Rev Lett 85:2
8. Dolan DH (2007) Characterizing the emissivity of materials under dynamic compression. SAND2007-6376, Sandia National Laboratory
9. Gupta YM, Turneaure SJ, Perkins K, Zimmerman K, Arganbright N, Shen G, Chow P (2012) Real-time, high-resolution X-ray diffraction measurements on shocked crystals at a synchrotron facility. Rev Sci Instrum 83:123905
10. Rigg PA, Schwartz CL, Hixson RS, Hogan GE et al (2008) Proton radiography and accurate density measurements: a window into shock wave processes. Phys Rev B 77:220101
11. Kalantar DH, Belak JF, Collins GW, Colvin JD, Davies HM et al (2005) Direct observation of the $\alpha - \epsilon$ transition in shock-compressed iron via nanosecond X-ray diffraction. Phys Rev Lett 95:075502
12. Schropp A et al (2015) Imaging shock waves in diamond with both high temporal and spatial resolution at an XFEL. Sci Rep 5:11089
13. Luo SN, Jensen BJ, Hooks DE, Fezzaa K, Ramos KJ, Yeager JD et al (2012) Gas gun shock experiments with single-pulse X-ray phase contrast imaging and diffraction at the advanced photon source. Rev Sci Instrum 83:073903
14. Jensen BJ, Luo SN, Hooks DE, Fezzaa K, Ramos KJ, Yeager JD et al (2012) Ultrafast, high resolution, phase contrast imaging of impact response with synchrotron radiation. AIP Adv 2:012170
15. Jensen BJ, Owens CT, Ramos KJ, Yeager JD et al (2013) Impact system for ultrafast synchrotron experiments. Rev Sci Instrum 84:013904
16. Wilkins SW, Gureyev TE, Gao D, Pogany A, Stevenson AW (1996) Phase-contrast imaging using polychromatic hard X-rays Nature 384:335
17. Wu X, Liu H (2003) Clinical implementation of X-ray phase contrast imaging: theoretical foundations and design considerations. Med Phys 30:2169
18. Zoofan B, Kim JY, Rokhlin SI, Frankel GS (2006) Phase-contrast X-ray imaging for nondestructive evaluation of materials. J Appl Phys 100:014502
19. Wang Y, Liu X, Im KS, Lee WK, Fezzaa K (2008) Ultrafast X-ray study of dense-liquid-jet flow dynamics using structure-tracking velocimetry. Nature 4:305–309
20. Fezzaa K, Wang Y (2008) Ultrafast X-ray phase contrast imaging of the initial coalescence phase of two water droplets. Phys Rev Lett 100:104501
21. Yeager JD, Luo SN, Jensen BJ, Fezzaa K, Montgomery DS, Hooks DE (2012) High-speed synchrotron X-ray phase contrast imaging for analysis of low-Z composite microstructure. Comput Part A 43:885
22. Ramos KJ, Jensen BJ, Iverson AJ, Yeager JD, Carlson CA, Montgomery DS et al (2014) In situ investigation of the dynamic response of energetic materials using IMPULSE at the advanced photon source. J Phys Conf Ser 500:142028

23. Ramos KJ, Jensen BJ, Yeager JD, Bolme CA, Iverson AJ et al (2014) Investigation of dynamic material cracking with in situ synchrotron-based measurements. In Song B, Casem D, Kimberley J (eds) Conference proceedings of the society for experimental mechanics series. Springer, New York, pp 413–420
24. Ramos KJ, Jensen BJ, Iverson AJ, Yeager JD et al (2014) In situ investigation of the dynamic response of energetic materials using IMPULSE at the advanced photon source. In Buttler WT, Evans WJ (eds) Journal of physics: conference series. IOP Publishing 500:142028
25. Jensen BJ, Ramos KJ, Iverson AJ, Bernier J et al (2014) Dynamic experiment using IMPULSE at the advanced photon source. In Buttler WT, Evans WJ (eds) Journal of physics: conference series. IOP Publishing 500:042001
26. Jensen BJ, Cherne FJ, Prime MB, Fezzaa K, Iverson AJ et al (2015) Jet formation in cerium metal to examine material strength. J Appl Phys 118:0195903
27. Brown EN, Furmanski J, Ramos KJ, Dattelbaum DM, Jensen BJ et al (2014) High-density polyethylene damage at extreme tensile conditions. In Buttler WT, Evans WJ (eds) Journal of Physics: Conference Series. IOP Publishing 500:112011
28. Hawreliak JA, Lind J, Maddox B, Barham M, Messener M, Barton N, Jensen BJ, Kumar M (2016) Dynamic behavior of engineered lattice materials. Nat Sci Rep 6:28094
29. Willey TM, Champley K, Hodgin R, Lauderbach L, Bagge-Hansen M, May C, Sanchez N, Jensen BJ, Iverson AJ, van Buuren T (2016) X-ray imaging and 3D reconstruction of in-flight exploding foil initiators. J Appl Phys 119:235901
30. Herrmann W (1969) Constitutive equation for the dynamic compaction of ductile porous materials. J Appl Phys 40:2490–2499
31. Carroll M, Holt AC (1972) Static and dynamic pore-collapse relations for ductile porous materials. J Appl Phys 43:759–761
32. Grady D, Winfree NA, Kerley GI, Wilson LT, Kuhns LD (2000) Computational modeling and wave propagation in media with inelastic deforming microstructure. J Phys IV France 10: 15–20
33. Studman CJ, Field JE (1984) The influence of brittle particles on the contact between rigid surfaces. J Phys D: Appl Phys 17:1631–1646
34. Lorenz A, Tuozzolo C, Louge MY (1997) Measurement of impact properties of small, nearly spherical particles. Exp Mech 37:292–298
35. Andrews EW, Kim KS (1999) Threshold conditions for dynamic fragmentation of glass particles. Mech Mater 31:689–703
36. ABAQUS (2011) ABAQUS Documentation. Dassault Systèmes, Providence, RI, USA
37. Addessio FL, Johnson JN (1990) A constitutive model for the dynamic response of brittle materials. J Appl Phys 67:3275
38. Snigirev A, Snigireva I, Kohn V, Kuznetsov S, Schelokov I (1995) On the possibilities of X-ray phase contrast microimaging by coherent high-energy synchrotron radiation. Rev Sci Instrum 66:5486
39. Pogany A, Gao D, Wilkins SW (1997) Contrast and resolution in imaging with a microfocus X-ray source. Rev Sci Instrum 68:2774
40. Montgomery DS, Nobile A, Walsh PJ (2004) Characterization of National Ignition Facility cryogenic beryllium capsules using X-ray phase contrast imaging. Rev Sci Instrum 75:3986
41. Montgomery DS, Gautier DC, Kozioziemski BJ, Moody JD, Evans SC et al (2006) Characterization of D-T cryogenic layer formation in a Beryllium capsule using X-ray phase contrast imaging. J Phys IV 133:869
42. Medlovic D, Zalevsky Z, Konforti N (1997) Computation considerations and fast algorithms for calculating the diffraction integral. J Mod Opt 44:407–413
43. Cloetens P, Ludwig W, Baruchel J, Guigay JP, Pernot-Rejmankova P, Salomé-Pateyron M et al (1999) Hard X-ray imaging using simple propagation of a coherent synchrotron radiation beam. J Phys D 32:10A
44. Dejus RJ, Sanchez del Rio M (1996) XOP: a graphical user interface for spectral calculations and X-ray optics utilities. Rev Sci Instrum 67:3356

45. Gerchberg RW, Saxton WO (1972) A practical algorithm for the determination of phase from image and diffraction plane pictures. Optik 35:237
46. Fienup JR (1982) Phase retrieval algorithms: a comparison. Appl Opt 21:2758
47. Teague MR (1985) Image formation in terms of the transport equation. J Opt Soc Am A 2:2019
48. Paganin D, Mayo SC, Gureyev TE, Miller PR, Wilkins SW (2002) Simultaneous phase and amplitude extraction from a single defocused image of a homogeneous object. J Micro 206:33
49. Gureyev TE (2003) Composite techniques for phase retrieval in the Fresnel region. Opt Commun 220:49
50. Mayo SC, Davis TJ, Gureyev TE, Miller PR, Paganin D et al (2003) X-ray phase contrast microscopy and microtomography. Opt Express 11:2289
51. Jensen BJ, Gupta YM (2006) X-ray diffraction measurements in shock compressed magnesium doped LiF crystals. J Appl Phys 100:053512
52. Rigg PA, Gupta YM (2006) Time-resolved X-ray diffraction measurements and analysis to investigate shocked lithium fluoride crystals. J Appl Phys 93:3291–3298
53. Rigg PA, Gupta YM (2001) Multiple X-ray diffraction to determine transverse and longitudinal lattice deformation in shocked lithium fluoride crystals. Phys Rev B 63:094112
54. Gupta YM, Zimmerman KA, Rigg PA, Zaretsky EB, Savage DM (1999) Experimental developments to obtain real-time X-ray diffraction measurements in plate impact experiments. Rev Sci Instrum 70:4008
55. Sarrao JL (2012) MaRIE 1.0: a flagship facility for predicting and controlling materials in dynamic extremes. LA-UR 12-00500. Los Alamos National Laboratory

Shock Compression of Porous Materials and Foams Using Classical Molecular Dynamics

J. Matthew D. Lane

1 Introduction

Computational modeling of shock compression in porous media has become a useful partner to experiment, augmenting difficult experiments, especially where direct experimental measurements of local temperature and flow are not yet possible. Molecular modeling can open new avenues of exploration, in cases where the internal structure of voids, powder packings or foams are difficult to fabricate, control, or even characterize experimentally. Over the past decade, molecular dynamics simulation has made significant strides in modeling extreme environments and heterogeneous materials. Some of that progress will be summarized here. However, there is still much to do to understand the particularly challenging overlap of extreme states and heterogeneous materials. One thing appears certain: that advances will require progress on every length scale from mesoscale and macroscopic material inhomogeneities (i.e., pores and void structure) to molecular-scale energy dissipation processes (deformational and chemical).

What makes classical molecular dynamics (MD) an attractive method for studying processes in porous collapse? One can think of MD as a bridge between two powerful tools. On the quantum scale, density functional theory (DFT) and quantum Monte Carlo (QMC) can produce highly-reliable first-principles calculations, but these are limited to time and length scales which are not well matched to porous materials. On the continuum scale, equation of state (EOS) calculations do not generally incorporate non-equilibrium processes and lack molecular scale processes (e.g., surface reconstructions). MD relies on quantum scale methods to develop powerful classical interatomic potentials which can then be applied to larger scale

J. M. D. Lane (✉)
Sandia National Laboratories, Albuquerque, NM, USA
e-mail: jlane@sandia.gov

T. J. Vogler, D. A. Fredenburg (eds.), *Shock Phenomena in Granular and Porous Materials*, Shock Wave and High Pressure Phenomena,
https://doi.org/10.1007/978-3-030-23002-9_8

problems such as those discussed here. On today's computers, a simulation of a few hundred atoms would be a large DFT calculation, due to poor scaling of the approach. MD, on the other hand, can model tens to hundreds of millions of atoms with reasonable efficiency, and even account for billions of atoms if massively parallel computing resources are available.

Moreover, MD is a capable tool to capture and understand the varying mechanisms in both hard and soft distended systems. In the area of soft materials, research has been driven by two primary application areas: hot spot initiation in energetic materials, and pulse shaping with polymer foams. In hard materials, work has focused on how dislocations and phase transitions nucleate near, or interact with, collapsing pores.

Figure 1 shows a simple depiction of the void collapse. A propagating compression wave enters from the left, and passes around and through the pore, causing a localized release of stored strain energy which can couple to increased kinetic energy, or any number of other dissipative structures available to a specific material. In this schematic image, material from the upstream side of the pore bursts into the void space, but the details of this collapse depend on the specific material and the strength of the wave. In high-strength materials, the collapse may be slow and lead to phase transition. In soft materials, the collapse may be rapid and lead to vaporization and chemistry. In every case, temperatures rise, and materials are pushed out of homogeneous equilibrium. The timescales for thermalization and a return to equilibrium depend on the mechanisms available.

In the development of molecular understanding, much work has focused on the processes in single voids. But, understanding the bulk response of porous materials requires modeling multiple voids and their interactions. The length scale of individual voids and scaling to a collective response are constant challenges in comparing MD results to experiment.

Recent work has identified several fundamental processes which appear in collapsing porous materials:

- Collapse and crush
- Flow, focussing, and Jetting
- Hot spot heating and thermalization
- Ejecta vaporization and free surface

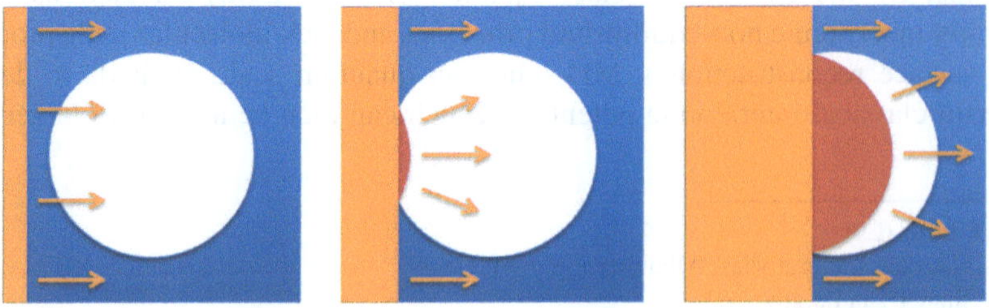

Fig. 1 Schematic of void collapse mechanisms

- Chemistry and detonation
- Fracture and fragmentation
- Dislocation and plasticity

Collapse and crush is the removal of free volume in the material. In pressure vs. volume plots of the porous Hugoniot, this is often seen as a nearly horizontal curve at relatively low pressures, indicating significant density increase without significant pressure increase. This crush phase is sometimes described as a preliminary stage, which occurs with little resistance, until the material comes to roughly full-density. At this point, the Hugoniot response either continues to compress or can reverse and expand depending on the degree of porosity and the amount of heat produced. Hot spot formation is a universal feature of pore collapse, while other processes are found only in certain classes of materials. For instance, chemistry and detonation would be seen only in organics such as polymers and explosives. Dislocations and plasticity would be seen in metals. And, fracture and fragmentation would be seen in brittle materials. At lower shock regimes, the response is more sensitive to the particular mechanisms available to a given material. In high shock regimes, all materials will exhibit flow, focussing, and ultimately jetting and vaporization. The molecular simulation community initially focused on the more universal response exhibited at high shock intensities, but more recently has focused on the material specific response seen at lower shock strengths.

This chapter reviews the findings of atomistic simulation of single void and collective porosity effects in uniaxial compression. It is organized into four sections, the first being this brief introduction. Section 2 is a summary of the methodology, which is divided into three subsections. These subsections cover, first, an introduction to the principles of molecular dynamics (MD); second, a summary of approaches used to produce nanoscale porosity; and third, issues particular to modeling porosity collapse and how to address them. Section 3 turns to discussion of particular material response, divided into subsections to address metals, high-strength covalently bonded materials, and soft and energetic materials. Section 4 summarizes the material in this chapter and proposes areas for future study.

2 Methods in Molecular Study of Porous Materials

Classical molecular dynamics (MD) is based on the assumption that the behavior of a material can be approximated by calculating the trajectories of its atomic mass points, interacting through a conservative energy potential, U, called an interatomic potential. The interatomic potential can be quite simple or very complex. In general U is a function of all the atom positions, $\mathsf{U}(\mathbf{r}_1, \mathbf{r}_2, \ldots, \mathbf{r}_N)$, and it incorporates in some average way the behavior dictated by quantum mechanics and the electronic degrees of freedom, which are not explicitly modeled in classical MD. The force on a particle i at a discrete time l, $\mathbf{F}_i^l$ is given by the gradient of the interatomic potential,

$$\mathbf{F}_i^l = -\nabla_i \mathsf{U}^l,$$

where U^l is calculated for the positions at time l.

The motion is constrained by Newton's Second Law,

$$\mathbf{F}_i = m_i \frac{d^2\mathbf{r}_i}{dt^2} = m_i \frac{d\mathbf{v}_i}{dt},$$

which can be discretized.

Therefore if we know the positions $(\mathbf{r}_1, \mathbf{r}_2, \ldots, \mathbf{r}_N)$ and velocities $(\mathbf{v}_1, \mathbf{v}_2, \ldots, \mathbf{v}_N)$ of all N atoms at some point in time, t_0, then we can write equations that relate the current positions and velocities to the positions and velocities of the atoms a short time, Δt later. These can be discretely formulated into what is termed a velocity Verlet algorithm,

$$\mathbf{v}_i^{(l+\frac{1}{2})} = \mathbf{v}_i^l - \frac{\mathbf{F}_i^l}{2m_i}\Delta t$$

$$\mathbf{r}_i^{(l+1)} = \mathbf{r}_i^l + \mathbf{v}_i^{(l+\frac{1}{2})}\Delta t$$

$$\mathbf{v}_i^{(l+1)} = \mathbf{v}_i^{(l+\frac{1}{2})} - \frac{\mathbf{F}_i^{(l+1)}}{2m_i}\Delta t$$

By iteratively integrating the equations of motion, one can discretely advance the atoms through space and time. From the velocity distributions, temperatures can be defined. For a more detailed description and derivation of the principles of MD, see one of the several excellent discussions [1–3], which describe other important aspects of MD, such as boundary conditions, thermostats, and barostats. The documentation of the extremely popular LAMMPS [4] code is another excellent starting point for exploration.

Modeling shocks in porous materials introduces several unique requirements on the simulation methodology. First, models must be able to handle significant volume change as free space is crushed out. As with all extreme environment simulations, the interatomic potentials and integration time step must be suitable for extreme pressures. But, collapsing voids also can produce much higher temperatures for a given pressure, by as much as an order of magnitude, over those seen in full-density shocks. In order to assure that the MD integration is stable in these conditions (i.e., conserves energy) the integration time step generally must to be shortened. Second, in order to compare with experimental results, simulations must be large enough to average over heterogeneous material and long enough in duration to allow equilibration or thermalization of the localized response. And, third, the structure of the initial porosity should be consistent with that seen in experiments.

2.1 Constructing Initial Porous States

Modeling the collapse of porous media assumes that a suitable initial material can be produced. A variety of approaches have been developed to produce model systems with appropriate initial porosity. Examples of each are represented in Fig. 2. In each case, distended systems are produced by introducing vacuum space or a low density molecular gas to reduce the overall density of a fully-dense phase. Ideally, the structure of these voids would match those in real materials, however, this is not always possible. This can be due to computing limitations or unknowns in experimental characterization of features such as pore size, pore shape, wall thickness, and pore interconnectedness. For example, it is common to model defects with idealized shapes, most commonly spheres; and, micron or larger pores are often modeled with nanoscale features, which match the overall density seen in experiments, but not the surface to volume ratios. Recent studies have begun to investigate sensitivity on these parameters, and will be discussed in Sect. 3. To this point, a thorough evaluation of pore size scaling from nanoscale to micron scale has not been demonstrated. However, in many cases, good agreement with experiments can be reached by matching the overall density and interconnectedness (open cell vs closed-cell), if not the average size of void features. This is an area for continued research, as the effect is likely to differ in different material classes. In cases such as nanoporous metallic foams, direct overlap between experimental and simulation pore length scales is possible.

Voids Cut from Pristine Material This straightforward approach simply cuts voids from an otherwise perfect crystal or uniform material. Voids are often spherical, but can be any shape. Multiple voids are often randomly placed, but can also be geometrically arranged. Cut voids are frequently used in crystalline systems to avoid introducing defect structures into the remaining material. It is believed that the artificially high long-range order which results from the remaining single-crystal material better represents an annealed porous metal than a material with high dislocation density which would be produced by a void growth approach (see below). Examples of cut voids are seen in Fig. 2a and b, which, respectively, depict a single spherical pore cut from crystalline explosive (PETN) from Shan et al. [5] and a 50% porous silicon with random arrangement of multiple cut voids from Lane et al. [7].

Grown Voids Grown voids are more commonly used to create pores in amorphous or polymeric systems. Here what is termed an indentor is used to gradually push material out of regions of the simulation box. Voids are most commonly spherical in shape, but cylinders and other shapes are possible. Growing voids in this manner, rather than cutting out material assures that molecular connectivity is preserved. For example, in polymer systems, chains are not severed by the void, but rather are pulled out or around the pores, preserving entanglements. In LAMMPS, `fix_indent` can be used to define spherical regions on which a soft harmonic potential can be used with a slowly increasing radius. The strength of the repulsion

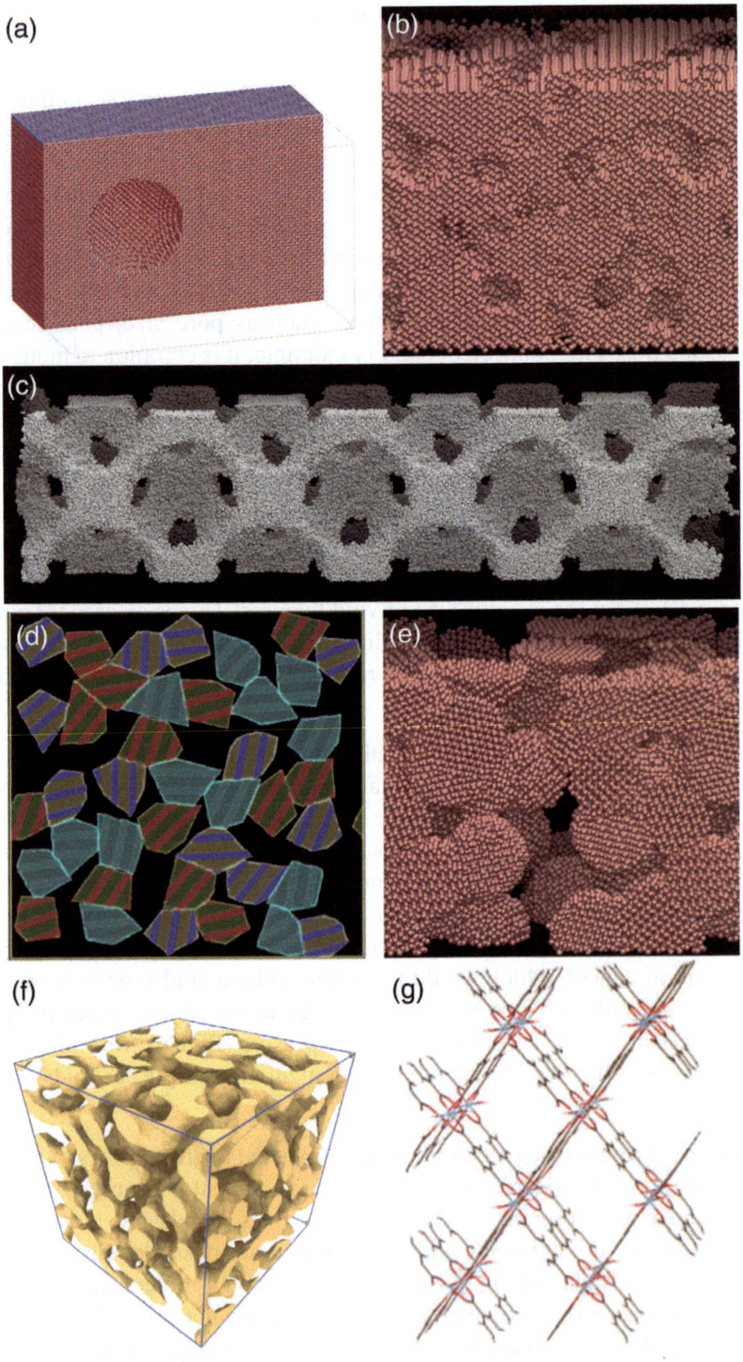

Fig. 2 Example of initial porous states (figure oriented left to right and down): (**a**) Cut single void from single crystal [5]; (**b**) Cut porous region [6]; (**c**) Grown voids in polymer foam [7]; (**d**) Accumulated powder grains of Ni/Al [8]; (**e**) Accumulated sintered silicon grains [6]; (**f**) Spinoidal phase-field [9]; and (**g**) An open pore metal-organic framework (MOF) [10]

and rate of expansion will depend on the system, but final structure should not depend on these parameters. These final porous systems need to be equilibrated to assure that they are stable against spontaneous collapse. Closed (non-percolating void) systems are easy to construct using this method, but open systems, especially ordered ones are possible by extending the growth until voids overlap. Such an example of a grown open-cell arrangement of voids in a hydrocarbon polymer foam [7] is shown in Fig. 2c.

Accumulated Grains As in experiments, porosity can be the result of void growth, or alternatively can be due to imperfect packing of extended rigid objects. To model the latter, one can produce simulation cells with a system of grain–grain contact points, to model force-chain evolution, contact deformation, and contact shear. Simulations of this type are currently limited to nanoscale grains. Two approaches can be used to produce aggregated grain systems. They can be constructed from a measured or calculated collection of non-overlapping large hard grains (usually spheres). The overall porosity of the packing is determined and often quite limited, by the packing density and required surface contact stability of these nanograins. Once the nanoscale grain packing is determined, each is filled with either arbitrarily rotated single crystal or pre-relaxed grains. This packing approach produced interconnected void space. Alternatively, a similar process can produce a structure more akin to sintered powders in which grains are randomly placed and filled. The overall system is constructed additively, by continuing to add grains until the desired porosity is reached (up to full density). When a new grain overlaps an existing grain, the new grain is only partially filled to avoid atomic overlaps. The sintered grain approach can produce open or closed void space. Two examples of systems constructed by accumulating grains are illustrated. In Fig. 2d we see a random sparse packing of layered Ni/Al grains incorporating approximately 41 million atoms from Cherukara et al. [8, 11], illustrating the contact-packed grain approach. In Fig. 2e we see an example of the sintered grain approach, for 50% porous silicon [6]. Note that Fig. 2b and e are equivalently porous materials with different microstructure. In cases, where underlying structure is unknown, sensitivity studies can be carried out.

Spinodal Phase-Field There has been significant recent experimental and theoretical interest in metallic nanoporous foams. In response, Sun et al. [12] developed a methodology to build a simulated open-cell foam which more realistically captures the spinodal structure of these materials. Experimentally, the foams are created by a dealloying process which is driven by phase separation, and resulting spinodal decomposition in a system. This process is mimicked with a phase-field approach to produce simulated nanoporous metals, with control of overall porosity and average ligament thickness. Figure 2f shows an example of nanoporous gold from Ngo et al. [9].

Other Specialized Structures Other more specialized approaches have been demonstrated to model specific systems, such as Metal-Organic Frameworks (MOFs). These are porous systems which are not constructed from a more dense phase. The molecular structure itself introduces the low density, rather than introduction of vacuum. These materials may be of great interest to modelers, but they are different in their construction and in their overall degree of homogeneity, compared to other porous media or foams. An example structure is illustrated in Fig. 2g from Yot et al. [10].

2.2 *Issues of Scale and Scaling*

As in non-porous molecular dynamics simulations, the most common approach to generating shock compression in molecular systems is to use non-equilibrium molecular dynamics (NEMD). In NEMD, a compression wave is created by driving a planar piston into a simulation cell, creating a 1D planar wave which traverses the system. Generally, the piston is an infinite-mass momentum mirror, although other boundaries are used to minimize noise and transients. NEMD is termed non-equilibrium because the equations of motion and integration algorithm do not incorporate any equilibrium assumptions. Generally a constant energy, constant particle number integrator is implemented (i.e., NVE ensemble), with periodic boundary conditions in directions transverse to the propagation direction. A variation of NEMD is to use a stationary wall rather than a moving piston, and instead throw the material toward the wall with a uniform average velocity. It can be shown that this is mathematically equivalent, but can make analysis more straightforward.

NEMD is a simple and powerful approach to modeling propagating compression waves, however, its computational cost scales poorly in situations in which long simulation times are needed. Because NEMD simulations model the propagating wave, longer simulations require larger systems to assure the wave does not run off the end of the system. Thus, extending simulation times by a factor m requires at least an m^2 increase in computational resources to model (i.e., the system must be m times larger, and be run for m times longer). Unfortunately, porous systems are particularly likely to require long simulation times. For instance, simulation times of hundreds of picoseconds to nanoseconds might be necessary to allow localized hot spots to thermalize and equilibrate—a process which is frequently necessary in order to compare to experimental measurements. In simulations of explosive initiation, chemical reaction timescales might require simulation durations of tens of nanoseconds or longer. With typical simulation timesteps of a fraction of a femtosecond, the total number of simulations timestep integrations can easily exceed 10^8. A daunting number of calculations, especially for large simulations.

For this reason, two methodologies have been employed to extend the modeling time (or reduce the computation expense) of long-duration compression simulations. The first of these is the Hugoniostat method of Ravelo et al. [13], and the second invokes the absorbing boundary condition of Bolesta et al. [14]. Both method geometries are shown in Fig. 3. While both of these methods have been used effectively to model compression in porous materials, both come with caveats and special considerations.

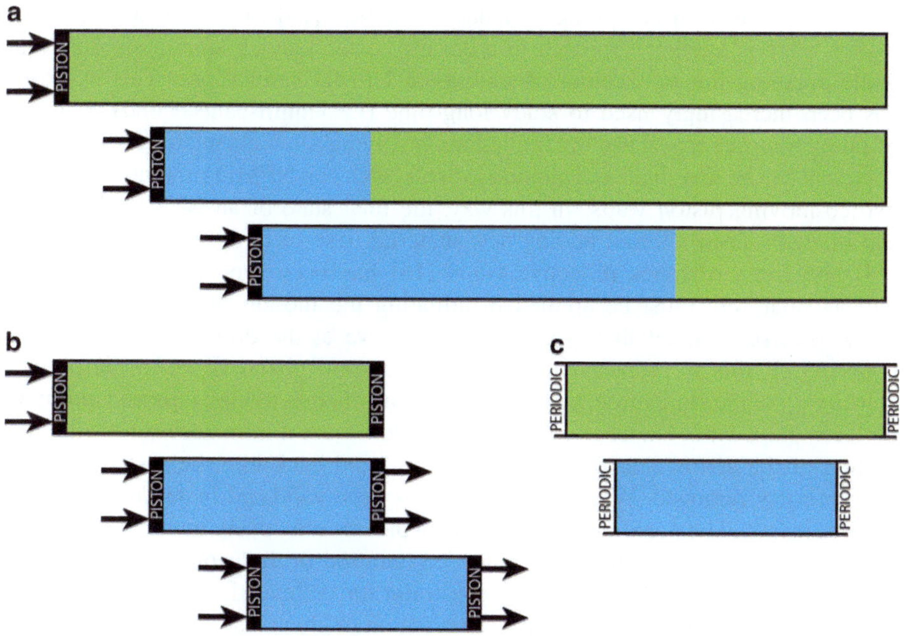

Fig. 3 Shock driver geometries: (**a**) Non-equilibrium molecular dynamics (NEMD); (**b**) Absorbing boundary condition [14]; (**c**) Constant-stress Hugoniostat [13]

The uniaxial constant-stress Hugoniostat [13] method homogeneously compresses a system while applying modified equations of motion, which guarantee the final state satisfies the Rankine-Hugoniot shock conditions. It is effectively a linked thermostat and barostat, which efficiently brings an initial system to the conditions which would be observed behind a steady-state shock at a given pressure. However, it does this without introducing a shock discontinuity or any spatial gradient in the system. The Hugoniostat is an efficient way to model a final homogeneous shock state, but does not capture the details of asymmetric void collapse, material flow, or stress/strain localization. For these reasons it has only recently been applied to shocks in porous systems [15–17], and even then only with significant care and validation. Jian et al. [16] argued that the Hugoniostat is likely to be useful in strong shock conditions where final states are highly homogenized, and intermediate processes do not significantly affect the final state. Lane et al. [15] showed that in a high-strength covalently bonded material, where void collapse is symmetric and slow relative to the shock propagation times, the Hugoniostat calculations predict strength and $P - V$ states which are consistent with NEMD calculations. Zhao et al. [17] also demonstrated only small differences ($\sim$1%) in $\rho - \sigma$ states of shock compressed 50% porous copper nanofoams. They showed that even this small difference disappeared at shock pressures above 10 GPa. However, care should be taken in applying the Hugoniostat methods in porous materials, especially when pore collapse processes are asymmetric and pressures are not sufficiently high.

Application of an absorbing boundary condition [14] to the downstream simulation edge is another approach to arbitrarily extend shock simulation durations while avoiding the m^2 scaling of traditional NEMD simulations. This approach has been increasingly used to study long-time (i.e., multi-nanosecond) evolution of pore collapse, especially in simulations involving slow chemistry [16–20]. The approach can be imagined as a method to "soft catch" an NEMD simulation between two co-moving piston walls. In this way, the final state of an NEMD simulation can be maintained without adding new material. Use of both momentum mirrors [14] and Lennard-Jones reflective walls [16] has been reported. Bolesta et al. indicate that care must be applied in initiating the motion of the second piston at the precise moment that atoms begin to move at the downstream simulation boundary. If this is properly timed, and the second piston is set to the particle velocity v_p of the simulation, then the wave is absorbed as though it passed through a perfectly impedance-matched interface. Reflections are minimized, but the thermal motion of the atoms near the boundary is reflected back into the system with v_p appropriately removed. While the method has been validated in several published reports, it should be noted that the co-moving pistons effectively maintain a constant volume simulation throughout the extended duration of the capture. It is not clear whether this would be an appropriate constraint for systems that undergo a delayed significant volume change due to phase transition or chemistry. This method is best applied to materials where void collapse to final density is abrupt, but slow volume-conserving processes persist. It has, for instance, been successfully applied to model the equilibration of hot spots, and ultimate recrystallization of local melting in nanoporous copper [17].

2.3 Interatomic Potentials and Other Considerations

Interatomic potentials are generally parameterized for ambient conditions and, therefore, are not necessarily quantitatively accurate at high pressures. Modelers must be thorough to test and validate potentials at pressure. A growing number of potentials are being developed specifically for high pressure, while some established potential forms are known to be better than others in capturing high-pressure dynamics. For soft materials, Buckingham-style (exponential-6) potentials, for instance, are generally better than Lennard-Jones (12-6) potentials, due to their softer core repulsion term. And, ReaxFF, while much more computationally expensive, has been shown to be accurate to nearly 2× compression [21]. For metals, EAM-based potentials have been parameterized to capture high-pressure response [22], and are frequently used to model porous metals.

At pressure, potential energy gradients are generally steeper, and therefore shorter timesteps are sometimes necessary to assure numerical stability. A method exists to estimate a timestep from the potential well depth and average maximum atomic velocity of the system. In practice, however, modelers will test a timestep by simulating in the NVE ensemble and looking for energy drift—a telltale signature of too long a timestep.

In addition to high-pressures, pore collapse produces extremely high temperatures which further test MD methodology. High temperatures indicate higher average velocities. And, as discussed above, high average velocities require shorter timesteps, in order to maintain numerical stability. In principle, one might think that simply continuing to decrease the timestep, one could go to ever higher temperatures. However, this is not the case. In highly porous materials, temperatures can reach 10,000 K at pressures as low as 15 GPa. At temperatures above an eV, one must assume that atoms will begin to ionize and that their interactions would be altered substantially. It is therefore generally thought that special care must be taken to validate simulations where temperatures exceed a few tens of thousands of Kelvin. Such temperatures are beyond the fidelity range of almost all potentials.

A final concern in interatomic potential selection is the large compression ratios in shock compression of porous materials. It is not unusual to reach 10× compression in extremely distended initial foams. Such compression can wreak havoc with electrostatics solvers and especially with long-range electrostatics. In some codes, such electrostatic interactions require periodic boundary conditions. Thus, NEMD, which requires open boundaries in the propagation direction, may not be an option when long-range electrostatic forces must be computed. In these cases, Hugoniostat methods can allow for charge calculations where other methods fail. It goes without saying that validation with experiment or DFT is necessary in these cases on the fringe of classical MD's applicability.

3 Application Areas

Early molecular dynamics studies of void collapse were motivated by a need to understand the initiation processes in energetic materials. Beginning in the early 1990s, researchers began to use molecular dynamics to probe whether, and how, material defects might contribute to the early stages of explosives detonation. It was generally believed that inhomogeneities in crystals might lead to energy concentrations which allowed for chemical initiation even in circumstances where a homogeneous material would be inert. However, there was little direct evidence for this proposition. Maffre and Peyrard [23] were one of the first to study the effects of shock on model grain boundaries, and vacancies, concluding that such defects did, in fact, lead to hot spot formation in some shock strength regimes. A series of papers followed, which used 2D simulations of non-reactive Lennard-Jones or Morse potentials to probe pore collapse in otherwise perfect 2D crystals. Phillips et al. [24] showed that voids caused heating in atomic lattices where planar gaps did not, concluding that there was an "essential physical difference between a gap and an included void." Tsai and Armstrong [25], using a simple molecular crystal, concluded that void collapse converted local strain energy into kinetic energy, which led to not only heating, but also additional defects such as kinks and slip plane formation. This was further supported by Bandak et al. [26] in atomic crystals who identified nanodislocation dipoles in the wake of void collapse. Fried and Tarver [27] used an early 2D model for TATB to conclude that 20% random vacancy was enough to significantly raise the overall temperature of an explosive, and increase equilibration times by an order of magnitude.

Mintmire et al. [28] were the first to use 3D models, studying void collapse in an fcc crystal of diatomic molecules. They showed that a threshold shock strength was necessary to drive "turbulent collapse" of voids. At lower shock strengths voids crushed to full density without producing hot spots, but above this threshold, significant local heating and excitation of vibrational modes were observed. These conclusions largely resolved the primary issue of whether void collapse could be a source of hot spots and early initiation of explosive detonation.

It was not until 2002 that Holian, Germann et al. [29, 30] reopened debate on the detailed processes involved in void collapse. They wrote "(i) A critical velocity for detonation exists for a perfect crystal. (ii) The critical velocity can be reduced by shock passage over sufficiently large voids, as long as the shock is strong enough to release into the vapor dome. (iii) The heating by recompression of the spalled (ejecta) gas depends upon the void size and asymptotes to a value considerably above the first-shock temperature. (iv) There is a delay time in the initiation of chemistry, because of the compaction process at the far side of the void. (v) Reactions occur first in the recompressed gas, not in the far wall." Holian et al. argued that vaporization and recompression of the resulting gas was the true source of hot spots and initiation, and that void shape was not a critical factor—demonstrating dramatic temperature overshoot even in a 2D planar gap geometry in an fcc lattice of unreactive Lennard-Jones particles. Hatano [31] responded by showing that, in fact, explosive initiation depends more on particle collision, rather than temperature, and that Holian et al. oversimplified the argument for initiation. Hatano argued that the details of the void shape, and the processes in void collapse matter, and that maximums in temperature, collision number, and chemical initiation may occur in different locations and at different times in the collapse of a porous material. Moreover, Hatano concluded that differently shaped voids maximize different elements of the response. For instance, he concluded that larger cross-section voids enhance the number of collisions within a collapsing void, while larger transverse voids maximize temperature. Later in this section, we will see that neither authors' generalizations hold for all materials. The importance of geometry is dependent on the material, the loading and on what measurements are of particular interest.

After 2004 the field shifted from general characterization of processes in model materials, to more quantitative and focused studies using realistic potentials for specific materials. At the same time, research has moved beyond the response of individual voids, to collective response of foams, and most recently toward realistically capturing porous structure, such as in the still emerging area of nanoporous metals. It is now clear that the details of porous materials, including chemistry, pore size, pore shape, wall thickness, and pore dispersion may all be factors in a material's shock compression properties, and that the relative importance of these factors may differ significantly between material classes.

The remainder of this section will explore the specific processes and mechanisms observed in different families of materials. We will discuss metals, high-strength covalent materials, soft materials, and explosives.

3.1 Shock Compression in Porous Metals

In porous metals, two void collapse regimes are seen. In the low to moderate shock strength regime, nanoporous metals accommodate local strain plastically with the emission of dislocation loops and/or stacking fault production. In the higher shock strength regime, metals exhibit local melting, and flow, to include jetting. The work of Zhao et al. (see Fig. 4) shows these two regimes for a number of columnar pore geometries in copper. The snapshots on the left show the low to moderate shock strength response, while the images on the right show high shock strength response. We see that the nature of the response shifts dramatically from defect production to flow and jetting. One might refer to these regimes as "material-specific strength" response at lower shock strength, and a more "universal flow" response at higher shock strengths. In copper, Zhao et al. [32] alternatively refer to these as "geometric" and "hydrodynamic" modes, respectively.

In the low shock strength regime, recent work has identified dislocation emission processes associated with closed-cell void collapse in fcc metals copper [33, 34] and aluminum [34], bcc tantalum [35–38] and more recently nanoporous gold [9, 12, 39, 40] and γ-TiAl alloy [41].

While dislocation and kink production were identified during void collapse in some of the earliest molecular simulations [25] of shocked pores, it is only recently that sophisticated analysis tools have been developed which can readily extract and visualize these extended defects. Tools such as the dislocation extraction algorithm (DXA) from Stukowski [42] and Stukowski et al. [43] have allowed for significant advances in understanding the localized deformation in metals, and how these defects interact to produce hardening.

Davila et al. [33] were the first to report detailed dislocation behavior associated with single pore collapse in copper. Subsequent work [34] extended these results to multiple void systems with porosity of approximately 4%, observing interaction of shear loops from separate voids. Tang et al. [35] did the same for single voids in tantalum. Figure 5 shows the emission of shear loops in tantalum under moderate shock state conditions of uniaxial compression. Such analysis allows for the identification of slip planes and testing of models of dislocation production and mobility, which can be quite complex and material specific. The single void tantalum work was extended to multiple voids by Ruestes et al. [36, 37]. These simulations, notably, identify shear loop production, but show little evidence of twinning in compression, unlike what is observed in tension.

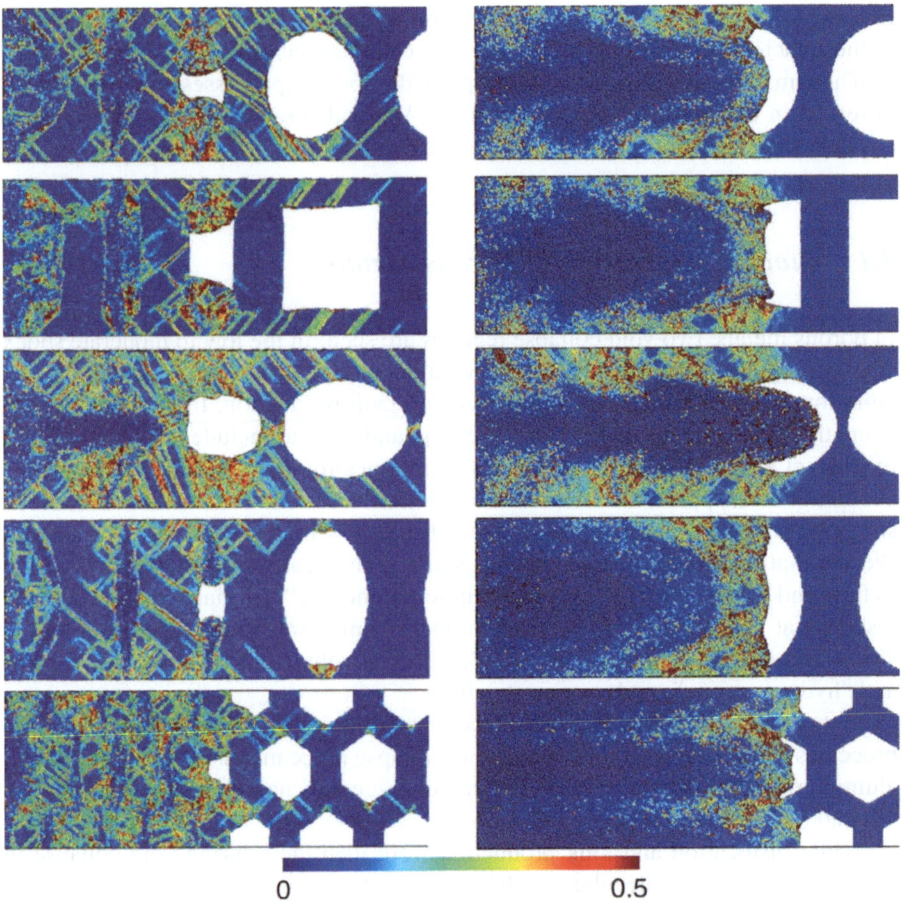

Fig. 4 Void collapse response for a variety of columnar pore shapes (circle, square, horizontal oval, vertical oval, and honeycomb, from top to bottom) in copper. For each pore shape, response is shown in both the low shock strength (left) and high shock strength (right) regimes. In copper, the threshold piston velocity u_p is approximately 0.7 km/s and these images are for $u_p = 0.625$ km/s and $u_p = 2.0$ km/s. The atoms are colored by local shear strain η_{vM}. Adapted with permission from Zhao et al. [32]. Copyright 2013 AIP Publishing LLC

At higher shock strengths, metals accommodate additional strain energy by local melting and flow processes. Work has focused on characterizing the effect of pore shape on flow dynamics and jetting, as well as characterizing local vs. global melting and subsequent recrystallization on thermalization and equilibration time scales. He et al. [20] studied melting and long-time recrystallization in copper at piston velocities over 2.75 km/s. Zhao et al. [17, 32, 44] also studied melting, and specifically determined a power law relation between incipient melt and porosity. Zhao also showed that the details of void collapse, jetting and the elastic-plastic transition depend strongly on void shape (or microstructure) in this regime.

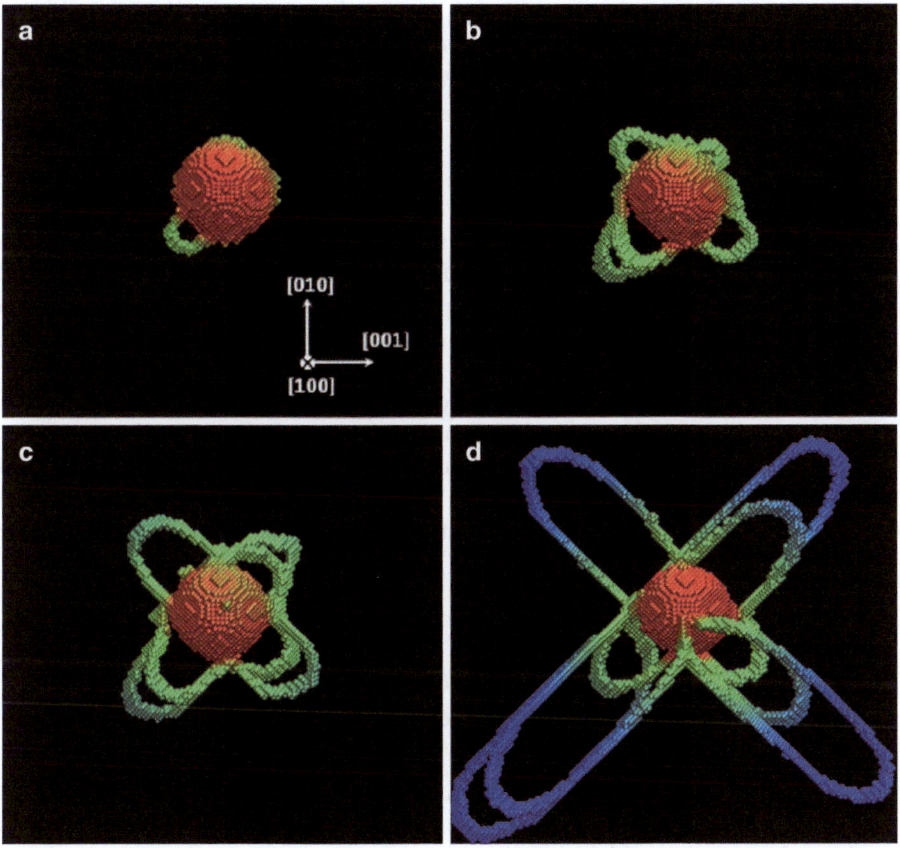

Fig. 5 A sequence showing the emission of shear loop dislocations from a collapsing 3.3 nm diameter void in Ta under uniaxial compression at 10^8 1/s, from Tang et al. [35]

However, the macroscopic shock Hugoniot state variables are not dependent on void microstructure. This distinction is interesting, and deserves more investigation in other materials.

Since 2013, modeling interest in higher-porosity open-cell metallic foams has grown significantly, largely due to congruency of the modeling and experimental scales. These structures were first modeled by Crowson et al. [45, 46], and refined by Sun et al. [12] to allow independent control of relative mass density and average ligament size. The construction methodology (described in Sect. 2) produces representative bicontinuous structures using a phase field model of spinodal decomposition in binary alloy. The methodology has been applied to study uniaxial compression of nanoporous gold [9, 39, 40], copper [16], and aluminum [47]. Neogi and Mitra [48] used spherical voids, to compare directly between open- and closed-cell geometries in copper.

Mechanical studies of open-cell foams demonstrate plastic response at extremely low loads, measured at below 0.1 GPa [9, 40], but also very high strain hardening. They also exhibit a propagating three-wave response of elastic, plastic, and “collapse” waves [47], and good agreement between experiments and atomic scale modeling.

Jian et al. [16] used open foams to quantitatively compare the macroscopic Hugoniot shock response in materials as a function of specific surface area. They found a strong dependence of shock pressure and temperature on porosity, and a weaker, but significant, dependence on surface area per volume. It should be noted that this dependence on specific surface area may be a nanoscale effect, due to nanoscale ligament thicknesses. Further study is needed to confirm this trend is sustained in microscale porosity.

Finally, Soulard et al. [49] recently conducted an atomistic simulation of gas-filled pores in copper. They found a significant reduction in heat production and melting in gas-filled pores, as compared to vacuum pores. It has long been assumed that gas densities in voids would have an insignificant effect on shock compressed porosity, and thus, gases are usually omitted. This work suggests that the issue should be more carefully investigated.

3.2 Shock Compression in Porous Ceramics, Covalent Materials, and Enhanced Densification

As we have discussed, heating and hot spot formation in pore collapse is a universal feature of shock compressed porous materials. In most materials, this heating leads to thermal expansion which drives up pressures (or, at constant pressure, drives down density). The net result is that the porous Hugoniot is generally above or to the left of the principal Hugoniot in $P - \rho$ plots of shock response. However, in some materials, the opposite is true. The introduction of initial porosity causes an enhanced densification as compared to the final states from full dense samples. This effect has been observed experimentally in silicon nitride, silicon dioxide, boron carbide, and uranium dioxide [50]. Several mechanisms have been proposed for this behavior [51]. One of these is that the densification is the result of a local solid–solid phase transition to a higher density state at pressures below the equilibrium transition pressure. This process is reminiscent of the pre-mature hot spot initiation in energetic materials, however here the local nucleation is driven by shear strain rather than temperature.

Recent molecular dynamics simulations have demonstrated that such a process is possible. Lane et al. [6, 15] showed that introduction of porosity from 5% to 50% in silicon reduces the pressure threshold for phase transition. Ribbons of phase transformed material nucleate at collapsing pores and spread into the bulk. The process relieve the driving shear stress, arresting the transition and leaves a metastable mixed phase state on simulations timescales. While these observations

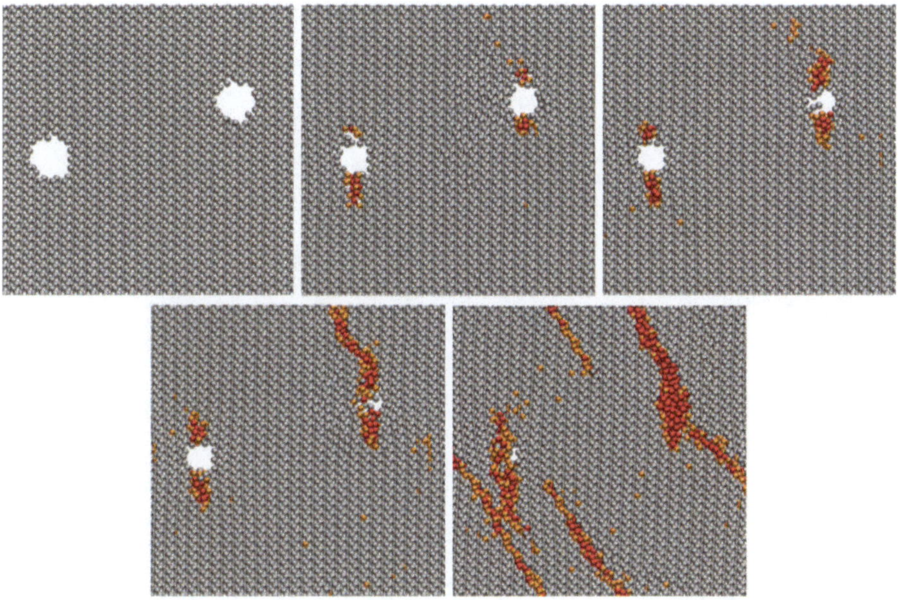

Fig. 6 Slices showing atoms near collapsing pores in 1% porous silicon. The 20 GPa shock nucleates a local phase transition (red atoms) below the pressure necessary to nucleate in homogeneous crystal. Atoms are colored by coordination number. Images from Lane et al. [6]

demonstrate the concept, simulations have not been conducted in materials where enhanced densification has been experimentally observed.

Figure 6 shows atomistic snapshots of void collapse in 1% porous silicon from Lane et al. [6]. The first image shows two cut voids in uncompressed material. Between the first and second snapshot an elastic shock moves through the region, and the pore closes over approximately 44 ps. As the pore collapses, shear stain nucleates local phase transition, which is illustrated by atomic coordination number. Gray indicated a diamond structure, and red, the phase transformed higher-density state.

Figure 7 shows the Hugoniot in $P - \rho$ for several porosities and void microstructures. As in Sect. 3.1, we see here only a weak dependence of the Hugoniot variables on the void structure, especially at shock intensities. The black curve is the principal Hugoniot for silicon, as measured by MD. The evidence of enhanced densification can be seen in all of the porous states.

While this was the first simulation to show solid–solid phase transition leading to enhanced densification, Wu et al. [52] have recently reported pore-enhanced phase transitions in single-crystal iron. They argue that stress assisted transformation mechanisms and strain induced transformation mechanisms in the local environment play a role in altering the homogeneous nucleation of high-pressure iron phases. Other studies have been carried out in silica [53] and ceria [54].

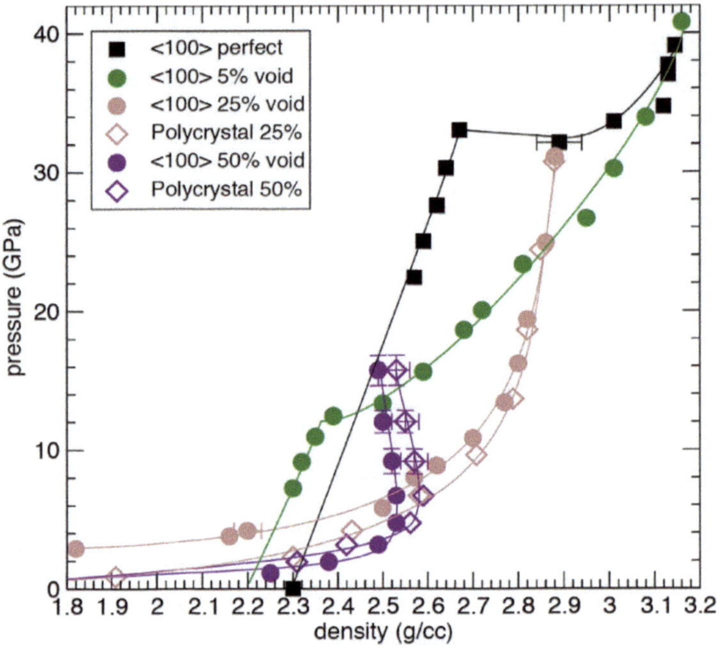

Fig. 7 Hugoniot curves for high-porosity silicon. Comparison with the principal Hugoniot (black) shows evidence of enhanced densification. Porous Hugoniots generally are shifted up and to the left of the principal Hugoniot due to increased heating and thermal expansion. Plot reprinted from Lane et al. [6]

This is an area which is ripe for study, especially with the advent of experimental apparatus which might more easily measure signatures of partial phase transitions in porous media. Such studies are possible [55] with facilities such as the Advanced Photon Source. See the chapter by Jensen et al. in this volume for details about such experiments.

3.3 Shock Compression in Porous Soft and Energetic Materials

Unlike hard materials, polymers and soft materials have low strength and generally exhibit plastic flow even at weak shock conditions. At higher shock pressures, organic materials are relatively quick to exhibit chemistry, dissociation, and vaporization. There are very few atomistic studies of shock compression of polymer foams. This is likely due to the dearth of suitable hydrocarbon potentials for extreme pressures and temperatures. Mattsson et al. [21, 56] compared DFT and experimental results with four classical MD interatomic potentials for full-density polyethylene and poly(4-methyl-1-pentene) (PMP) under shock loading up to 100s of GPa. Using $P-V$ Hugoniot data from Z-machine experiments, supplemented by

DFT calculations, they were able to show that ReaxFF [57, 58] was the most reliable of the four up to a compression ratio of approximately 2:1, or density of 1.8 g/cc. The other potentials, especially those based on a Lennard-Jones core repulsion, were generally too stiff at higher pressures.

Shock compression of polymeric foams has been used to study temperatures and thermalization in hot spots [7]. In simulations of polymeric foam compression, hot spot temperatures were shown to reach tens of thousands of degrees locally, but quickly thermalizes on picosecond timescales with their local environments. These local extreme temperatures, even short-lived are enough to vaporize polymeric systems, and significantly raise average temperatures on Hugoniot states by more than an order of magnitude. Figure 8 shows final average temperature behind shock fronts for several initial density foams, as well as full density polymer. While experimental temperature measurements were not possible, corresponding Hugoniot $P-V$ response confirmed that atomistic results agreed with Z-machine experiments conducted by Root et al. [59].

While initiation and explosives applications are not the focus of this book, it is appropriate to briefly summarize the latest atomistic simulation work in this area. A significant number of papers have recently been published which take the study of initiation to much larger system sizes and to much higher fidelity in interatomic potentials. The ReaxFF potential [57, 58] has been particularly successful in these applications. Studies have continued to investigate the general properties of hot spots in model energetic materials [60, 61], but have also moved to properties of specific systems, including RDX [18, 62, 63], HMX [64, 65], PETN [5, 19], and ETN [66]. Many of these studies include the calculation of specific initiation

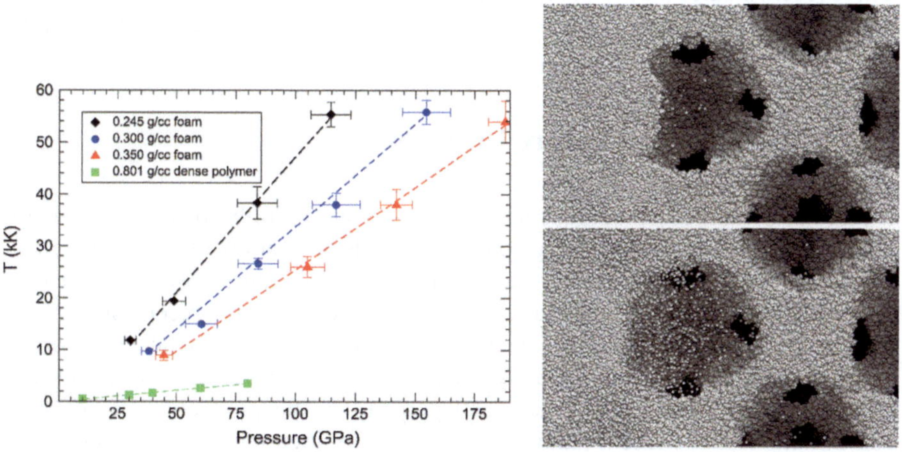

Fig. 8 (Left) Plot of temperature for several initial porosities of hydrocarbon foam from Lane et al. [7]. Final shock state temperatures can be more than an order of magnitude higher in foams than in full dense polymers for a given pressure. (Right) Snapshots of vaporized polymer propagating ahead of the shock front, through the open foam structure at (top) 10 km/s and (bottom) 25 km/s impact velocities in 0.3 g/cc foam [59]

thresholds and evolution properties, which is beyond the scope of this review. However, several address the fundamental issues of the role of void scale in the production of sustained deflagration.

Wood et al. [63] and Shan et al. [19] have recently reported simulations with pores sizes of 10–40 nm, and system sizes and durations which push the envelope of what is currently possible at even the largest computing scales. In addition to large sizes, these reactive simulations require high-fidelity potentials which are on the order of 10–30 times more costly than typical LJ or EAM potentials. With these larger systems, researchers have begun to differentiate between initiation of chemistry and sustained exponential growth of a detonation wave. This requires much longer simulations out to hundreds of picoseconds, even as the integration time step is held to a fraction of a femtosecond. The absorbing wall method discussed in Sect. 2 has been applied to these systems to extend simulations times [14, 19].

Interestingly, fundamental insight is continuing to be challenged with these more quantitative simulations. For instance, Wood et al. [63] recently compared dynamically-produced hot spots from collapsing voids with artificially-produced hot shots which mimicked the pressure, density, and temperature states found in void collapse, but not the flow dynamics. They showed that the artificial hot spots did not always lead to sustained detonation, where the dynamically produced hot spots did. They concluded that, contrary to common assumption, it is not simply the temperature within the "hot spot" which drives detonation—an idea which harkens back to Hatano's earlier assertions [31]. Eason and Sewell [18] similarly use tracer particles to investigate the mixing and rotation of flow in the nodes of dynamic pore collapse. They show strong dependence of these properties on the shock strength. These recent studies may invigorate the argument that non-linear processes and non-equilibrium dynamics drive chemical initiation, more than simple heat production.

4 Conclusions and Future Directions

Molecular dynamics simulations have been used successfully to study the specific mechanisms and processes during collapse of voids in a variety of porous materials. While the approach is currently limited to nanoscale features, we see increasing interest in this class of materials. We have surveyed, here, recent simulations in a wide variety of materials, including metals, ceramics, covalent materials, soft materials, and explosives. Early simulations were instrumental in confirming the importance of material defects and porosity in the production of hot spots and initiation. More recent studies have effectively identified the molecular scale relaxation mechanisms which are explored in weaker shock strengths.

From early work, with only thousands of atoms, to modern simulations with billions of atoms interacting with high-fidelity potentials, the scale of problems has grown with the computational capacity to efficiently solve larger and larger problems. We expect that this trend will continue, with future simulations striving to address mesoscale processes, which overlay the atomistic processes. In particular grain–scale interactions of powder compaction are currently on the threshold of what is possible. We expect future work will begin to explore and identify molecular mechanisms at contacts and fracture in nanoscale grains, especially at low shock strengths. These problems are currently in the mesoscale realm, where atomistic processes are difficult to incorporate. Similarly, as accessible timescales grow, new areas of research will open. Simulations on multi-nanosecond timescales and beyond will allow for slow chemistry to evolve and for homogenization of the inherently local and therefore heterogeneous response around voids. There is still much to learn about the role and interplay between equilibrium and non-equilibrium processes.

As system sizes grow larger and more complex, improved analysis tools and techniques become crucial. Data handling and extraction research will become key to advances in understanding. This has been recently demonstrated in the extraction and visualization of dislocations in metals. Improved techniques have driven rapid development in the understanding of strength mechanisms in nanoporous metals. Continued improvement of these approaches to higher compressions will be necessary to quantify the role of atomic disorder on hardening mechanisms in strong shocks.

Other areas of future research are likely to include rigorous study of strain-rate effects over wider ranges of strain rate. The role of gas filled, vs vacuum, void spaces is due for more rigorous study, in light of recent work by Soulard et al. [49]. The emerging area of anomalous densification and solid–solid phase transition studies are an intriguing area, which is also currently accessible to molecular studies. Current research points toward the unexplored possibility of using void collapse to achieve molecular scale mixing and/or sintering in heterogeneous materials.

Finally, realistic modeling of experimentally achievable nanoscale porous geometries must be expanded, especially to soft materials and powder beds. This will require improved experimental characterization of pore structure, and for modelers to utilize this improved characterization to build more relevant geometries, beyond the spherical pore. Doing so will move the molecular study of porous materials from the realm of physics to the growing field of materials science.

Acknowledgment Sandia National Laboratories is a multi-mission laboratory managed and operated by National Technology and Engineering Solutions of Sandia, LLC, a wholly owned subsidiary of Honeywell International, Inc., for the U.S. Department of Energy's National Nuclear Security Administration under contract DE-NA-0003525.

References

1. Allen MP, Tildesley DJ (1989) Computer simulation of liquids. Oxford University Press, New York
2. Rapaport DC (2004) The art of molecular dynamics simulation, 2nd edn. Cambridge University Press, New York

3. Frenkel D, Smit B (2001) Understanding molecular simulation: from algorithms to applications, vol 1. Academic, San Diego
4. Plimpton S (1995) Fast parallel algorithms for short-range molecular dynamics. J Comput Phys 117(1):1–19. LAMMPS website: http://lammps.sandia.gov
5. Shan TR, Thompson AP (2014) Shock-induced hotspot formation and chemical reaction initiation in PETN containing a spherical void. J Phys Conf Ser 500:172009
6. Lane JMD, Thompson AP, Vogler TJ (2014) Enhanced densification under shock compression in porous silicon. Phys Rev B 90(13):134311
7. Lane JMD, Grest GS, Mattsson TR (2013) Hot spot and temperature analysis of shocked hydrocarbon polymer foams using molecular dynamics simulation. Comput Mater Sci 79:873–876
8. Cherukara MJ, Germann TC, Kober EM, Strachan A (2016) Shock loading of granular Ni/Al composites. Part 2: Shock-induced chemistry. J Phys Chem C 120(12):6804–6813
9. Ngô BND, Stukowski A, Mameka N, Markmann J, Albe K, Weissmüller J (2015) Anomalous compliance and early yielding of nanoporous gold. Acta Mater 93:144–155
10. Yot PG, Boudene Z, Macia J, Granier D, Vanduyfhuys L, Verstraelen T, Van Speybroeck V, Devic T, Serre C, Férey G, et al (2014) Metal–organic frameworks as potential shock absorbers: the case of the highly flexible MIL-53 (Al). Chem Commun 50(67):9462–9464
11. Cherukara MJ, Germann TC, Kober EM, Strachan A (2014) Shock loading of granular Ni/Al composites. Part 1: mechanics of loading. J Phys Chem C 118(45):26377–26386
12. Sun XY, Xu GK, Li X, Feng XQ, Gao H (2013) Mechanical properties and scaling laws of nanoporous gold. J Appl Phys 113(2):023505
13. Ravelo R, Holian B, Germann T, Lomdahl P (2004) Constant-stress hugoniostat method for following the dynamical evolution of shocked matter. Phys Rev B 70(1):014103
14. Bolesta AV, Zheng L, Thompson DL, Sewell TD (2007) Molecular dynamics simulations of shock waves using the absorbing boundary condition: a case study of methane. Phys Rev B 76(22):224108
15. Lane JMD, Thompson AP, Vogler TJ (2017) Enhanced densification, strength and molecular mechanisms in shock compressed porous silicon. AIP Conf Proc 1793(1):120010. https://doi.org/10.1063/1.4971692
16. Jian W, Li B, Wang L, Yao X, Luo S (2015) Shock response of open-cell nanoporous Cu foams: effects of porosity and specific surface area. J Appl Phys 118(16):165902
17. Zhao F, Li B, Jian W, Wang L, Luo S (2015) Shock-induced melting of honeycomb-shaped Cu nanofoams: effects of porosity. J Appl Phys 118(3):5904
18. Eason RM, Sewell TD (2015) Molecular dynamics simulations of the collapse of a cylindrical pore in the energetic material α-RDX. J Dyn Behav Mater 1(4):423–438
19. Shan TR, Wixom RR, Thompson AP (2016) Extended asymmetric hot region formation due to shockwave interactions following void collapse in shocked high explosive. Phys Rev B 94(5):054308
20. He A, Duan S, Shao JL, Wang P, Qin C (2012) Shock melting of single crystal copper with a nanovoid: molecular dynamics simulations. J Appl Phys 112:074116
21. Mattsson TR, Lane JMD, Cochrane KR, Desjarlais MP, Thompson AP, Pierce F, Grest GS (2010) First-principles and classical molecular dynamics simulation of shocked polymers. Phys Rev B 81(5):054103
22. Ravelo R, Germann TC, Guerrero O, An Q, Holian BL (2013) Shock-induced plasticity in tantalum single crystals: interatomic potentials and large-scale molecular-dynamics simulations. Phys Rev B 88:134101
23. Maffre P, Peyrard M (1992) Molecular-dynamics investigations of shock-induced detonations in inhomogeneous energetic crystals. Phys Rev B 45(17):9551
24. Phillips L, Sinkovits RS, Oran ES, Boris JP (1993) The interaction of shocks and defects in Lennard-Jones crystals. J Phys Condens Matter 5(35):6357
25. Tsai D, Armstrong R (1994) Defect-enhanced structural relaxation mechanism for the evolution of hot spots in rapidly compressed crystals. J Phys Chem 98(43):10997–11000

26. Bandak F, Tsai D, Armstrong R, Douglas A (1993) Formation of nanodislocation dipoles in shock-compressed crystals. Phys Rev B 47(18):11681
27. Fried LE, Tarver C (1996) Molecular dynamics simulation of shocks in porous TATB crystals. In: AIP conference proceedings, vol 370. AIP Publishing, New York, pp 179–182
28. Mintmire J, Robertson D, White C (1994) Molecular-dynamics simulations of void collapse in shocked model-molecular solids. Phys Rev B 49(21):14859
29. Holian BL, Germann TC, Maillet JB, White CT (2002) Atomistic mechanism for hot spot initiation. Phys Rev Lett 89:285501
30. Germann TC, Holian BL, Lomdahl PS, Heom A, Gronbech-Jensen N, Maillet JB (2002) Molecular dynamics simulations of detonation in defective explosive crystals. In: Technical papers of 12th international detonation symposium, pp 11–16
31. Hatano T (2004) Spatiotemporal behavior of void collapse in shocked solids. Phys Rev Lett 92:015503
32. Zhao F, Wu H, Luo S (2013) Microstructure effects on shock response of Cu nanofoams. J Appl Phys 114(7):073501
33. Davila L, Erhart P, Bringa E, Meyers M, Lubarda V, Schneider M, Becker R, Kumar M (2005) Atomistic modeling of shock-induced void collapse in copper. Appl Phys Lett 86(16):161902
34. Erhart P, Bringa EM, Kumar M, Albe K (2005) Atomistic mechanism of shock-induced void collapse in nanoporous metals. Phys Rev B 72(5):052104
35. Tang Y, Bringa EM, Remington BA, Meyers MA (2011) Growth and collapse of nanovoids in tantalum monocrystals. Acta Mater 59(4):1354–1372
36. Ruestes C, Bringa E, Stukowski A, Nieva JR, Bertolino G, Tang Y, Meyers M (2013) Atomistic simulation of the mechanical response of a nanoporous body-centered cubic metal. Scr Mater 68(10):817–820
37. Ruestes C, Bringa E, Stukowski A, Nieva JR, Tang Y, Meyers M (2014) Plastic deformation of a porous bcc metal containing nanometer sized voids. Comput Mater Sci 88:92–102
38. Tramontina D, Ruestes C, Tang Y, Bringa E (2014) Orientation-dependent response of defective tantalum single crystals. Comput Mater Sci 90:82–88
39. Farkas D, Caro A, Bringa E, Crowson D (2013) Mechanical response of nanoporous gold. Acta Mater 61(9):3249–3256
40. Ruestes CJ, Farkas D, Caro A, Bringa EM (2016) Hardening under compression in Au foams. Acta Mater 108:1–7
41. Xu XT, Tang FL, Xue HT, Yu WY, Zhu L, Rui ZY (2015) Molecular dynamics simulations of void shrinkage in γ-TiAl single crystal. Comput Mater Sci 107:58–65
42. Stukowski A (2012) Structure identification methods for atomistic simulations of crystalline materials. Model Simul Mater Sci Eng 20(4):045021
43. Stukowski A, Bulatov VV, Arsenlis A (2012) Automated identification and indexing of dislocations in crystal interfaces. Model Simul Mater Sci Eng 20(8):085007
44. Zhao F, An Q, Li B, Wu H, Goddard III W, Luo S (2013) Shock response of a model structured nanofoam of Cu. J Appl Phys 113(6):063516
45. Crowson DA, Farkas D, Corcoran SG (2007) Geometric relaxation of nanoporous metals: the role of surface relaxation. Scr Mater 56(11):919–922
46. Crowson DA, Farkas D, Corcoran SG (2009) Mechanical stability of nanoporous metals with small ligament sizes. Scr Mater 61(5):497–499
47. Gunkelmann N, Rosandi Y, Ruestes CJ, Bringa EM, Urbassek HM (2016) Compaction and plasticity in nanofoams induced by shock waves: a molecular dynamics study. Comput Mater Sci 119:27–32
48. Neogi A, Mitra N (2014) On shock response of nano-void closed/open cell copper material: non-equilibrium molecular dynamic simulations. J Appl Phys 115(1):013504
49. Soulard L, Pineau N, Clérouin J, Colombet L (2015) Molecular dynamics simulations of shock compressed heterogeneous materials. I. The porous case. J Appl Phys 117(11):115901
50. Grady D, Fenton G, Vogler T (2013) Equation of state and evidence of enhanced phase transformation for shock compression of distended compounds. Int J Impact Eng 56:19–26

51. Medvedev AB, Trunin RF (2012) Shock compression of porous metals and silicates. Phys-Uspekhi 55(8):773–789
52. Wu L, Wang K, Xiao S, Deng H, Zhu W, Hu W (2016) Atomistic studies of shock-induced phase transformations in single crystal iron with cylindrical nanopores. Comput Mater Sci 122:1–10
53. Jones K, Lane JMD, Vogler TJ (2018) Molecular dynamics study of shock compression in porous silica glass. AIP Conf Proc 1979:050010
54. Weck PF, Cochrane KR, Root S, Lane JMD, Shulenburger L, Carpenter JH, Sjostrom T, Mattsson TR, Vogler TJ (2018) Shock compression of strongly correlated oxides: a liquid-regime equation of state for cerium(IV) oxide. Phys Rev B 97:125106
55. Turneaure SJ, Sinclair N, Gupta YM (2016) Real-time examination of atomistic mechanisms during shock-induced structural transformation in silicon. Phys Rev Lett 117:045502
56. Mattsson TR, Cochrane KR, Lane JMD, Root S (2019) Simulations of hydrocarbon polymers related to compression experiments on Sandia's Z machine. Springer International Publishing, Cham, pp 1–23
57. Van Duin AC, Dasgupta S, Lorant F, Goddard III WA (2001) ReaxFF: a reactive force field for hydrocarbons. J Phys Chem A 105(41):9396–9409
58. Chenoweth K, van Duin ACT, Goddard III WA (2008) ReaxFF reactive force field for molecular dynamics simulation of hydrocarbon oxidation. J Phys Chem A 112:1040
59. Root S, Haill TA, Lane JMD, Thompson AP, Grest GS, Schroen DG, Mattsson TR (2013) Shock compression of hydrocarbon foam to 200 GPa: experiments, atomistic simulations, and mesoscale hydrodynamic modeling. J Appl Phys 114(10):103502
60. Shi Y, Brenner DW (2008) Jetting and detonation initiation in shock induced collapse of nanometer-scale voids. J Phys Chem C 112(16):6263–6270
61. Herring SD, Germann TC, Grønbech-Jensen N (2010) Effects of void size, density, and arrangement on deflagration and detonation sensitivity of a reactive empirical bond order high explosive. Phys Rev B 82(21):214108
62. Nomura K, Kalia RK, Nakano A, Vashishta P (2007) Reactive nanojets: nanostructure-enhanced chemical reactions in a defected energetic crystal. Appl Phys Lett 91(18):183109
63. Wood MA, Cherukara MJ, Kober EM, Strachan A (2015) Ultrafast chemistry under nonequilibrium conditions and the shock to deflagration transition at the nanoscale. J Phys Chem C 119(38):22008–22015
64. Duan XH, Li WP, Pei CH, Zhou XQ (2013) Molecular dynamics simulations of void defects in the energetic material HMX. J Mol Model 19(9):3893–3899
65. Zhou T, Lou J, Zhang YG, Song HJ, Huang FL (2016) Hot spot formation and chemical reaction initiation in shocked HMX crystal with a nanovoid: a large-scale reactive molecular dynamics study. Phys Chem Chem Phys 18(26):17627–17645
66. Furman D, Kosloff R, Zeiri Y (2016) Effects of nanoscale heterogeneities on the reactivity of shocked erythritol tetranitrate. J Phys Chem C 120(50):28886–28893

Additively Manufactured Cellular Materials

Ron Winter and Graham McShane

1 Introduction

It is well known that porous metals have potential application as absorbers of energy generated by high velocity impact or explosive loading. Lattice materials in which the porosity arises from a regular array of nominally identical cells consisting, for example, of planar walls or cylindrical struts are an important class of porous materials. The choice of a lattice structure to meet the requirements demanded by a particular loading scenario requires a careful balance of strength, stiffness and cost in addition to optimised response to high-rate loading. The state of the art on manufacturing, mechanical properties and applications of metallic micro-lattice materials was reviewed by Rashed et al. [1]. One of the manufacturing techniques highlighted by Rashed was additive manufacturing (or AM). AM has emerged as a versatile technique to closely define the structure and properties of lattice, or cellular, materials. In particular the technique allows components to be manufactured in which the properties vary with position in order, for example, to tailor the shape and strength of a compression wave passing through the component. A brief review of additive manufacturing processes, with particular emphasis on cellular metals, is presented in Sect. 2. Clearly, to optimise cellular materials for energy absorbing applications it is important to determine how their energy-absorbing capability in the loading regime of interest depends on the architecture and material of the lattice. It will be argued here that, in studies aiming to optimise the properties of lattice materials, AM technology offers significant

R. Winter (✉)
Physics, AWE, Reading, UK
e-mail: ron.winter@awe.co.uk

G. McShane
Department of Engineering, University of Cambridge, Cambridge, UK

T. J. Vogler, D. A. Fredenburg (eds.), *Shock Phenomena in Granular and Porous Materials*, Shock Wave and High Pressure Phenomena,
https://doi.org/10.1007/978-3-030-23002-9_9

advantages relative to traditional manufacturing methods. One of the tools available to understand and quantify the impact response of lattice materials are computer simulations which have been validated against experiments. Two regimes, which will be termed here, "intermediate" rate loading and "high" rate loading, will be discussed in Sects. 3 and 4, respectively. Intermediate-rate loading is loosely defined as situations in which the elastic and plastic stress waves generated by the initial impact propagate through the full thickness of the specimen before it is fully densified. It is assumed that if intermediate-rate loading were to be maintained for a long period it would eventually lead to full densification as a result of multiple wave reflections within the sample. In our intermediate-rate studies, samples are loaded at velocities ranging from 105 to 157 ms^{-1} using a direct impact Hopkinson bar apparatus. In high-rate loading, by contrast, the material tends to be fully densified by the first plastic shock. It is assumed that at the shock front there is a density discontinuity with uncompressed (or only elastically compressed material) ahead of the front and fully densified (solid) material behind it. In the studies reported in Sect. 4 a 100 mm light gas gun is used to load cellular samples at velocities in the range 300–700 ms^{-1}. Note that Sect. 4 is based on the work presented in [2, 3]. Conclusions and possible future applications of additively manufactured cellular materials are presented in Sect. 5.

2 Additive Manufacture Processes

2.1 Outline

Additive manufacturing (AM) came to the industrial market in the 1980s and was a revolutionary way of making components. This technology made it possible to produce parts directly from 3D model data which dramatically reduced manufacturing times. Over the next three decades the number of processes increased and material properties improved and by the early 2000s AM was being used for the production of tooling and end use parts. The adoption of AM has steadily increased over the last 10 years but the reader should note that a considerable amount of qualification work is still required for high consequence applications, e.g. load bearing safety equipment. A significant new design space is the ability to manufacture cellular structures that could not be produced using conventional processes. The broad range of AM techniques currently in use have been comprehensively reviewed by Gibson et al. [4], Srivatsan [5], Thompson et al. [6] and Herzog et al. [7]. Therefore only those techniques which have been used to manufacture cellular metals will be outlined here. Since the experiments described in the body of this paper were manufactured using selective laser melting (SLM), this technique is described in detail in Sect. 2.2. Alternative techniques, which also appear suitable for the manufacture of cellular metal structures, are described briefly in Sects. 2.3, 2.4, and 2.5. Note that all the processes share a common manufacturing principal—

additive manufacturing of cross sectional layers one at a time to build up a 3D physical component. It is convenient to classify the structures as either 3D or 2.5D. The descriptor "3D" refers to a three-dimensional topology which fits within a unit cell; an array of 3D cells generates an array in which the structure varies in all three Cartesian directions. By contrast "2.5D" refers to a cellular pattern, like honeycomb, for example, that is drawn in one direction; a 2.5D structure varies in two directions but not in the third direction. In general 2.5D structures are easier to make than 3D structures.

2.2 *Selective Laser Melting*

In SLM a metal powder is spread on to a horizontal build platform. A numerically controlled laser beam is then used to selectively melt the metal to a density that is normally above 99% of solid. A second layer of powder is then deposited on top of the first and this, in turn, is melted in a pattern prescribed by the electronic file. A three-dimensional structure is built up in this way, resulting in a solid metal component or structure surrounded by un-melted powder. The finished component is obtained by removing the structure from the base plate and then shaking out the unfused powder. The process is not conducted at an elevated temperature but sometimes uses heat in the build platform to reduce stress generation. The process requires build supports for overhanging features that exceed a certain angle which sets restrictions of the shapes that can be created. Further limits on the structures that can be created are set by the need to be able to remove the powder from the finished component. Thus structures with fully enclosed cells are not possible. The process is used routinely to produce both 3D cellular and 2.5D cell structures. Final components often need annealing to alleviate residual stress. SLM has been used to make components from many metals including steels, titanium alloys, nickel alloys, cobalt chrome, copper and refractory metals.

The technique is well suited for the fabrication of cellular materials. For example, low density lattices, below ~15% relative density, have been manufactured by SLM in stainless steel [8–14] and in titanium alloy [15–22]. For reasons which will be explained later, Winter and co-workers chose to study stainless steel 316L lattices of significantly higher density (64%) than previous workers. The reasons for this choice and the results of the corresponding experimental study are presented in Sect. 3.

2.3 *Electron Beam Melting*

Electron beam melting (EBM) has been used to fabricate cellular metal material. Much of this work has been based on the titanium alloy Ti6Al4V (Ti-64). See, for example, novel research by Martin et al. [22], work by Li et al. [21] who studied the compressive fatigue behaviour of Ti-64 lattices and Ozedemir who studied the quasi-static and impact response of a range of Ti-64 cellular structures [25]. As with SLM in this method a layer of powder is spread on a base plate. An electron beam is then used to preheat the layer to an elevated temperature. The same electron beam is then used to selectively fuse the metal in a pattern controlled by an electronic file. As with SLM the process requires build supports for some overhanging features. Unfused powder tends to be more difficult to remove than with laser melting. The process has been used to produce both 3D cellular and 2.5D metal cell structures.

2.4 Binder Jetting

Examples of the use of binder jetting are found in the work of Meteyera et al. [26] and Tang et al. [27]. As in the SLM and EBM methods a layer of powder is deposited on a base plate. An inkjet print head is then used to jet an adhesive binder onto the surface of the powder in the desired cross sectional pattern. The process repeats until the part is completed. Components do not need build support as the powder bed provides sufficient support to structural features. The component is removed from the powder bed where the un-adhered powder is removed. The part is then placed into a furnace where material is infiltrated with the same or different material. The process has been used to produce both 3D cellular and 2.5D cell structures in steels, nickel alloys and tungsten.

2.5 Direct Laser Deposition (DLD)

In a typical DLD process, a high power laser and blown powders are used to create a melt pool that subsequently solidifies for generating tracks/layers [4, 28–30]. Parts are built track-by-track and layer-by-layer upon a build plate (or substrate) that moves relative to the laser source via a predefined tool path, or scanning pattern, realised via computer numerical control. DLD has been demonstrated to effectively fabricate a wide range of materials such as titanium alloys, tool steels, austenitic steels, martensitic steels, nickel-base super-alloys and cobalt-base alloys. The process does not use build supports and can achieve a range of features by changing the position of the work piece. The process can also be used to produce graded materials. This is achieved by having two material feedstocks that are blown into the energy source using two nozzles. The amount of material blown into the melt pool can be varied allowing control of the material density gradient. The process is not usually used to produce small-scale 3D cellular structures. However, small 2.5D cell structures can be achieved using this process. The technique has been applied to steels and alloys of titanium, nickel and aluminium.

3 Intermediate-Rate Loading

3.1 Introduction

The dynamic compressive deformation of cellular materials at intermediate loading rates has been investigated in the context of applications such as energy absorbing structures for vehicle crash protection, or (at higher loading rates) blast protection, for example, as the cores for blast mitigating sandwich panels. Here, intermediate-rate loading is considered to span the range for which dynamic effects are present (such as inertial stabilisation of buckling phenomena and material strain rate effects), but for which shock deformation conditions are not established.

The compressive collapse response of cellular materials in this intermediate loading regime—bridging the quasi-static and shock responses—is strongly dependent on the cellular topology. Calladine and English [31] identified two classes of response, applicable to cellular structures in this regime. Type I structures crush quasi-statically at a near constant compressive strength, as exhibited by foam-like materials or honeycombs compressed transverse to their prismatic direction. Type II structures exhibit softening on compression, due to the buckling of structural members. Calladine and English [31] argue that the latter are more sensitive to the rate of compression, due to inertial stabilisation of the buckling process. A number of experimental and numerical studies have been carried out into the dynamic compressive collapse of cellular structures of both types. Amongst metallic cellular materials demonstrating Type I characteristics, aluminium foams have been investigated for their energy absorbing potential [32–34]. Tan et al. [34] propose that inertial stabilisation of the foam cell walls against buckling (a “micro-inertial” mechanism) leads to an increase in the compressive strength, relative to the quasi-static value. At higher impact velocities, deformation of the foam localises at the impacted face. The shock propagation analysis for cellular solids of Reid and Peng [35] provides a model for this regime of response, predicting a quadratic dependence of the dynamic strength elevation with impact velocity. A practical example of a Type II cellular material is a honeycomb loaded parallel to the prismatic direction. Studying the dynamic compressive response of aluminium hexagonal honeycombs at impact velocities up to around 30 ms^{-1}, Zhao and Gary [36] confirmed the greater inertial enhancement of the strength for loading in this direction. Radford et al. [37] considered the dynamic compression of square honeycombs, manufactured by brazing together stainless steel sheets, at impact velocities up to around 300 ms^{-1}. For velocities above around 50 ms^{-1}, elastic–plastic wave propagation effects begin to dominate the dynamic strength elevation at the impacted face of the honeycomb, a contribution shown to increase linearly with impact velocity. Lattice materials, periodic arrays of straight bars, represent another category of cellular materials displaying Type II characteristics. The dynamic buckling of individual bars has been studied numerically by Vaughan and Hutchinson [38] and McShane et al. [39]. As the impact velocity is increased, the buckle wavelengths are shown to reduce, a consequence of wave propagation effects. McShane et al. [39] show that a shock-type regime is seen for inclined struts at impact velocities approaching the speed at which the plastic wave can propagate away from the impacted face. Buckling is almost entirely suppressed, with the inclined strut folding flat against the impactor. In this regime, the dynamic elevation in strength at the impacted face is dominated by inertia, and shows a quadratic dependence on impact velocity (similar to that seen for foams in the shock regime).

Additive manufacturing (AM) presents a number of opportunities to support the optimal design of cellular materials for impact energy absorption. Firstly, there is the traditional rapid prototyping role of AM, which makes it feasible to study a wider range of specimen configurations. This is particularly important in a response regime that is sensitive to topology. Secondly, AM enables novel material-geometry combinations to be realised, such as complex lattice structures (otherwise only accessible through casting routes) fabricated from high strength, ductile wrought alloys, such as stainless steel. This presents greater scope for performance optimisation compared to traditional fabrication routes. Thirdly, AM gives scope to miniaturise cellular test specimens, supporting dynamic experimental methods such as the split and direct impact Hopkinson bar techniques and flyer plate impact experiments. However, AM routes, with their wide range of processing parameters, also present challenges in these applications, including microstructure control, geometry resolution and defects.

There have been relatively few studies published to date on the dynamic deformation of additively manufactured cellular materials. Research has concentrated on lattice materials (periodic arrays of straight bars), fabricated from either stainless steel or titanium alloy via a selective laser melting (SLM) process. The process–property interactions for 316L stainless steel lattices fabricated via this route have been reported by Tsopanos et al. [8]. The sensitivity to the laser power and laser exposure time are shown, with insufficient heat input resulting in higher porosity in the SLM steel, and a lattice material with lower compressive strength. Cherry et al. [9] show a clear optimum in input energy density that both maximises the hardness and minimises the porosity of SLM-processed 316L stainless steel. Laser processing also results in a fine sub-grain structure, a consequence of rapid cooling, that contributes to a high material strength. The characterisation and modelling of the quasi-static mechanical properties of these SLM lattices has been the focus of the majority of research to date; see, for example, Gumruk and Mines [10] and Ushijima et al. [11]. However, the blast loading of SLM stainless steel lattice materials with relative densities in the range 4–16% has been investigated by McKown et al. [12] and Smith et al. [13]. The consistency and repeatability of the dynamic collapse mechanisms is noted, and identified as a potential advantage of AM lattices compared to, for example, metallic foams, which may be susceptible to a distribution of defects. Ozdemir et al. [25] use a Hopkinson bar apparatus to obtain the transient stress–time response during the dynamic compression of titanium alloy lattice materials (with relative densities of approximately 15%), at impact velocities up to around 190 ms^{-1}. At low compression velocities, deformation is distributed uniformly through the height of the specimen (of side length 25 mm), with local perturbations in strength dictating the sequence of buckling of each layer of cells.

For impact velocities above around 100 ms^{-1}, collapse of the cells concentrates at the impacted face, and higher peak stresses are measured at the impacted face compared to the distal face.

Each of these investigations of the dynamic deformation of AM lattice materials at intermediate rates of loading consider a limited number of lattice configurations. The importance of the lattice topology and processing conditions for energy absorption and impact mitigating performance emerges from this work. However, the optimisation of performance, and the identification of the best strategies to exploit the capabilities of AM to achieve this, remains to be resolved.

Dynamic mechanical testing of additive manufactured cellular materials has an important role to play in supporting this optimisation process. First, the experiments are essential to identify any practical constraints and limitations imposed by the AM on the impact performance. Of particular importance is feature resolution, and differences in dimensions and added mass compared to the nominal specimen geometry. Also important are defects introduced by the manufacturing process, for example, porosity at the microstructure scale and cell-wall defects at the level of the cellular topology. Second, experimental measurements support numerical modelling, which is an efficient route to performance optimisation. The outcome of the experiments informs the most appropriate modelling strategy (e.g. Lagrangian or Eulerian finite element analysis), helps to identify the required complexity of the material constitutive models (e.g. strain rate dependent plasticity, damage and fracture) as well as providing data for model validation. The relative lack of available material data for materials processed via AM routes makes it particularly important to target key gaps in material characterisation.

In the following, we present some experiments that aim to measure the dynamic compressive response of additively manufactured metallic cellular (i.e. porous) materials at intermediate loading rates: impact velocities up to 150 ms^{-1}. A range of cellular topologies will be considered, which span a range of minimum feature size and relative density. This will demonstrate any limitations of the AM process that might constrain the useful parameter space for optimisation. The outcome of the dynamic compression tests will then be used to inform the most appropriate modelling strategies in this loading regime.

3.2 Experimental Samples

The cellular material illustrated in Fig. 1 is loosely based on the "pillar-octahedral" architecture studied by McKown and co-workers [12]. This structure was a periodic array which incorporated both vertical bars or struts and 45° angular struts. The bars of diameter d form a triangulated structure, arranged in planar walls, which are repeated periodically with a wall separation L. Cells of side 1.5 mm and 2.5 mm were studied with relative densities varying between 0.033 and 0.166. The architecture described in [12] is convenient for our study because varying

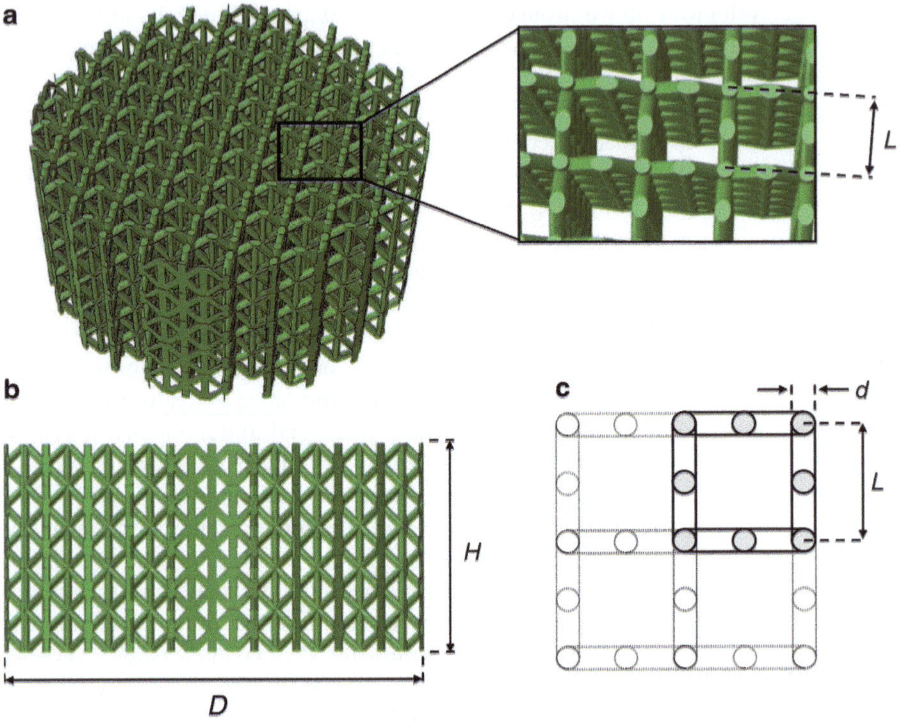

Fig. 1 The experimental samples were manufactured as cylinders of height 10 mm and diameter 20 mm. Electronically generated images of specimen D are shown here as (**a**) an angled view, (**b**) a side view and (**c**) a plan view of four cells, showing the cell size *L* (2 mm), and lattice bar diameter, *d* (0.4 mm)

the two geometric parameters *d* and *L* allows resolution issues and the effect of relative density to be probed independently. For modelling reasons, to be described in more detail in Sect. 4, the relative density of our specimens was increased relative to those in [12]. To produce a practical test specimen, the lattice structure was specified to conform to a circular cylindrical envelope of diameter $D = 20$ mm and height $H = 10$ mm, obtained by trimming the periphery in the CAD model, prior to manufacturing. Five lattice configurations were considered, with geometric parameters as listed in Table 1. Specimens A–C have the same wall space $L = 1$ mm, but increasing values of the bar diameter *d*, and therefore increasing relative density. Specimens D and E have the same ratio *d*/*L* as specimens A and C (and therefore the same nominal relative density), but the values of *d* and *L* are doubled. This allows any resolution limitations affecting the small-celled specimens to be assessed. Small cell sizes are attractive for laboratory-scale experimental investigations, and so the scope to miniaturise without significantly altering the characteristics is of practical interest.

Table 1 Cellular specimen configurations

Specimen	Nominal geometry		Nominal relative density, $\overline{\rho}$	Measured relative density, $\overline{\rho}$
	Cell size, L (mm)	Bar diameter, d (mm)		
A	1.0	0.20	0.22	0.32
B	1.0	0.28	0.37	0.42
C	1.0	0.40	0.57	0.64
D	2.0	0.40	0.22	0.22
E	2.0	0.80	0.57	0.58

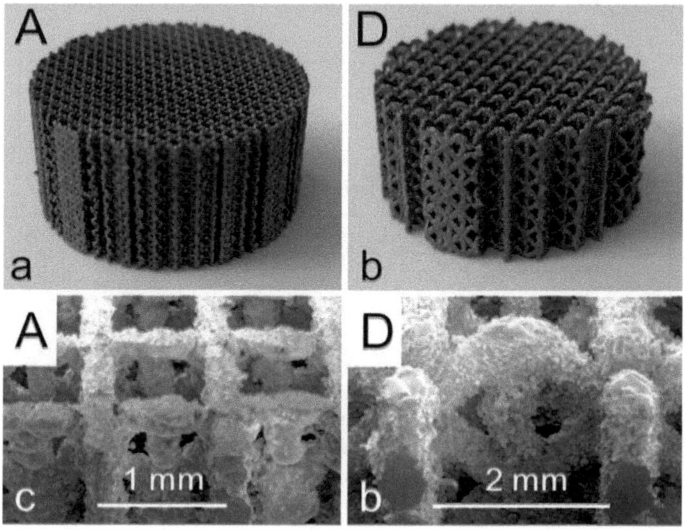

Fig. 2 (**a**, **b**) As-manufactured 316L stainless steel specimens; the cylinders' height and diameter were 10 mm and 20 mm, respectively. (**c**, **d**) are scanning electron microscope (SEM) images of the manufactured specimens. Capital letters indicate specimens in Table 1

Metallic test specimens were additively manufactured from 316L stainless steel using an EOS M 280 selective laser melting (SLM) process. A 316L stainless steel powder of particle size 20–40 μm was used, and the laser spot size was approximately 100 μm. Examples of the as-manufactured test specimens are shown in Fig. 2, and the measured relative densities are given in Table 1. The height and diameter of the cylinders in Fig. 2a, b are 10 mm and 20 mm, respectively. Figure 2c, d shows scanning electron microscope (SEM) images. For specimen A, the minimum feature size is approaching the laser spot size, which affects the ability of the process to resolve this geometry. The reproduction of the lattice structure is improved for the larger cell-sized variant, specimen D. Also apparent in the SEM images is the adhesion of un-melted powder particles to the lattice struts, which will increase the relative density of the specimens compared to the nominal value.

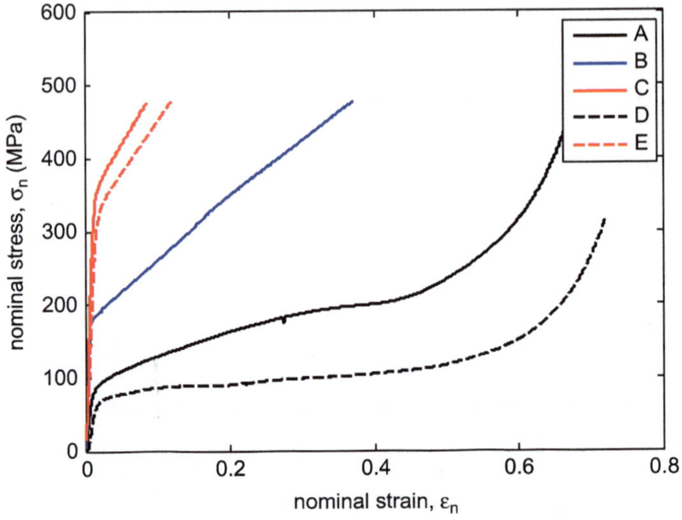

Fig. 3 Quasi-static compression of the five lattice specimens

3.3 *Quasi-Static Testing*

The cellular specimens are first tested quasi-statically in compression using an Instron screw-driven materials testing machine. The force acting on the specimen (F) was obtained from the load cell of the test machine. The nominal stress is defined as

$$\sigma_n = \frac{4F}{\pi D^2}. \tag{1}$$

The relative displacement of the platens (u) was measured using a laser extensometer, and the nominal strain $\varepsilon_n = u/H$. The compressive response of the five specimens is shown in Fig. 3. The lower density specimens (A, D) demonstrate a stress plateau after the onset of plastic collapse of the cell walls, followed by a sharp rise in stress at the onset of densification. The smaller cell-sized variant (A) shows a greater degree of strain hardening during the plateau phase. However this effect is probably due to the fact that, as a consequence of the AM resolution limitations, the relative density of specimen A is higher than that of D. The higher density specimens (B, C and E) don't display a plateau phase, but instead harden progressively after the onset of yielding.

3.4 Dynamic Testing

Dynamic compressive testing was performed using a direct impact Hopkinson bar apparatus, as described by Radford et al. [37]. A maraging steel Hopkinson bar of diameter 28.5 mm was used. Strain gauges mounted diametrically opposite each other at a distance of 10 bar diameters from the impacted end were used to measure the force versus time acting on the end of the bar during compression of the specimen. The cellular specimen was impacted by a steel projectile of diameter 28.5 mm and mass 0.1 kg, fired using a gas-gun apparatus. The impact velocity V_0 was measured using laser velocity gates and verified using high speed photography. Two impact configurations were used, in order to obtain the force–time history at the impacted and distal faces of the specimen. In the distal face configuration, the specimen was attached to the face of the Hopkinson bar (using a thin layer of adhesive tape) before impact by the projectile. In the impacted face configuration, the specimen was mounted to the projectile, and impacted with it onto the face of the Hopkinson bar. Note that the mass added to the projectile by the specimen is of the order of 5%. These two measurements—impacted and distal face—allow the degree of stress non-uniformity through the height of the specimen to be assessed.

The results are shown in Fig. 4 for two cellular materials, specimens A and B, and two different impact velocities $V_0 = 100$ and $150\ \mathrm{ms}^{-1}$, for both impacted and distal face impact configurations. These impact velocities correspond to initial nominal strain rates $\dot{\varepsilon}_n = 10 \times 10^3$ and $15 \times 10^3\ \mathrm{s}^{-1}$, respectively (though deceleration of the projectile means that the rate of compression will reduce during the impact). For both specimens, the features of the measured stress–time histories from the dynamic experiments show similarities to the quasi-static stress–strain response. Specimen A shows a plateau in stress, followed by a rise in strength at the onset of densification. Specimen B shows approximately linear hardening after the onset of yielding. However, both specimens show a significant increase in compressive strength as a result of dynamic loading, when compared to the quasi-static case. This can be attributed to the strain rate sensitivity of the 316L stainless steel and the influence of inertia on the buckling resistance of the cellular structure. Slightly higher initial compressive strengths are measured at the impacted face compared to the distal face, and this is more pronounced at the higher impact velocity, $V_0 = 150\ \mathrm{ms}^{-1}$. This indicates that wave propagation effects are beginning to have an effect at these impact velocities. As analysed by Radford et al. [37] for stainless steel square honeycombs, the propagation of plastic deformation from the impacted to the distal face can account for the higher impact velocity sensitivity of the front face stress. However, the relatively small differences in the measurements at the impacted and distal faces indicate that strong gradients in deformation through the height of the specimen have yet to develop at these intermediate impact velocities.

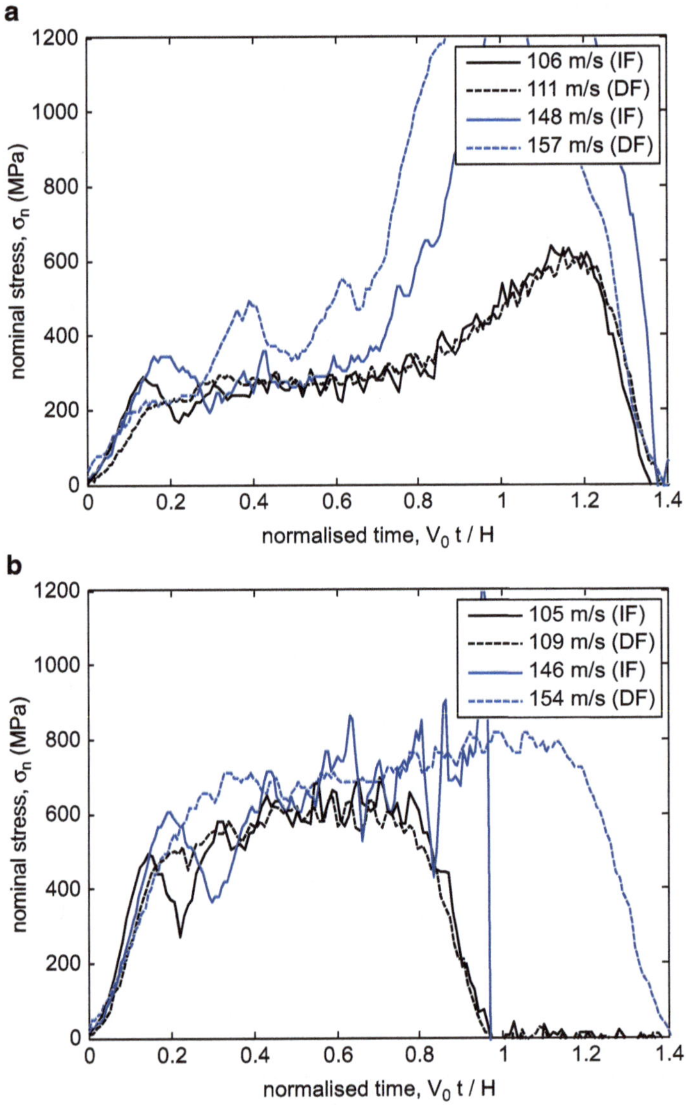

Fig. 4 Dynamic compression measurements for (**a**) specimen A and (**b**) specimen B. Impacted face and distal face measurements are shown for the impact velocities indicated

3.5 Conclusions from Intermediate-Rate Study

The quasi-static and intermediate rate testing reveals the value of additive manufacturing as a tool to support the optimisation of cellular materials for impact mitigation. Miniaturisation of lattice topologies is possible, producing small-scale specimens which perform comparably with their larger-scale equivalents, even when feature sizes are reduced to the point where processing resolution limits begin to have an effect. The AM manufacturing process also facilitates the testing of a wide range of different lattice topologies and relative densities, to support the shortlisting of candidates for detailed optimisation. The specimens show a significant increase in compressive strength as a result of dynamic loading when compared to the quasi-static case, an effect which can be attributed to the strain rate sensitivity of the 316L stainless steel and the influence of inertia on the buckling resistance of the cellular structure. The results are available for the validation of simulation codes in the intermediate loading regime. It is envisaged that the experiments will be modelled using Lagrangian mesh schemes and the whole specimen will be meshed. The validated code can then be used to optimise the materials under investigation for applications in the scenarios of interest.

4 High-Rate Loading

4.1 Introduction

Ideally, in order to provide an insight into the mechanisms of energy absorption in alternative lattice structures simulations should resolve the fine detail of the deformations of the struts or walls which make up the lattice. This is especially true in the very high velocity loading regime. For example, the simulations presented in Figs. 13 and 18 of this article suggest that at impact velocities above ~500 ms^{-1} large increases in plastic work are generated as a result of the intense concentrations of flow associated with jets or other microkinetic processes. These features typically have dimensions ~1/10 of the struts themselves. In addition an Eulerian meshing scheme is needed if the code is to withstand the severe deformations associated with jetting and other microkinetic flows. It follows that a typical strut should be meshed with at least 20 computational cells through its thickness if the expected flow concentrations are to be clearly resolved. However implementing Eulerian meshing makes a heavy demand on computing resources because, in general, the complete cell, including the voids, must be finely meshed. As a result, in practice relatively coarse meshing has to be used, especially in 3D simulations. The requirement for fine meshing in the high velocity loading regime strongly encourages the use of large diameter plate impact configurations. On a macroscopic scale this loading mode is termed "one-dimensional (1D)" or "uniaxial strain". Typically 1D conditions are generated when a large diameter gun is used to impact a flat projectile plate against a target disc specimen whose diameter is large compared with its thickness. 1D loading is maintained near the centre of the specimen during the early stages of experiment before release waves from the outer perimeter reach the central region. Alternatively a reverse configuration is often used in which a flat plate of the specimen material is accelerated to impact a flat plate target. Although the plate impact configuration can be used to impose 1D loading of a lattice material on a macroscopic scale, at the scale of the cell itself the loading is, of course, not 1D. However (in simulations at least) adjacent cells in planes perpendicular to the load direction experience identical load histories. In this situation it is only necessary to compute the response of one cell, or small group of cells, or one column of cells stacked in the direction of the load. The presence of identical adjacent cells can be accounted for by setting up the simulation with rigid, or reflective, lateral boundaries. It is important to note here that the combination of high relative density and high impact velocity leads to high velocity longitudinal waves and therefore very short times for waves to transit the thickness of the specimen. It follows that there is little time for lateral waves to affect the deformation of the sample. Of course, it is likely that in as-manufactured specimens, slight differences between adjacent cells will lead to flow across cell boundaries but for the reasons just explained we would not expect this to play a significant role in the pattern of deformation.

An important advantage of AM over more conventional techniques for manufacturing lattice materials relates to the need, described above, to maintain 1D conditions on a macroscopic scale in validation experiments. AM technology allows components consisting of arrays of cells as small as 1 mm to be manufactured. With a typical gas gun of diameter 70–100 mm, say, a useful 1D region can be generated in the central region of a specimen disc of thickness 5–10 mm. This range would allow up to 5–10 cells through the thickness of the sample enabling an experimental study of the successive compression of multiple cells stacked in the direction of the impact shock. Experiments will be described here in which the high-rate shock response of cellular structures manufactured by selective laser melting (SLM) was studied using velocity interferometry. Simulations of the experiments, run using the Sandia code CTH [40], were shown to predict the material response well. The validated CTH model was then used to evaluate the energy absorbing potential of a series of alternative lattice configurations of different structures but fixed average density. Note that the need to maintain 1D conditions during the period of the transition of the initial plastic densification shock sets a lower limit on the relative density of the sample. The reason is that a low density material takes longer to densify, and therefore the region of interest would be vulnerable to releases from the perimeter of the sample at an earlier stage.

4.2 *Experimental Samples*

A graphical depiction of the lattice structure which forms the basis of the experimental study is shown in Fig. 5. The cells which make up the structure use the same electronic definition as those in Type C used for the intermediate-rate experiments described in Sect. 3. Samples consisting of 99.2 mm diameter porous discs of thickness 6 mm were prepared for plate impact studies. The experimental components were weighed allowing the average density of the structure to be

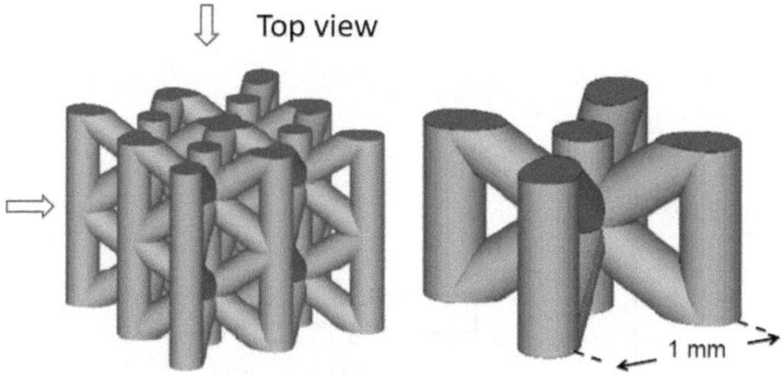

Fig. 5 The structure defined by the CAD-generated stl file. The structure consists of a regular array of intersecting rods of diameter 0.4 mm, with a repeated unit spacing of 1 mm. The "top" and "side" views are indicated by the vertical and horizontal arrows, respectively. The density is 64% of solid

determined. At a density of 64% of solid the structure is significantly denser and the cells are significantly smaller that the similar structure studied by McKown et al. [12]. One reason for our move towards higher density is that, as explained above, an important aim of the work was to compare the experimental results with the predictions of 3D computer simulations. The maximum mesh size it is reasonable to use in the simulations is determined, to a large extent, by the smallest features in the structure; in other words, narrow struts require finer meshes. This led us to choose a nominal strut diameter of 0.4 mm, which was an appreciable fraction of the cell size (1 mm). Another reason is that, as explained in Sect. 4.1, it is easier to maintain 1D conditions in shocked high density than in low density materials. Note that the need to maintain 1D conditions restricts the relative densities of the lattices to a rather narrow range.

In order to characterise the SLM-processed 316L stainless steel, specimens for tensile testing were manufactured using the same SLM process parameters as the lattice specimens. Harris et al. [41] showed that the microstructure of dogbone and cellular structures generated using the same SLM process were very similar suggesting that the mechanical properties of the lattice material can be estimated from tests on dogbone specimens. Specimens dimensioned as shown in Fig. 6 were loaded in tension in an Instron screw-driven test machine. The tensile stress–strain curve for the material is also shown in Fig. 6. The stress at yield, σ_y, was estimated from these tests to be ~640 MPa.

Further information was obtained by examining structures which had been sectioned and etched. It was estimated from optical micrographs that the grain diameter in the SLM material was ~5 μm. According to the well-known Hall–Petch analysis [42, 43] a material with a small grain size tends to have a high flow stress. Kashyap and Tangri [44] measured the flow strength of 316L stainless steel in which grain sizes from 3.1 to 85.7 μm were generated by rolling at room temperature. The strength value of ~640 MPa measured for SLM stainless steel is consistent with the

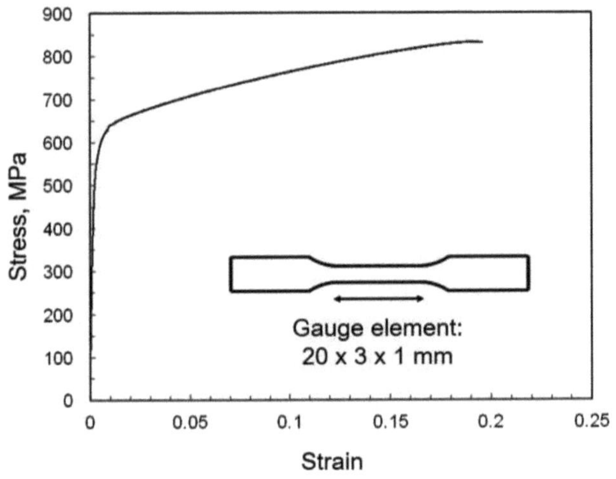

Fig. 6 Tensile stress–strain curve for SLM material

values observed by Kashyap and Tangri for material with a grain size of ~5 μm. We assume that the small grain size is a result of the very rapid sequential cooling experienced by the melt pools as the laser spot passes across them.

4.3 High-Rate Experiments

Three plate impact experiments were conducted using the 100 mm diameter single-stage gas gun at The Institute of Shock Physics, Imperial College London. A reverse impact configuration was employed in this study. The experiment is illustrated in Fig. 7. A 6 mm thick disc of the structure was manufactured from 316L stainless steel using SLM. As shown the disc incorporates a protective steel cylinder of outer diameter 99.2 mm and wall thickness 0.5 mm. After manufacture by SLM a lathe was used to machine a cutaway at the leading edge of the disc, as shown in Fig. 7, to provide a light reflecting surface for velocity measurement using a heterodyne velocimetry (Het-V) probe [45]. The sample was mechanically attached to a polycarbonate sabot. This projectile impacted an instrumented 6 mm thick × 120 mm diameter 316L solid steel target, or "velocity plate", at nominal velocities of 300, 500 and 700 ms^{-1}. The rear-surface velocity history of the steel velocity plate was monitored using an array of three Het-V probes mounted on a diameter of 40 mm. A hole in the target plate allowed a fourth Het-V probe to be used to measure the projectile velocity. Note that since the diameters of the flyer and target are much greater than their thicknesses, uniaxial strain conditions persist in the centre of the system for the duration of the experiment. Based on computer simulations the time at which release waves from the edges of the experiment would be expected to reach the central probe was 4.6 μs from impact time. The schematic

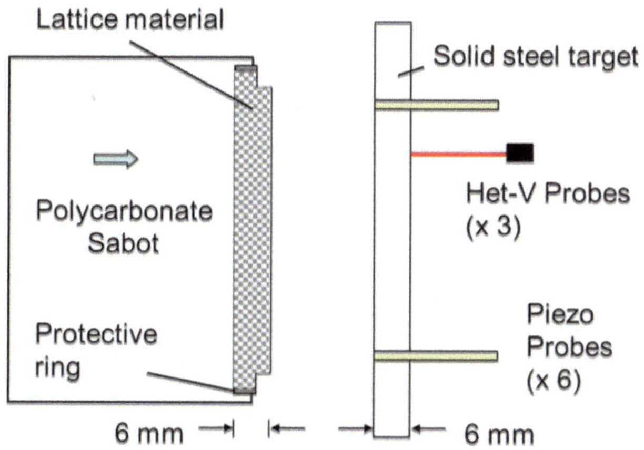

Fig. 7 Experimental setup. The lattice material is mounted on a sabot which is accelerated using a gas gun. The velocity of the back surface of the target is measured using three Het-V probes. A fourth probe measures the velocity of the flyer via a hole near the edge of the target. A series of three shots were fired at nominal velocities of 300, 500 and 700 ms^{-1}

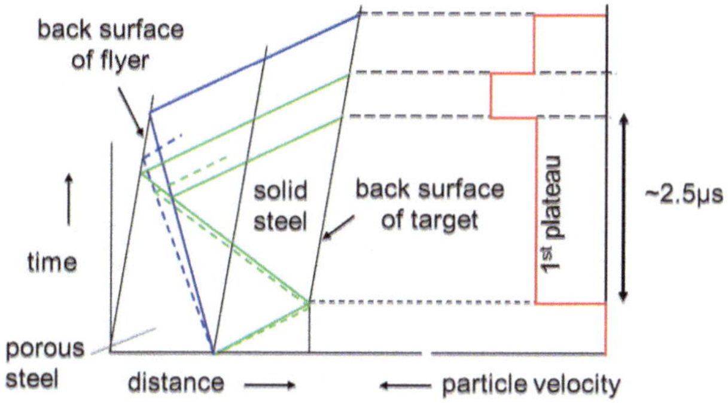

Fig. 8 Schematic distance vs. time and velocity vs. time plots showing the main features of the expected velocity profile at the free surface of the steel target. For simplicity the elastic waves are omitted from this diagram

distance vs. time and surface velocity plots in Fig. 8 show that the back surface of the target will be accelerated by the wave generated by the initial impact, resulting in a velocity plateau (labelled "first plateau" in Fig. 8). The wave reflected from the interface between the compressed and uncompressed flyer, shown by a green line, then generates a second, shorter-lived velocity plateau. Eventually the surface velocity is reduced by a rarefaction from the interface between the back surface of the flyer and the polycarbonate sabot. It will be noted in experiments and simulations presented later that, as the impact velocity increases, the length of the first plateau increases. A qualitative explanation for this effect is illustrated in Fig. 8 where wave trajectories for low and high velocity impacts are depicted by solid and dashed lines respectively. The blue lines show wave trajectories into the porous material. It is

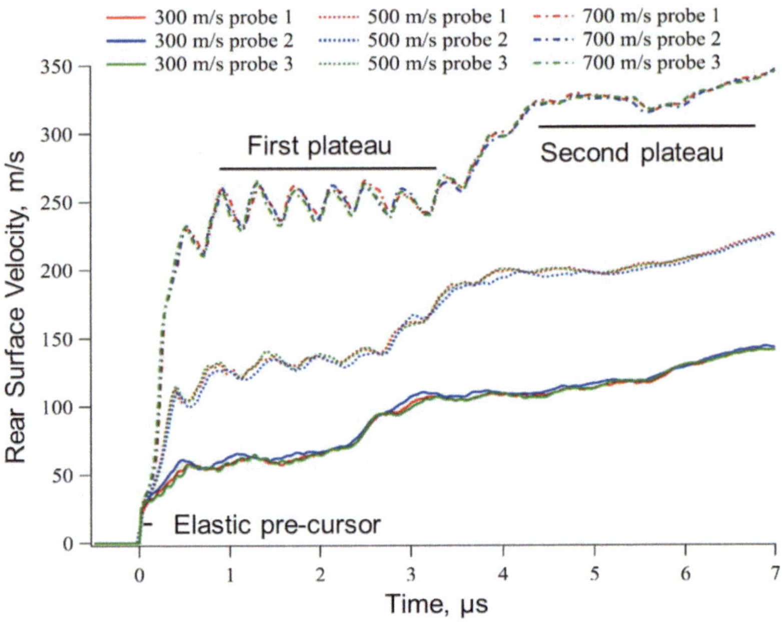

Fig. 9 Particle velocity records measured using Het-V for the three experiments. The time $t = 0$ represents first detectable motion of the target rear surface

assumed that the material is fully densified by the shock. Because the material is highly porous (relative density 0.64) the shock wave propagates much faster for high velocity loading than for low velocity loading. The green lines show the waves propagating into the stainless steel. At the relatively low pressures in the regime considered in this article the waves travel only slightly faster in the higher than in the lower velocity experiment. It is clear from the x-t plot that the distance travelled by the waves that generate the second plateau increases as the impact velocity increases. Therefore the length of the first plateau increases with impact velocity.

Unfolded data for the three experiments (at impact velocities of 300, 500 and 700 ms^{-1}) are shown in Fig. 9. The traces have been positioned along the time axis so that they all start at the same time. Typically each trace consists of an initial steep rise in velocity corresponding to the arrival of the elastic wave. The subsequent reduction in slope of the velocity trace indicates the transition to the plastic wave. This is followed by a plateau corresponding to the first velocity plateau in Fig. 8. Oscillations in velocity during the period of the first plateau are believed to be caused by the successive collapse of the cells in the impactor. Simulations supporting this contention are presented in Sect. 4.7. The amplitude of the "cell-collapse" waves decreases as the impact velocity decreases and the frequencies are higher at higher impact velocities. Clearly, samples with smaller or larger cells sizes would generate correspondingly higher or lower frequencies. After 2–3 μs the trace

Table 2 State variables for plateau 1 and plateau 2 deduced from Fig. 10

Impact velocity	$mm\mu s^{-1}$	0.30	0.5	0.7
Plateau 1 particle velocity	$mm\mu s^{-1}$	0.03	0.065	0.125
Plateau 2 particle velocity	$mm\mu s^{-1}$	0.052	0.09	0.17
Plateau 1 stress	GPa	1.25	2.55	4.83
Plateau 2 stress	GPa	2.06	3.48	6.55

Table 3 Constants used for the constitutive model of SLM 316L

Solid steel density	ρ_{s0}	7.90 gcm^{-3}
Solid steel specific volume	v_{s0}	0.127 gcm^{-3}
Room temperature	T_0	298.1 K
Sound speed	C_0	4.569 ms^{-1}
Slope of Hugoniot	S	1.49
Specific heat at constant volume	C_v	0.446 Jg^{-1} K^{-1}
Density of porous material	ρ_0	5.056 cm^3 g^{-1}
Specific volume of porous material	v_0	0.198 gcm^{-3}
Elastic pressure of porous material	P_E	0.44 GPa
Crush pressure of porous material	P_C	1.9 GPa
Poisson's ratio	ν	0.283

ramps up to a second plateau before lateral release waves arrive. We assume this ramp and second plateau arise as illustrated schematically in Fig. 8. The particle velocities for the first and second plateaus were measured from Fig. 9. Assuming that the in situ particle velocities are half the free surface velocities allows in situ particle velocities in the plateau regions to be estimated and these are recorded in Table 2. Further, assuming a linear Gruneisen equation of state for the solid steel target with parameters as listed in Table 3 allows the stress states in the first and second plateaus to be estimated; they are listed for the three experiments in Table 2. Assuming equality of stress and particle velocity across the boundary between the solid steel target and the porous flyer, the states in Table 2 also describe the states reached by the porous material.

It can be seen that the length of the first plateau increases as impact velocity increases. As illustrated in Fig. 9, we assume that this effect arises because as the impact velocity increases so does the length of the path travelled by the wave which generates the second plateau.

4.4 CTH Simulations of High-Rate Experiments

In order to gain an insight into the processes occurring during the dynamic collapse of the lattice structure both 2D continuum and 3D "structural" simulations in CTH were employed. CTH is a multi-dimensional, multi-material Eulerian-based code developed by Sandia National Laboratories [40]. It is suitable for the simulation of large deformations and material distortions loaded under shock and high strain rate conditions. The code incorporates many popular equations-of-state (e.g. Mie–Gruneisen) and strength models (e.g. Steinberg–Guinan [46], Johnson–Cook [47]) used to describe the behaviour of materials subjected to high-rate loading. It has well developed continuum models for porous materials, namely the porous alpha ($P(\alpha)$) model [48] (used in the current study) and the $P(\lambda)$ [49]. It also has a 3D capability together with the ability to read in descriptions of shapes defined using the surface triangle representation (stl) generated with computer aided design (CAD) software. Note that, in principle, the ability to import stl files into CTH offers considerable savings in the time and effort spent on setting up the simulations. However it was found that in practice differences are often found between the manufactured structure and the structure defined by the stl file. For example, in our study it was found that the density of the actual components tended to be higher than that corresponding to the stl file. In this situation it was decided to increase the diameter of the strut in the stl file such that the file matched the measured density.

CTH was used to perform both $P(\alpha)$ and 3D structural simulations of the experiments.

The fundamental assumption in the $P(\alpha)$ model is that the pressure in the porous material is expressed as

$$P = \frac{1}{\alpha} F\left(\frac{v}{\alpha}, E\right), \tag{2}$$

where $P_s = F(v_s, e)$ is the equation of state of the fully dense material and $\alpha = v/v_s$ is the distention ratio (the ratio of the porous material volume to that of the fully dense material at the same pressure and energy). E and v are the internal energy and volume per unit mass, respectively.

The compaction behaviour is then governed by a functional form assumed to describe the variation of the distention ratio with pressure between its initial value $\alpha_0 = v_0/v s_0$ and the value 1.0 when the material is fully compacted. In the CTH simulations we have used the simple form

$$\alpha(p) = 1 + (\alpha_0 - 1)\left[\frac{P_C - P}{P_C - P_E}\right]^2, \tag{3}$$

where P_E is the pressure at which plasticity first occurs and the crush strength, P_C, is the pressure at which complete compaction occurs.

An estimate of P_E is provided by the work of Carrol and Holt [49] and Wu and Jing [50]

$$P_E = \frac{2}{3}\sigma_Y \ln \frac{\alpha_0}{\alpha_0 - 1}. \tag{4}$$

Entering α_0 as 1.56 and using the yield stress, σ_y, determined from the dogbone test described earlier, (0.64 GPa) gives $P_E = 0.44$ GPa. We have made the simplifying assumption here that the structure may be treated as a solid matrix containing spherical voids.

The crush strength of the porous steel was estimated using the simplified Fischmeister–Arzt model [51]:

$$P_C = 2.97\sigma_y, \tag{5}$$

where σ_y is the yield stress of the constituent particles. Again a simplifying assumption has been made since the structure considered in this paper does not contain discrete particles, as assumed in [51], but instead consists of a regular connected structure. However, as a starting point it seemed justified to apply the Fischmeister–Arzt factor to the measured quasi-static yield strength of 0.64 GPa (see results in Fig. 6) giving a crush strength, P_c, of 1.9 GPa.

The stresses in the porous materials in the first and second plateaus are listed in Table 2. In most cases the estimated stress exceeds the crush pressure. However the stress in the first plateau in the 0.3 $\mathrm{mm\mu s^{-1}}$ experiment was 1.25 GPa, significantly less than the crush pressure estimated using [51] which suggests that this material might not be fully densified.

Since the experimental configuration is radially symmetric it was modelled as two-dimensional (2D). Note that in $P(\alpha)$ simulations of a porous material the shapes of the pores in an irregular structure or the cell shapes in a regular structure are not explicitly modelled. Instead, the material is represented as a homogeneous region throughout which the porosity is uniformly distributed. As described above the effect of porosity is approximated using a macroscopic constitutive description which is judged to represent the bulk behaviour of the distended material.

In both the $P(\alpha)$ and 3D simulations described here the 316L stainless steel which forms the base material of the cellular flyer was modelled using a Mie–Gruneisen equation of state:

$$p = p_h(v) + \Gamma_0\rho_0\,(E - E_h(v)) \tag{6}$$

in conjunction with a linear shock velocity, U_s, vs. particle velocity, u_p, relationship:

$$U_{s=}C_0 + su_p \tag{7}$$

with constants as listed in Table 3. Note that a value for C_v, the specific heat at constant volume, is required by CTH. The value listed in Table 3 is taken from the CTH materials database.

The mesh size for the $P(\alpha)$ calculations was 0.02 mm in all regions. The density of the cellular material was set to 64% of solid (5.056 $\mathrm{gcm^{-1}}$).

The response of the solid stainless steel target was modelled using the Mie–Gruneisen EoS and the Steinberg–Guinan strength [46] together with the default constants in the CTH package.

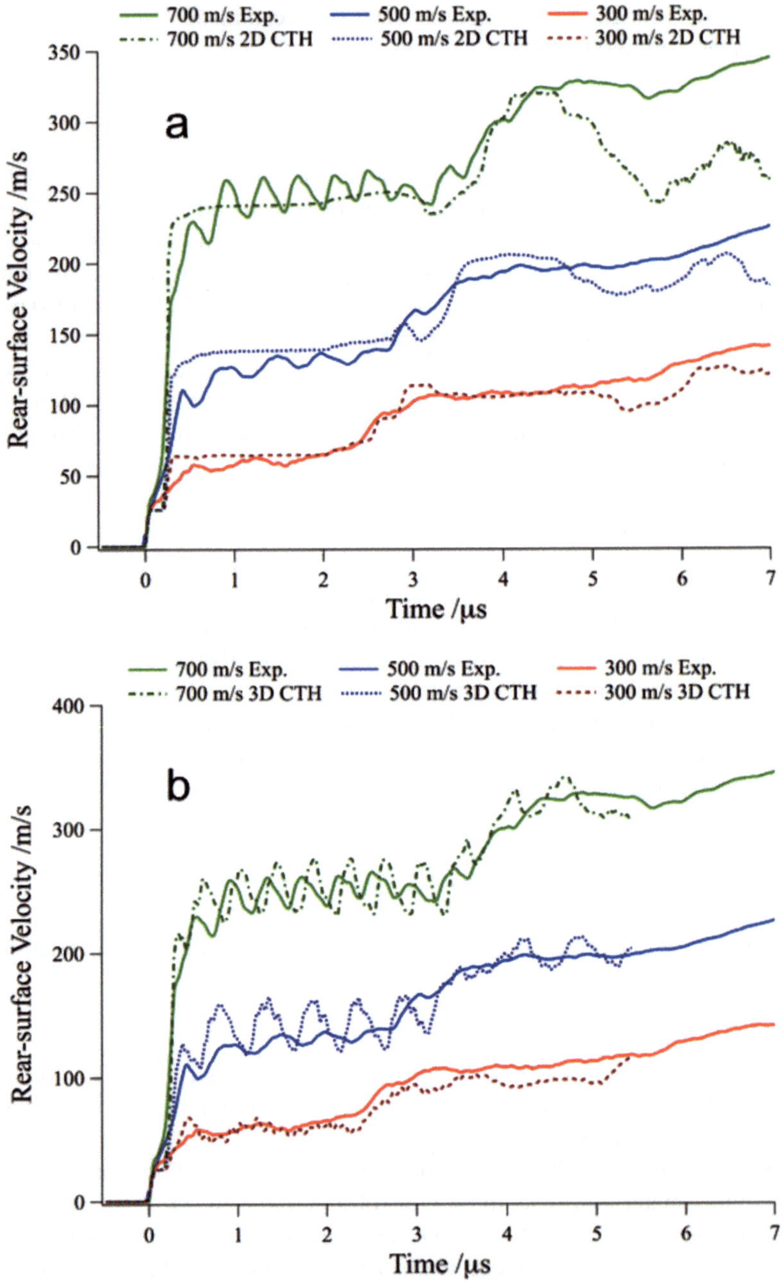

Fig. 10 (**a**) Comparison between the experiments and 2D continuum CTH simulations run with the p-alpha model with elastic and crush pressures of 0.44 GPa and 1.9 GPa, respectively; (**b**) Comparison between experiments and 3D CTH

4.5 Results of CTH Continuum Simulations

The velocity profile measured at the back surface of the solid steel target is shown for comparison with the 2D continuum $P(\alpha)$ calculations in Fig. 10a. Since the individual cells of the structure are not modelled in the continuum simulations the oscillations observed in the experimental records are not reproduced in the simulations. The levels and relative timings of the initial shock, and the shock generated by the second interaction with the flyer/target interface (as illustrated in the distance vs. time plot in Fig. 8), are well matched by CTH. However in the 700 ms^{-1} experiment the $P(\alpha)$ model exhibits a release at ~4 μs which is not observed in the experiment. This suggests that, for the highest velocity experiments, the release wave (see Fig. 8) travels faster in the simulation than in the experiment.

One-dimensional and two-dimensional $P(\alpha)$ simulations were run to estimate the time for which the Het-V measurement remains one-dimensional. Surface velocities were monitored at a radius of 20 mm and the diameters of both the flyer and the target plate were varied. It was found that with a 49.6 mm radius plate (as used in the experiment) the trace begins to diverge slightly from true 1D behaviour at ~3 μs after the shock reaches the rear surface of the target. However the observation that CTH tends to overestimate the velocity of release waves suggests that 3 μs may be an underestimate of the "1D time window". This assessment suggests that waves from the edges do not affect the first plateau but do have a very small effect on the second plateau.

It is clear that the $P(\alpha)$ continuum simulations have captured the first order response of a typical cellular structure reasonably well. This finding has important implications for research into the potential of cellular materials for application as energy absorbers. For example, consider a scenario in which a cellular structure formed from a base material whose equation of state is known is to be used for a given application requiring specific energy absorption characteristics. The results presented in Fig. 10a suggest that a good indication of the system performance can be gained without specifying the lattice structure; that is, only the initial porosity and the flow stress of the base material need to be specified. Three-dimensional calculations can then be used to optimise the detailed structure of the lattice.

4.6 3D Structural Simulations

For the 3D structure simulations, hereinafter referred to as "3D" simulations, the stl file used in the SLM manufacturing process was read into CTH, providing a direct geometrical representation of the lattice configuration including the detailed architecture of the individual cells. The stainless steel in the porous flyer was modelled using the Mie–Gruneisen EoS (and based on the strength measurement described in Sect. 2) a constant flow stress of 0.64 GPa. The solid steel target was modelled using the same constitutive model that was used for the $P(\alpha)$ simulations described in the previous section.

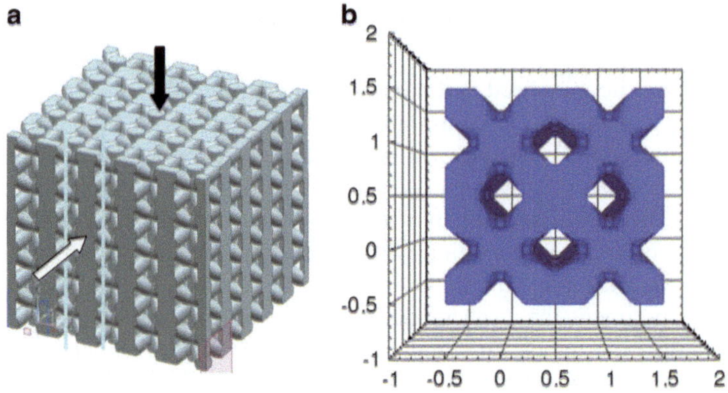

Fig. 11 (**a**) Computer generated image of a 5 × 5 × 5 cell region of the full structure. (**b**) The region used in "cut down" 3D simulations after import into CTH

Two types of 3D simulations, designated "full" and "cut down", were run. In both cases a uniform 0.05 mm mesh was employed giving ~8 meshes through the 0.4 mm struts. Clearly finer meshing would allow better resolution of the flow concentrations expected in high velocity impacts. However, in 3D simulations, reducing the mesh size below 0.05 mm would increase the computational load to an impractical level. A CAD package was used to define the structure as an array of triangles which was stored as an stl file. The cutaway at the edge of the disc seen in Fig. 7, which was used in the experiment to provide a reflective surface for velocity measurement, was not modelled as this would have required an additional stl file to be generated. Figure 11a shows a computer-generated view of a 5 cell × 5 cell × 5 cell block of the "full" structure.

To save computing time some simulations were run using a cut-down model of the lattice. Figure 11b shows a view in the impact, or axial, direction of the region of the cellular flyer modelled in the cut-down variant. Effectively it is made up of a group of four cells bounded by reflective boundaries. Comparison of output obtained using the full and cut-down models showed only minor differences. The simulations whose results are presented in this article were run using the cut-down model.

4.7 Results of 3D Structural Simulations

The 3D CTH simulations are shown for comparison with the experiments in Fig. 10b. It is seen that for all three experiments the levels of the first and second plateaus are reasonably well matched. The simulations also match the frequencies of the oscillations superimposed on the experimental traces. However, at the highest velocity the oscillations generated by sequential collapse of the cells appear out of phase with the experiment. In the 500 ms^{-1} experiment the calculated amplitude of the oscillations is significantly greater than that observed.

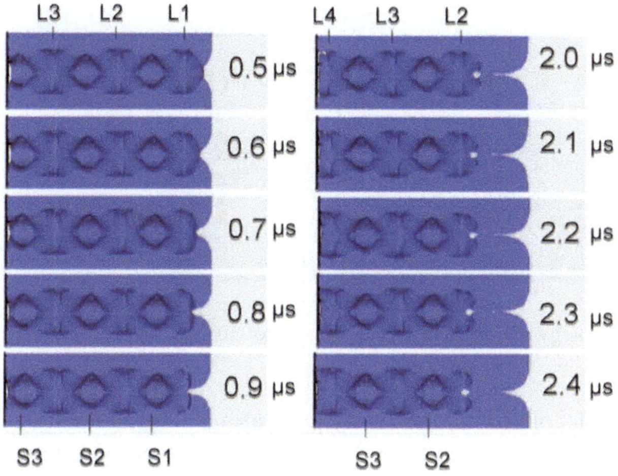

Fig. 12 A sequence of material plots for a 300 ms^{-1} impact, illustrating the sequential collapse of the voids in the structure under study. Note that as the incident shock wave reaches each cavity a pressure trough followed by a peak is propagated into the target. In this sequence the time between successive collapses is ~1.5 μs

The good match to experiment achieved by CTH justifies using the code as a tool to understand the mechanisms of energy absorption and, thereby, to contribute to the optimisation of energy absorbing structures. We also believe the work to be relevant to other areas in which the microscopic flows occurring in impacted heterogeneous materials are important. Examples are the shock consolidation of porous materials [52] and the generation of hot spots in heterogeneous explosives [53].

The sequence in Fig. 12 provides insight into the likely mechanisms at play during the shock loading of cellular structures and clearly shows that the time frequency of the velocity oscillations depends on the linear frequency of the unit cells in the shock propagation direction. Note that the time between the peaks of the oscillations is actually *half* of the time for the incident shock to propagate across a complete cell. This is because, when viewed in the direction indicated by the white arrow in Fig. 11a, the structure effectively contains two types of cavities, designated here "large" and "small". The sections in Fig. 12 have been chosen such that both the small (S1–S3) and large (L1–L4) are evident. The sequence of material plots illustrates the evolution of the structure for an impact velocity of 300 ms^{-1}. Note that in this sequence, and those that follow, an initially stationary solid target is impacted by a porous flyer approaching from the left. Therefore the crush wave in the flyer propagates to the left. In Fig. 12 the numbers (0.5–2.4) give the time in microseconds from impact. It can be seen that, viewed in this section, each 1 mm cell has two cavities, labelled large (L1–L4) and small (S1–S3). The sequences on the left and right of Fig. 12 show times from 0.5 to 0.9 μs and 2.0–2.4 μs, respectively. The partial collapse of cavity L1 is shown in the sequence on the left; a similar process, occurring ~1.5 μs later, collapses cavity L2. A complete time sequence would show that the small cavities S1–S2 undergo a similar collapse process. The

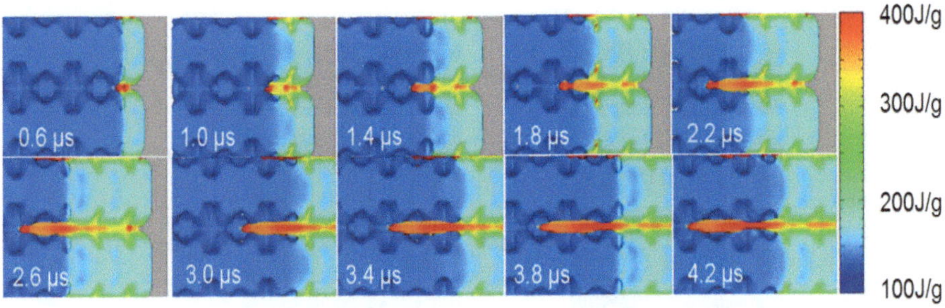

Fig. 13 Internal energy maps for an impact velocity of 500 ms^{-1}

process by which collapse of the cavity generates a pulse in the transmitted wave is as follows. As the compression wave meets the surface of the cavity a radial release wave is generated which propagates both laterally and back in the direction from which the wave has come. Eventually this release wave reaches the back surface of the target where it produces a dip in the velocity field. On full collapse of the cavity a compression wave is generated which propagates radially outward resulting in an increase in pressure. For the 300 ms^{-1} sequence illustrated, the successive collapse of the large and small cavities generates a roughly sinusoidal "primary" wave into the solid target with a peak-to-peak time of $\sim$0.75 μs. Note that high magnification photographs of the SLM structure show that the surface of the specimens are not sharply defined but instead tend to be coated by un-melted or partially melted particles of the base material. One of the effects of this surface unevenness will be to cushion the impact following cavity collapse. Therefore the compression wavelets generated by the collapse will be smoothed relative to the predicted waveform. This phenomenon probably explains the observation in Fig. 10b that the oscillations generated by cavity collapse are more pronounced in the simulations than in reality.

Figure 13 shows a series of internal energy maps corresponding to the 500 ms^{-1} experiment. As in the material plot of Fig. 12, large and small cavities are visible. Since the impact velocity ($\sim$500 ms^{-1}) is greater than that in the experiment depicted in Fig. 12, the cavities collapse in a different way. In the higher velocity case, as each cavity collapses it generates a jet which propagates in the direction of the shock in the lattice. The role of jetting in energy absorption will be discussed further in Sect. 4.9. It will be argued that jetting processes increase the amount of plastic flow associated with densification of a porous material.

4.8 Conclusions from High-Rate Study

Computer simulation is needed to assess the relative merits of different cellular structures and materials as energy absorbers in different loading environments. In high-rate experiments the deformations tend to be so severe that Lagrangian meshing is not an option. Instead, the relatively mesh-hungry Eulerian schemes are needed. This leads us towards one-dimensional or uniaxial strain systems. Further, we believe that for high-rate experiments the most useful comparisons between simulation and experiment are those in which nominally one-dimensional, or uniaxial strain, conditions persist, at least until maximum (or full) densification of the porous material is reached. We note that 1D conditions are most readily achieved in high density (low porosity) high velocity systems.

Selective laser melting was used to fabricate a cellular structure, consisting of intersecting 316L stainless steel rods. The shock delivered to a solid target by a gas-gun driven flyer with the chosen structure was measured experimentally using a Heterodyne Velocimetry (Het-V) technique. It was found that all of the traces consisted of a small elastic precursor followed by a plastic wave which took the velocity to a first plateau of duration about one microsecond. A second reflection from the flyer–target interface then took the trace up to a second velocity plateau. In all of the experiments the first plateau exhibited roughly sinusoidal oscillations.

Continuum simulations of the experiments were run using the $P(\alpha)$ model [48, 49] in conjunction with the Sandia National Laboratories code CTH. Values for the crush pressure and elastic pressure required by the model were estimated from the measured yield strength of a test piece of solid SLM steel. Although the experimentally observed oscillations were not reproduced by the continuum model the average levels and timing of the plateaus were well matched. This finding justifies the use of continuum models, in which only the porosity and flow stress of a structure formed from a particular base material are specified, to assess energy absorbing potential.

Three-dimensional "structural" CTH calculations were run in which stl files used to manufacture the experimental samples were imported into CTH. In this case the detailed, dimensioned, structure of the lattice was modelled. The approach is similar to that adopted by Borg in an earlier chapter in this book. A 0.05 mm mesh was used giving $\sim$8 cells through the rods which made up the cell. These "3D" simulations gave a reasonable match to all of the features observed in the experiments including the oscillations which were superimposed on the first and second velocity plateaus at the back surface of the solid steel target. The 3D simulation provided insights into the processes at play in the shocked lattice. As the compression wave reaches a cavity, a release wave followed by a compression wave is generated. The array of cavities in the structure thereby generates stress oscillations which propagate in all directions. As these waves reach the back surface of the target they cause the velocity oscillations observed in the experiment. It was found that jetting process plays an increasingly important role as impact velocity increases. The jetting phenomenon has the effect of increasing the plastic flow associated with densification of the lattice.

4.9 *Two-Dimensional CTH Study of Alternative Cell Architectures*

The observation that the experiments were well matched by simulations encouraged us to use CTH to assess the effect of structural parameters on energy absorption. Because the structure in the experimental investigation just described was rather complex it was not considered ideal as a basis for a systematic study of the effects of cell architecture on energy absorption. It was felt that a study of the effect of cell shape on energy absorption needed to be based on a simpler structure. In [3] we describe the use of the validated CTH models to study a simplified lattice structure in which the simulations were confined to a single 6 mm × 6 mm two-dimensional cell allowing the distortions of the structure to be imaged at significantly higher resolution than previously. The aim was to provide a clearer picture of the energy absorbing processes occurring in metal-based cellular samples during impact than was provided by the relatively complex structures in the earlier study and, thereby, to support the optimisation of cellular structures for application as energy absorbers.

Adopting a two-dimensional rather than a three-dimensional configuration yields significant savings on computer resources. Figure 14 shows the idealised two-dimensional configuration, chosen as the basis for the study. The impactor is directed vertically upwards. The energy absorbing elements (EAE) and the impactor and target were considered to consist of infinitely long bars extending in and out of the paper. The outlines shown in Fig. 14 represent sections through these bars. For example, the grey rectangle illustrates the orientation of the "rod" variant (also shown on the extreme right of Fig. 14). The flyer and the target are rectangles of dimensions 6 mm × 18 mm, again effectively extending an infinite distance in and out of the paper. In order to provide an improved understanding of the effects of lattice structure on energy absorption, it was necessary to evaluate a systematic

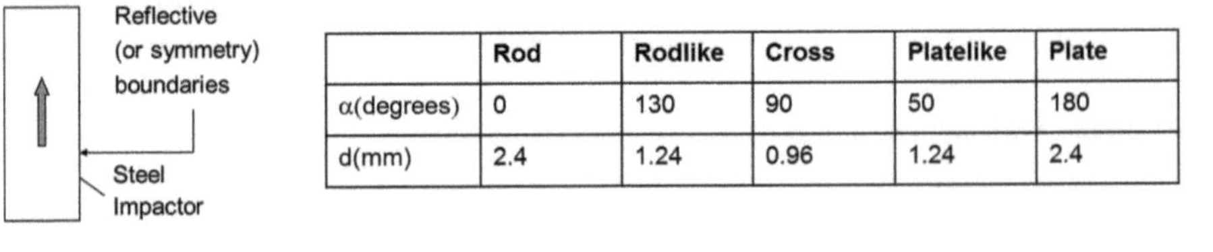

	Rod	Rodlike	Cross	Platelike	Plate
α(degrees)	0	130	90	50	180
d(mm)	2.4	1.24	0.96	1.24	2.4

Fig. 14 Schematic of the two-dimensional (2D) configuration used a basis for CTH study of energy absorption; the range of structures in the energy absorbing element (EAE) are as depicted in diagrams **a**–**e**

range of potential strut orientations. Therefore simulations of the five alternative EAE structures depicted in Fig. 14 were performed. In all cases the solid occupies 40% of the total volume of the cell. The flyer impacts from below and the target, or acceptor, is above the EAE. For convenience the EAE on the extreme right (Fig. 14e) is labelled "rod". However, since this is a 2D setup the "rod" element actually consists of a plate extending into the paper. The structure on the extreme left (Fig. 14a) also consists of a plate extending into the paper but it is rotated by 90° relative to the structure on the extreme right. The structure in the centre, labelled "cross" consists of two plates at 90° to each other. The inclusion of the rod-like, cross and plate-like structures is intended to provide a simple transition from rod to plate structure. Since the shape and mass of the unit cell is held constant the strut thickness, d, depends solely on the angle α. It can be shown that the porosity, $\varnothing$, defined as the fraction of the cell occupied by solid material, is given by:

$$\varphi = d/L \tag{8}$$

for options a and e

$$\varphi = \left[\frac{2dL}{\text{Sin}\,(\alpha/2)} - \frac{d^2}{2\text{Cos}\,(\alpha/2)\,\text{Sin}\,(\alpha/2)}\right]\frac{1}{L^2} \tag{9}$$

for options b and d and

$$\varphi = \left[d\sqrt{8L^2} - 2d^2\right]\frac{1}{L^2} \tag{10}$$

for option c. where α, d and L are as defined in Fig. 14.

It is envisaged that, theoretically, each of the unit cells depicted in Fig. 14 could be repeated horizontally and vertically to provide a macroscopic structure. However, note that, with the exception of the "cross", the macroscopic structure that would be formed by repeating the unit cells studied here would not form practicable, interconnected, material. Structural parameters for the five EAE designs are tabulated in Fig. 14.

The steel components were modelled using a Mie–Gruneisen equation of state and the Steinberg–Guinan (SG) strength model, the same approach used in the 3D structural simulations described in Sect. 4.6. The material parameters were the same as those used in the earlier investigation apart from the yield strength. Note that in the previous study an elevated value of Y_0 of 0.64 GPa, based on the measured strength of SLM steel was used. However, since the simulation described here does not currently form the basis of an experimental study, the CTH default value for Y_0 (0.34 GPa) was used. We note that changing the yield strength in the simulations to that used in the earlier high-rate study (0.64 GPa) would change the absolute, but not the relative, values of the energies calculated by CTH.

The mesh size in the study reported here was 0.02 mm as compared with 0.05 mm in the earlier work. Therefore the number of cells through the thickness of each strut in Fig. 14 varied between 48 and 120 depending on which EAE was being investigated; this compares with ~8 cells through each strut in the earlier study [2].

Key objectives of the CTH study were to obtain insights into the mechanisms at play during the interaction between the flyer and the cushioned target and to estimate from simulations which of the chosen variants maximises the irreversible energy generated in the EAE and thereby minimises the energy transferred to the target.

CTH allows the kinetic and internal energies of each cell to be calculated. The kinetic energy of the cell is calculated from its known mass and velocity. To calculate the internal energy the code uses the known elastic–plastic constitutive model to separately evaluate the elastic and plastic contributions to the internal energy. The code then adds the two energies together to yield the total increase in internal energy. Knowledge of the internal energies in each cell allows the distribution of energy through the impact configuration to be represented by a series of energy maps as shown, for example, in Figs. 15 and 18.

Note that in [3] the balance between kinetic and internal energy following impact of the configurations in Fig. 14 was assessed using a simple analytic model in addition to CTH simulations. Further the relative contributions to internal energy of hydrodynamic compression energy, distortional elastic energy and plastic energy were estimated. Table 4 lists the specific internal energies generated in the EAEs at 0.3 and 0.9 mmμs^{-1}. Also listed in Table 4 are the elastic and compression energies estimated for shock waves generated by steel-on-steel impact without an EAE present [3]. We assume that these latter values represent upper limits on the elastic and compression energy generated in an EAE. It can be seen that, at both impact velocities and for all EAE variants, the elastic components are small compared

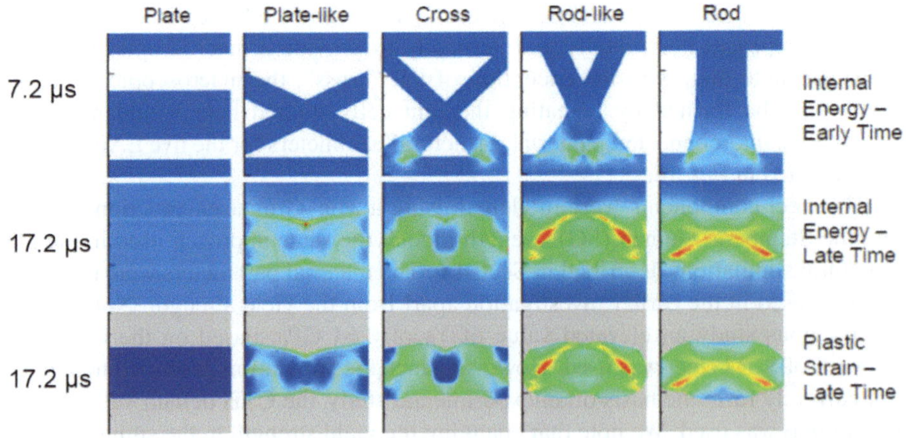

Fig. 15 The effect of cell geometry on plastic strain, internal energy and temperature during a 300 ms^{-1} impact. Internal energy is shown at both early and late times. Plastic strain is shown at late times only. Note that plastic strain, energy and temperature all increase as we go from plate and plate-like geometry to rod-like and rod geometry

Table 4 CTH simulations: peak volume average specific internal energies generated in the EAE (Jg^{-1})

Impact velocity ($mm\mu s^{-1}$)	Plate (Jg^{-1})	Plate-like (Jg^{-1})	Cross (Jg^{-1})	Rod-like (Jg^{-1})	Rod (Jg^{-1})	Elastic energy (Jg^{-1})	Compression energy (Jg^{-1})
0.3	15	40	74	122	126	<1.7	10
0.9	117	225	324	316	323	<1.7	75

with plastic energy. For the plate configuration the estimated compression energy is an appreciable fraction of the total internal energy, which is not surprising since the plate configuration is geometrically similar to a simple steel-on-steel impact. However it is clear that in the other EAE configurations the total internal energy is dominated by the plastic component of the internal energy. It follows that, in most of the simulations run in support of this investigation, the total internal energy may be taken as an approximate measure of the plastic energy. A possible exception to this trend is the plate EAE where, as might be intuitively expected, most of the internal energy is in the form of hydrodynamic compression energy.

The energy maps provide insights into the mechanisms at play during the loading of cellular structures. For example, Fig. 15 shows internal energy and plastic strain for the five EAE variants just after full densification for a simulation at 300 ms^{-1}. It is clear that the plastic strain and, therefore, internal energy absorbed by the EAE tends to increase as the configuration is varied from plate to rod but that there is little difference between the rod-like and rod variants.

The energies in each of the components in our impact configuration can be calculated. CTH allows the energies in each computational cell of each of the three regions (flyer, EAE, target) to be summed to give the total energy for a 1 cm length (measured into the paper) of each region or component as a function of time. Figure 16a shows the total internal energy delivered to the different EAEs by a 300 ms^{-1} impact. It is seen that the internal energy in all of the EAEs initially ramps up and then reaches a rather uneven "plateau". The onset of the plateau corresponds roughly to the time at which the EAEs reach full densification. The greatest energy absorption is achieved with the rod and rod-like EAEs and the least energy is absorbed by the plate and plate-like EAEs. The cross structure lies between these two. It was suggested in [2] that the small hills and troughs in the plateau regions are generated as waves reverberate through the test system in the impact direction. One of the effects of the reverberations is to cause the component of internal energy due to hydrodynamic compression to oscillate. Corresponding results for a 900 ms^{-1} impact are shown in Fig. 16b. It can be seen that, again, the energy rises roughly linearly to a "perturbed plateau". Again the onset of the plateau corresponds to full densification. However there are two significant differences between the 300 and 900 ms^{-1} results. First, the internal energies dissipated in the 900 ms^{-1} simulation are at least twice the corresponding energies in the 300 ms^{-1} simulations. Second, in the higher velocity simulation the energy generated in the cross structure is similar to that generated in the rod and rod-like structures, whereas at 300 ms^{-1}

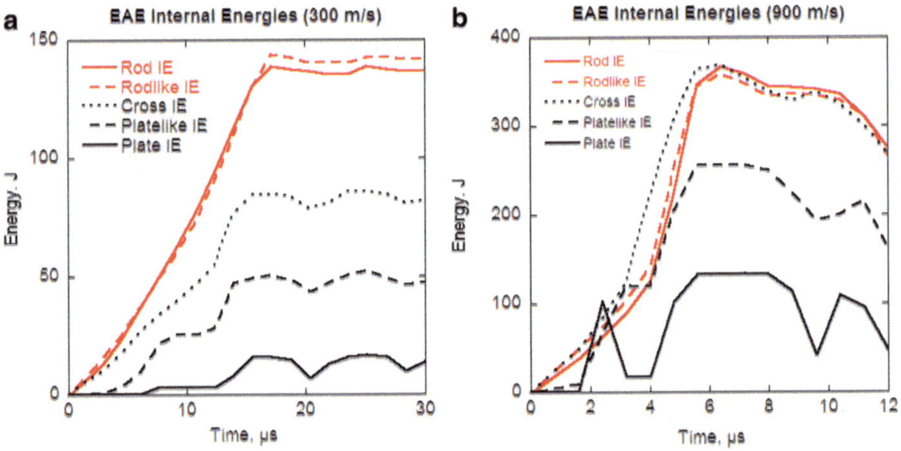

Fig. 16 Internal energies generated in the five alternative energy absorbing elements by impacts at (**a**) 300 ms^{-1} and (**b**) 900 ms^{-1}. The periodic dips in internal energy correspond to releases in hydrodynamic pressure as the impact wave reverberates through the system

the cross is intermediate between the rod and plate structures. It is argued in [3] that the differing effectiveness of the cross structure as the impact velocity changes may be explained by comparing the deformation mechanisms of the cross and rod-like structures at 300 and 900 ms^{-1}. As shown in Fig. 17a, according to CTH, at 300 ms^{-1} the cross structure closes by axial movement of the parts of the cell driven by the impactor. By contrast, as shown in Fig. 17b, at 300 ms^{-1} the rod-like cell closes predominantly by lateral flow, a process accompanied by significantly more plastic deformation than with the cross variant. Therefore, at 300 ms^{-1}, more energy is absorbed by the rod-like than by the cross variant. The simulations show that as the impact velocity is increased beyond 500 ms^{-1} there is a transition from a mechanism dominated by roughly uniform lateral flow to one dominated by jetting. At 900 ms^{-1}, both the cross and the rod-like variants deform by a concentration of lateral flow in the region of the cell adjacent to the impactor. This produces a mushroom feature which occupies the whole width of the cell and intense lateral flow and jetting predominate. The result is that at 900 ms^{-1} the cross and rod-like structures absorb similar amounts of energy.

The growing importance of jetting processes as velocity increases is clearly illustrated for the rod EAE in the internal energy sequence in Fig. 18. It is important to note that as jetting mechanisms begin to play an increasing role in the collapse of the cell it would be expected that the total amount of plastic flow associated with the densification would increase.

A particularly useful metric for estimating the energy-absorbing capability of a given cushion (or EAE) is illustrated in Fig. 19 which shows simulations run using the interface separation feature in CTH. The effect of using this feature is to allow two components to separate when the axial stress at the interface between them

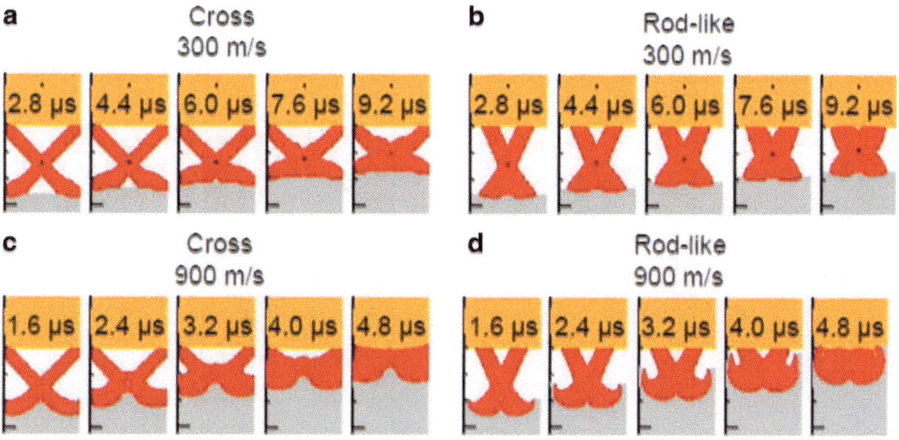

Fig. 17 CTH simulations for 300 m/s impacts onto cross and rod-like structures are shown in (**a**) and (**b**) and, correspondingly, in (**c**) and (**d**) for 900 m/s impacts

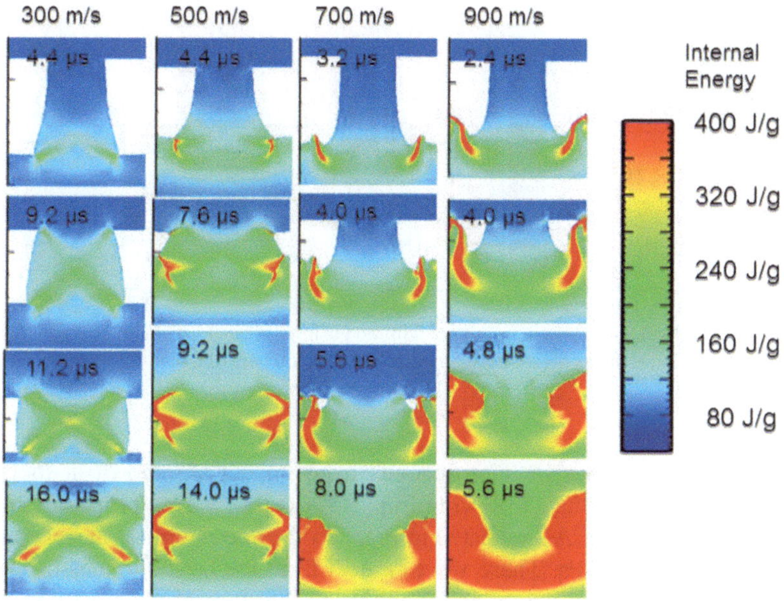

Fig. 18 Microkinetic processes increase as velocity increases

becomes negative. The upper sequence in Fig. 19 shows a CTH time sequence of a velocity map for a simple, un-cushioned, 300 ms^{-1} impact between an 18 mm long stainless steel projectile and an initially stationary 18 mm target. As with the previous simulations, in this "baseline" simulation the steel was modelled using a Gruneisen EoS and a Steinberg–Guinan strength model. The lateral boundaries were modelled as reflective, thereby imposing in effect 1D or uniaxial strain conditions.

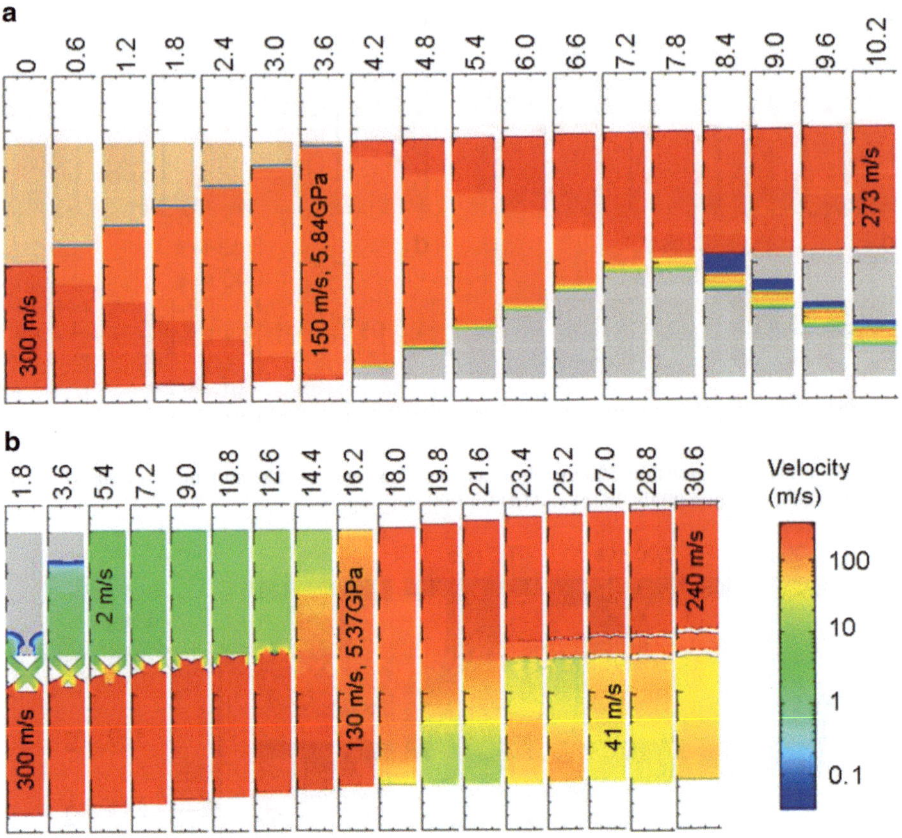

Fig. 19 (**a**) Axial velocity following impact with no cushion present. (**b**) With cross cushion. It is seen that the cushion reduces the velocity imparted to the target and the maximum stress developed in the target

The times of the frames in microseconds from impact are shown at the top of each frame. It is seen that, following impact (at 0 μs), waves propagate into the impactor and the target. As indicated in Fig. 19a the maximum stress generated by this impact was 5.84 GPa. These waves reflect from the back surfaces of the flyer and the target as release waves and meet at the interface at ~8 μs from impact. The tension resulting from the interaction between the two release waves causes the interface to separate. As indicated in Fig. 19a the residual velocity of the target is 273 ms^{-1}. Note that this is less than the velocity (300 ms^{-1}) which would have been obtained in a perfectly elastic collision. The reason is that some energy has been lost as a result of plastic flow of the components. The elasto-plasticity also explains the residual wave which continues to oscillate through the components following their separation. The peak stress developed in the target as a result of the impact is estimated from the simulation to be 5.84 GPa and to occur ~3.6 μs from impact.

By contrast Fig. 19b shows a sequence generated by a 300 ms^{-1} impact on a cross EAE. From this sequence we judge that full densification of the EAE occurs at ~14 μs. It is seen that an elastic wave propagates through the EAE into the target and reaches the back face of the target at ~5.4 μs. Note that a similar wave propagates into the impactor but this is not visible because the velocity scale is logarithmic. The average velocity of the target, following separation, was estimated from a velocity print-out along the centre of the system. As indicated in Fig. 19b the velocity in the system cushioned by a cross EAE was 240 ms^{-1}, significantly less than the velocity imparted to the un-cushioned system. Further the peak stress imparted to the target has been reduced and the time at which the peak occurs has been significantly delayed.

It was noted in [3] that, particularly at the higher velocities, the internal energy generated in the EAE is only a very small fraction of the total energy delivered to the system by the flyer. Consequently it is expected that, in the test configuration studied here, the different EAE designs will have only a very small effect on the energy imparted to the target. This expectation was confirmed by studies similar to that depicted in Fig. 19b in which the EAE type and impact velocities were varied. At 300 ms^{-1} the residual velocity of the target varied from ~250 ms^{-1} for the plate EAE to 180 ms^{-1} for the rod. However for 900 ms^{-1} impact there was no detectable difference in the velocity imparted to the target. We note that it is likely that the difference in effectiveness between the different EAE designs would become more evident if the EAE occupied a greater fraction of the test system. One obvious route would be to reduce the length of the flyer and target in our chosen system. Another option would be to stack EAEs in the axial direction.

4.10 Conclusions from Cell Architecture Study

Simulations run at the relatively low velocity of 300 ms^{-1} showed that all EAE variants were compressed to solid density in about the same time. However, significantly more plastic flow and internal energy were generated by rod and rod-like EAEs than in the plate and plate-like EAEs. It was shown that, at the low impact velocity, for most cell types the cells densify by simple broadening of the elements of the EAE. By contrast, as impact velocity is increased to 500 ms^{-1} and beyond, jetting processes play an increasingly important role. In effect, jetting increases the amount of plastic flow associated with the full densification of the porous material. The increased plastic flow is accompanied by an increase in the irreversible energy absorbed during the densification process. Quantitative comparison of the effectiveness of the different EAEs can be provided by using CTH to calculate the energy distribution in the system. For example, the pictorial evidence indicated that more energy was generated in the rod and rod-like EAE than in the plate-like versions, implying that the former variant would provide a more effective cushion than the latter. This conclusion was supported by CTH calculations in which the energies generated in each cell of each of the three components of the system were summed. For example, simulations at 300 ms^{-1} showed that, relative to a plate-like EAE, the rod-like EAE led to a lower gain in total energy of the target.

CTH simulations of a representative cushion system were run in which separation of the components was allowed. In this configuration the residual velocity of the target following separation provides a useful metric through which to judge the effectiveness of alternative cushions. That is, in systems where the cushion absorbs more energy, a lower velocity will be imparted to the target and vice versa.

Future investigations will consider how to compare and verify the simulations described here with experimental data. One promising method of accessing this data is dynamic X-ray imaging of the impacted lattice. This technique, which has been utilised by Eakins and Chapman [54], has the potential to furnish line of sight mass images at selected viewing directions and times. Corresponding deformation profiles can then be extracted from 3D simulations for comparison with experiment.

5 Overall Conclusions and Future Directions

Specimens of 316L stainless steel were manufactured as tensile test specimens using selective laser melting (SLM). The quasi-static stress–strain response was measured and the yield stress evaluated. The parameters used for the tensile tests specimens were also used to manufacture cellular structures with measured relative densities ranging from 0.22 to 0.64. The cell sizes were either 1 mm or 2 mm.

Lattices were tested quasi-statically using an Instron screw-driven testing machine. As expected the shapes of the stress–strain relationships obtained depended on the scale and density of the lattices.

A Hopkinson bar configuration was used to test the lattices dynamically at impact velocities ranging from 105 ms^{-1} to 154 ms^{-1}, a regime in which the lattice is not likely to be fully densified by the impact. The specimens showed a significant increase in strength relative to the quasi-static case. This effect was attributed to the strain rate sensitivity of the stainless steel and the influence of inertia on the buckling resistance of the struts which formed the cellular structure.

High-rate loading of the cellular structures was achieved using a 100 mm light gas gun. Samples consisting of discs of diameter 99.2 mm and thickness 6 mm were mounted on the end face of a polycarbonate sabot. The specimens impacted a solid steel target at velocities ranging from 500 to 700 ms^{-1}. Surface velocities at the back face of the target were monitored using Het-V. The experiments were modelled using both 2D continuum and 3D structural simulations in the SNL code CTH. The continuum simulations matched the macroscopic features of the observed velocity profiles well. For example, all three of the observed records were divided into an elastic region, a first plateau and a second plateau. The continuum simulations gave a good match to the average amplitude and duration of the plateaus in all the experimental records. The analysis was further supported by a schematic x-t construction. The x-t diagram was used to explain qualitatively why the length of the first plateau increased as the impact velocity increased.

CTH was also used to run 3D structural simulations of the high-rate loading experiments. The 3D simulations matched the general form of the experimental records reasonably well. In particular the frequencies of the oscillations which were superimposed on the first and second plateaus were well matched by the code. Detailed examination of the interaction of the shock front with the internal structure of the lattice provided an explanation for the observed velocity oscillations. However although the frequencies of the oscillations were well matched the simulations tended to overestimate their amplitude. The probable reason for this discrepancy is that the surfaces of the samples tended to be coated with un-melted particles and these would have had the effect of cushioning the impact generated by the collapse of the included voids. This phenomenon would reduce the amplitude of the waves generated by the collapse.

Following the observation that experiments were well matched by CTH, further simulations were run with the aim of developing a better understanding of the way in which energy absorption depended on the structure of the cell. Simulations were set up in which five energy absorbing elements (EAEs) all with the same volume and mass were positioned to "cushion" the impact between a steel projectile and a steel anvil. The simulations showed that the energy absorbed by the EAE depended on the configuration of the EAE and on the impact velocity. When rod and rod-like EAEs were compressed to full densification more plastic flow, and therefore more irreversible energy, was generated than when plate or plate-like EAEs were densified. Increasing the impact velocity also increased the amount of plastic flow associated with full densification. At relatively low velocities the elements compressed via a simple shape change but at high impact velocities the flow became complex, exhibiting intense localised flows. A potentially useful additional metric for assessing the effectiveness of a chosen EAE is to compute the velocity imparted to the target. However for most of the configuration in this study this velocity was rather insensitive to the EAE type. Options for increasing the sensitivity of this approach include increasing the volume occupied by the EAE relative to the flyer and target.

By applying the principles presented in the preceding text when considering the merits of alternative energy absorbing structures, it should be possible to assess which is most effective in a specific impact regime. In conjunction with the close control over cellular material design which is possible using SLM, this approach offers the opportunity to manufacture and test cellular configurations which maximise generation of irreversible internal energy.

Finally, an attractive feature of AM cellular structures is the ease with which the properties such as density, cell shape, cell size and even base material could be varied with position in a component. This ability offers the potential to tailor the pressure and velocity of waves delivered to a component by shock or impact. For example, electronic or other delicate components are often mounted in locations which may be subject to shocks or other high pressure waves. In such situations it may be possible to avoid or delay damage by mounting the sensitive item in cellular material whose properties have been optimised to deliver a benign and/or delayed loading profile to the component. Similar principles may be applied to a range of scenarios where protection against blast and/or impact is required. Examples of areas where highly optimised energy absorbing cellular structures might be of benefit include protective helmets and body armour, protecting military vehicles against mines and bombs and protecting experimental firing enclosures from fragment and blast damage.

Acknowledgements Some of the experiments described in this article were performed using the 100 mm light gas gun at the Institute of Shock Physics, Imperial College London. We thank Matthew Cotton and Mark Swan, AWE, and Jonathan Harris, Department of Engineering, University of Cambridge, for their significant contributions.

References

1. Rashed MG, Ashraf M, Mines RAW, Hazell PJ (2016) Metallic microlattice materials: a current state of the art on manufacturing, mechanical properties and applications. Mater Des 95:518–533
2. Winter RE, Cotton M, Harris EJ, Maw JR, Chapman DJ, Eakins DE, McShane G (2014) Plate-impact loading of cellular structures formed by selective laser melting. Model Simul Mater Sci Eng 22:025021
3. Winter RE, Cotton M, Harris EJ, Eakins DE, McShane G (2016) High resolution simulations of energy absorption in dynamically loaded cellular structures. Shock Waves 27:221–236. https://doi.org/10.1007/s00193-016-0651-2
4. Gibson I, Rosen D, Stucker B (2015) Additive manufacturing technologies,3D printing, rapid prototyping, and direct digital manufacturing. Springer, New York
5. Srivatsan TS (2016) Additive manufacturing – innovations, advances and applications. CRC Press, Boca Raton. ISBN: 9781498714778
6. Thompson MK, Moroni G, Vaneker T, Fadel G, Campbell RI, Gibson I, Bernard A, Schulz J, Graf P, Ahuja B, Martina F (2016) Design for additive manufacturing: trends, opportunities, considerations, and constraints. CIRP Ann Manuf Technol 65:737–760
7. Herzog D, Seyda V, Wycisk E, Emmelmann C (2016) Additive manufacturing of metals. Acta Mater 117:371–392
8. Tsopanos S, Mines RAW, McKown S, Shen Y, Cantwell WJ, Brooks W, Sutcliffe CJ (2010) The influence of processing parameters on the mechanical properties of selectively laser melted stainless steel microlattice structures. J Manuf Sci Eng 132:041011
9. Cherry JA, Davies HM, Mehmood S, Lavery NP, Brown SGR, Sienz J (2014) Investigation into the effect of process parameters on microstructural and physical properties of 316L stainless steel parts by selective laser melting. Int J Adv Manuf Technol 76:869–879. https://doi.org/10.1007/s00170-014-6297-2
10. Gumruk R, Mines RAW (2013) Compressive behaviour of stainless steel micro-lattice structures. Int J Mech Sci 68:125–139
11. Ushijima K, Cantwell WJ, Chen DH (2013) Prediction of the mechanical properties of micro-lattice structures subjected to multi-axial loading. Int J Mech Sci 68:47–55
12. McKown S, Shen Y, Brookes WK, Sutcliffe CJ, Cantwell WJ, Langdon GS, Nurick GN, Theobald MD (2008) The quasi-static and blast loading response of lattice structures. Int J Impact Eng 35:795–810
13. Smith M, Cantwell WJ, Guan Z, Tsopanos A, Theobald MD, Nurick GN, Langdon GS (2010) The quasi-static and blast response of steel lattice structures. J Sandw Struct Mater 13:479–501

14. Yan C, Hao L, Hussein A, Young P, Raymont D (2014) Advanced lightweight 316L stainless steel cellular lattice structures fabricated via selective laser melting. Mater Des 55:533–541
15. Mines RAW, Tsopanos S, Shen Y, Hasan R, McKown ST (2013) Drop weight impact behaviour of sandwich panels with metallic micro lattice cores. Int J Impact Eng 60:120–132
16. Van Bael S, Kerckhofs G, Moesen M, Pyka G, Schrooten J, Kruth JP (2011) Micro-CT-based improvement of geometrical and mechanical controllability of selective laser melted Ti6Al4V porous structures. Mater Sci Eng A 528:7423–7431
17. Brenne F, Niendorf T, Maier HJ (2013) Additively manufactured cellular structures: impact of microstructure and local strains on the monotonic and cyclic behavior under uniaxial and bending load. J Mater Process Technol 213:1558–1564
18. Wauthlea R, Vranckenc B, Beynaertsc B, Jorissena K, Schrootenc J, Krutha J-P, Van Humbeeck J (2015) Effects of build orientation and heat treatment on the microstructure and mechanical properties of selective laser melted Ti6Al4V lattice structures. Addit Manuf 5:77–84
19. Merkt S, Hinke C, Bültmann J, Brandt M, Xie YM (2015) Mechanical response of TiAl6V4 lattice structures manufactured by selective laser melting in quasistatic and dynamic compression tests. J Laser Appl 27:S17006
20. Tancogne-Dejean T, Spierings AB, Mohr D (2016) Additively-manufactured metallic micro-lattice materials for high specific energy absorption under static and dynamic loads. Acta Mater 116:14–28
21. Li SJ, Murr LE, Cheng XY, Zhang ZB, Hao YL, Yang R, Medina F, Wicker RB (2012) Compression fatigue behavior of Ti–6Al–4V mesh arrays fabricated by electron beam melting. Acta Mater 60:793–802
22. Martin G, Fabreque D, Mercier F, Chafino-Axia JA, Dandievel R, Blandin JJ (2016) Coupling electron beam melting and spark plasma sintering: a new processing route for achieving titanium architectural microstructures. Scr Mater 122:5–9
23. Maskery I, Aremu AO, Simonelli M, Tuck C, Wildman RD, Ashcroft IA, Hague RJM (2015) Mechanical properties of Ti-6Al-4V selectively laser melted parts with body-centred-cubic lattices of varying cell size. Exp Mech 55:1261–1272
24. Carlton HD, Lind J, Messner MC, Volkoff-Shoemaker NA, Barnard HS, Barton RB, Kumar M (2017) Mapping local deformation behaviour in single cell metal lattice structures. Acta Mater 129:239–250
25. Ozdemir Z, Hernandez-Nava E, Tyas A, Warren JA, Fay SD, Goodall R, Todd I, Askes H (2016) Energy absorption in lattice structures in dynamics: experiments. Int J Impact Eng 89:49–61
26. Meteyera S, Xub X, Perry N, Zhao YF (2014) Energy and material flow analysis of binder-jetting additive manufacturing processes. Procedia CIRP 15:19–25
27. Tang Y, Zhou Y, Hoff T, Garon M, Zhao YF (2016) Elastic modulus of 316L stainless steel lattice structures fabricated via binder jetting process. Mater Sci Technol 32:648–656
28. Thompson SM, Bianc L, Shamsaei N, Yadollahi A (2015) An overview of direct laser deposition for additive manufacturing; part I: transport phenomena, modeling and diagnostics. Addit Manuf 8:36–62
29. Shamsaei N, Yadollahi A, Bianc L, Thompson SM (2015) An overview of direct laser deposition for additive manufacturing; part II: mechanical behavior, process parameter optimization and control. Addit Manuf 8:12–35
30. Yadollahi A, Shamsaei N, Thompson SM, Seely DW (2015) Effects of process time interval and heat treatment on the mechanical and microstructural properties of direct laser deposited 316L stainless steel. Materials Science & Engineering A 644:171–183
31. Calladine CR, English RW (1984) Strain-rate and inertia effects in the collapse of two types of energy absorbing structures. Int J Mech Sci 26:689–701
32. Ashby MF, Evans AG, Fleck NA, Gibson LJ, Hutchinson JW, Wadley HNG (2000) Metal foams: a design guide. Butterworth Heinemann, Boston
33. Deshpande VS, Fleck NA (2000) High strain rate compression behaviour of aluminium alloy foams. Int J Impact Eng 24:277–298

34. Tan PJ, Harrigan JJ, Reid SR (2002) Inertia effects in uniaxial dynamic compression of a closed cell aluminium alloy foam. Mater Sci Technol 18:480–488
35. Reid SR, Peng C (1997) Dynamic uniaxial crushing of wood. Int J Impact Eng 19:531–570
36. Zhao H, Gary G (1998) Crushing behaviour of aluminium honeycombs under impact loading. Int J Impact Eng 21:827–836
37. Radford DD, McShane GJ, Deshpande VS, Fleck NA (2007) Dynamic compressive response of stainless steel square-honeycombs. J Appl Mech 74:658–667
38. Vaughn DG, Hutchinson JW (2006) Bucklewaves. Eur J Mech A/Solids 25:1–12
39. McShane GJ, Deshpande VS, Fleck NA (2012) Dynamic buckling of an inclined strut. Int J Solids Struct 49:2830–2838
40. McGlaun JM, Thompson SL, Elrick MG (1990) CTH a three dimensional shock wave physics code. Int J Impact Eng 10:351–360
41. Harris JA, Winter RE, McShane GJ (2017) Impact response of additively manufactured metallic hybrid lattice materials. Int J Impact Eng 104:177–191
42. Hall EO (1951) The deformation and ageing of mild steel: III discussion of results. Proc Phys Soc Sect B 64:747
43. Petch NJ (1953) The cleavage strength of polycrystals. J Iron Steel Inst 174:25
44. Kashyap BP, Tangri K (1995) On the Hall-Petch relationship and substructural evolution in type 316L stainless steel. Acta Metall Mater 43:3971–3981
45. Strand OT, Goosman DR, Martinez C, Whitworth TL, Kuhlow WW (2006) Compact system for high-speed velocimetry using heterodyne techniques. Rev Sci Instrum 77:083108
46. Steinberg DJ, Cochran SG, Guinan MW (1980) A constitutive model and data for metals applicable at high strain rates. J Appl Phys 51:1498–1504
47. Johnson GR, Cook WH (1983) A constitutive model and data for metals subjected to large strains, high strain rates and high temperatures. In: Proceedings of Seventh International Symposium on Ballistics. American Defense Preparedness Association, The Hague, pp 541–547
48. Herrmann W (1969) Constitutive equation for the dynamic compaction of ductile porous materials. Journal of Applied Physics 40:2490–2499
49. Carroll MM, Holt AC (1972) Suggested modification of the P-alpha model for porous materials. Journal of Applied Physics 43:759–761
50. Wu Q, Jing F (1996) Thermodynamic equation of state and application to Hugoniot predictions for porous materials. Journal of Applied Physics 80:43434349
51. Fischmeister HF, Arzt E (1983) Densification of powders by particle deformation 1983. Powder Metallurgy 26:82–88
52. Benson DJ, Nesterenko VF, Jonsdottir F, Meyers MA (1997) Quasistatic and dynamic regimes of granular material deformation under impulse loading. Journal of the Mechanics and Physics of Solids 45:1955–1999
53. Bowden FP, Yoffe AD (1952) Initiation and growth in explosives in liquids and solids. Cambridge University Press, Cambridge
54. Eakins DE, Chapman DJ (2014) X-ray imaging of subsurface dynamics in high Z materials at the diamond light source. Rev Sci Instrum 85:123708. https://doi.org/10.1063/1.4904275

MIX
Papier aus verantwortungsvollen Quellen
Paper from responsible sources
FSC® C105338

If you have any concerns about our products,
you can contact us on
ProductSafety@springernature.com

In case Publisher is established outside the EU,
the EU authorized representative is:
Springer Nature Customer Service Center GmbH
Europaplatz 3, 69115 Heidelberg, Germany

Printed by Libri Plureos GmbH
in Hamburg, Germany